Internet Performance Survival Guide

QoS Strategies for Multiservice Networks

Geoff Huston

Wiley Computer Publishing

John Wiley & Sons, Inc.

New York ◆ Chichester ◆ Weinheim ◆ Brisbane ◆ Singapore ◆ Toronto

Publisher: Robert Ipsen
Editor: Carol Long
Managing Editor: Marnie Wielage
Text Design & Composition: Pronto Design & Production, Inc.

Library of Congress Cataloging-in-Publication Data:

ISBN 0-471-37808-9

Printed in the United States of America.
10 9 8 7 6 5 4 3 2

Wiley Networking Council Series

Scott Bradner
Senior Technical Consultant, Harvard University

Vinton Cerf
Senior Vice President, MCIWorldCom

Lyman Chapin
Chief Scientist, BBN/GTE

Books in the Series

Geoff Huston, *ISP Survival Guide: Strategies for Running a Competitive ISP* 0-471-31499-4

Elizabeth Kaufman & Andrew Newman, *Implementing IPsec: Making Security Work on VPNs, Intranets, and Extranets* 0-471-34467-2

For more information, please visit the Networking Council Web site at www.wiley.com/networkingcouncil.

contents

networking council foreword

The Networking Council Series was created in 1998 within Wiley's Computer Publishing group to fill an important gap in networking literature. Many current technical books are long on details but short on understanding. They do not give the reader a sense of where, in the universe of practical and theoretical knowledge, the technology might be useful in a particular organization. The Networking Council Series is concerned more with how to think clearly about networking issues than with promoting the virtues of a particular technology; that is, how to relate new information to what the reader knows and needs, so that he or she can develop a customized strategy for vendor and product selection, outsourcing, and design.

In *Internet Performance Survival Guide* by Geoff Huston, you'll see the hallmarks of Networking Council books: examination of the advantages and disadvantages, strengths and weaknesses of market-ready technology; useful ways to think about options pragmatically; and direct links to business practices and needs. Disclosure of pertinent background issues necessary to understand who supports a technology and how it was developed is another goal of all Networking Council books.

The Networking Council Series is aimed at satisfying the need for perspective in an evolving data and telecommunications world that is filled with hyperbole, speculation, and unearned optimism.

In *Internet Performance Survival Guide: QoS Strategies for Multiservice Networks*, you'll get clear information from experienced practitioners.

We hope you enjoy the book. Let us know what you think. Feel free to visit the Networking Council Web site at www.wiley.com/networkingcouncil.

Scott Bradner
Senior Technical Consultant, *Harvard University*

Vinton Cerf
Senior Vice President, *MCIWorldCom*

Lyman Chapin
Chief Scientist, *BBN/GTE*

from the author

This is book is both a second edition of an earlier book, and a new work. The previous book, *Quality of Service: Delivering QoS on the Internet and in Corporate Networks,* written in collaboration with Paul Ferguson, studied Quality of Service tools, and the way they may be deployed in Internet networks. The book looked at the state of the various tools as they existed in 1997.

Much has happened in the industry since then, and the engineering agenda has broadened considerably. This book will be looking at some of the engineering issues that arise when engineering Internet networks to deliver defined levels of performance as well as examining the array of service performance tools that can be used within networks today.

The order in which the material is presented has been restructured from the first work to show how performance profiles are constructed upon the foundations of a small number of simple actions. The book will also be introducing recent approaches to QoS and looking at the broader area of Internet performance from an engineering perspective. The result is that the revision of the material includes a clear statement of focus on service performance. The book's title, *Internet Performance Survival Guide: QoS Strategies for Multiservice Networks,* reflects this changed emphasis. I hope that with these revisions, this book will continue to be a useful and relevant resource for network engineers tasked with the design of high performance and highly efficient Internet service networks.

Geoff Huston

acknowledgments

Writing a book is never an easy task and there are a number of individuals who have helped me to turn this idea into the completed project.

First I'd like to acknowledge Paul Ferguson, the coauthor of the first edition of this book. The material in this book builds upon the original approach adopted by Paul of being a straightforward and offering no nonsense commentary on the state of service differentiation mechanisms. I trust I've done justice to Paul's original concept by attempting to keep the marketing-inspired smoke, mirrors, and pixie dust out of this book. Where something appears to be unworkable, I've followed Paul's lead and simply called it broken.

Also, I would like to particularly acknowledge the assistance of Scott Bradner whose careful reading and helpful comments have been invaluable in shaping the material into a readable whole. I would also like to acknowledge Carol Long who has kept this project on track as well as editing my rough drafts into decent prose.

And, most importantly, I wish to acknowledge the patient support of my wife Michele and my children, Chris, Sam, and Alice, who have a much better idea of what "just a little project" means than I do.

Geoff Huston

introduction

Any sufficiently advanced technology is indistinguishable from magic.

—Arthur C. Clarke

For some years now, those in the communications industry have been hearing the call for convergence. Surely, they argue, we don't need a panoply of service networks to meet all these diverse communications requirements. At the heart of all of today's communications networks is a collection of flows of bits, of 1s and 0s. At this level of detail, you cannot tell whether the 1s and 0s carry voice, video, Web fetches, or electronic mail. Surely, if all these networks are just streams of 1s and 0s, we should be able to converge all these services into a common digital carriage network that can carry every requirement on a single switching and transmission platform. After all, think of the benefits that such a single network would provide: reduced cost of operation, economies of scale, and enhanced functionality.

The call for convergence is by no means recent. It was voiced in connection with the Asynchronous Transfer Mode (ATM) network architecture, the Integrated Services Digital Network (ISDN), and the Broadband cable network architecture. After all these convergence efforts, the industry appears to have now landed on one common theme: If there is to be convergence, the platform for it will be the Internet. According to this theme, what we will see is a *multiservice Internet*, where all forms of communications—voice, video, data, and signaling and all forms of transmission, including high-capacity fiber optic cable, broadband coaxial cable plant, copper pairs, and various forms of wireless systems—are bonded in a single-service platform through the technology of the Internet.

However, convergence is not just a case of joining bit streams. Each application has an associated set of service characteristics. To support the operation of a call, a voice conversation needs a reliable, constant rate, low jitter, symmetric bit flow, and a call control signal structure is required to frame the start and finish of the bit flow. A Web page retrieval operation generates a set of interactions, starting with a number of single data-packet exchanges with various domain name servers, followed by a data transfer from the selected Web server, where the transfer is capable of adapting to various levels of delay, jitter, and packet loss within the network. In this case, the underlying data is framed into packets, and the interaction with the network is viewed as a packet-based interaction. Convergence implies using a single-bit transmission and switching architecture for the network and adapting the service profile of the network to suit the requirements of each application.

So far, the Internet has been deployed with a single model of network service: that of unreliable packet delivery where the unit of interaction with the network is a packet and each packet is delivered to the destination address independently of any previous or following packets. On top of this has been layered a reliable transfer protocol, TCP, whose objectives are, first, reliability of the data transfer and, second, maximum efficiency of the transfer. The Internet performs no active resource management. If there is available capacity, the network will make it available to application traffic. If there is no available capacity, but there are additional demands, then the network congests. The reaction of TCP applications to such congestion is to reduce the flow rates to accommodate the additional demand. In other words, TCP delivers a variable response, faster under light load and slower under heavy load. The Internet protocols do not intrinsically support constant bit rate data flows, whereby the transmission is controlled by a synchronous clock and the constant sending bit rate is precisely matched by the arrival bit rate. Nor does the Internet intrinsically support reproducible transactions in which the network transaction can be reliably repeated in terms of elapsed time and reliability.

This raises a fundamental question: Can the Internet be managed in such a way as to deliver particular service performance responses in response to particular signals? If so, then it may be possible to emulate a synchronously clocked circuit, at least to an acceptable level of accuracy. It also may be possible to support constant responses so that a transaction can be provided with a constant level of resource, independent of other traffic levels imposed on the network. With such resource management tools, it should be possible to control the performance response from the network. This in turn should allow a second wave of service applications to enter the Internet application space, supporting high speed multimedia applications, constant quality voice and video applications, as well as allowing the efficient use of low-bandwidth wireless links supporting mobile personal digital assistants, which appear to form the next wave of mobile communications devices.

In all of these potential futures for the Internet as a common convergent service platform, there is one constant need: To realize many of these potential applications and fulfill the role of a multiservice platform, the Internet network needs to engage active resource management so that it can deliver an accurate response to signaled requirements. This may be in the form of a premium service response or a consistent service response, depending on the requirements of the application.

Managing the service response of an Internet network is often referred to as *Quality of Service (QoS)*, assuming that the major motivation for performance management is one of improving the service response of the network in either a general or specific fashion. While this is certainly a large part of the motivation for performance management, there is a broader issue of managing the *service performance* of a network, enabling the network to respond accurately to a broad range of service requirements.

A Service Performance Reality Check

This book is intended to introduce some semblance of uniformity into the understanding of service performance in Internet networks, and the related role of quality of service in managing performance—a needed reality check, if you will. There is a need to articulate the scope of the problem we are trying to solve before we throw multiple solutions into the networking landscape. Concomitant with that need is a need to define the parameters of the service management environment and then to describe a toolkit of available options to approach service management problems. This book attempts to provide the necessary tools to define the requirements of multiservice Internet networks, examine the problems, and to offer an overview of the possible solutions.

This book is intended for all types of readers: the network manager who is trying to understand the technology and its capabilities; the network engineering staff, who are designing methods to provide differentiated service outcomes; and the consultant or systems integrator, who is matching requirement to proposed solution.

Managing Expectations

Service quality and performance management are terms that mean different things to different people in different roles. Empirical evidence indicates that service differentiation mechanisms are in greater demand in corporate, academic, and other private campus intranets (private networks that interconnect portions of a corporation or organization) than they are on the global Internet (the Internet proper, which is the global network of interconnected computer networks), and the ISP (Internet service provider) community. Not only are there unique expectations within each of these communities, but there are different expectations within each organization.

It is not unusual for the engineering and marketing areas of an organization to have conflicting views on many matters. The marketing department of an ISP, for example, may be persistent about implementing some type of differentiated quality of service with the expectation that it will create new opportunities for additional revenue. Conversely, the engineering group may want to implement a managed service environment to provide predictable behavior for a particular class of applications, or to impose some graduated response to resource contention within the network, regardless of any particular market outcomes. An expectation common among engineering managers is that a multiservice network requires a managed performance environment, and that such an environment is an essential precondition to a converged service platform on which all forms of communications services can coexist and interact within the confines of a single network.

These are the types of issues that service performance and quality of service conjures in the imagination—*automagic* mechanisms that provide this managed differentiated service quality. This is not a simple issue, nor one that generates immediate industry consensus, especially when agreement cannot be reached within the network engineering community or, worse, within a single organization. Furthermore, it introduces a fair amount of controversy over exactly what the problem is that must be solved. Definitions often are crafted to deliver functionality to meet short-term goals. Is it technical? Economic? Market-driven? All of these things?

It is as important to clearly describe what is not part of service performance management as it is to attempt to describe what is within the scope of this study. The subject of study here is the Internet Protocol layer, or IP, and the network on which IP is carried; it does not include the various encoding and signaling technologies that form part of the support for voice-over-IP, nor does it include the various efforts to bridge remote segments of the telephone network with voice over IP services. The study also precludes descriptions of the multimedia gateway controllers and their related signaling requirements. This study is undertaken from the perspective of the network manager who is tasked with supporting a variety of services on a base IP platform, and, as part of that task, must determine how to support voice-over-IP, how to support multimedia gateway controllers, and how to provide differentiated service responses to different clients and different applications.

This book explores the general topic of how to tune a network to support a collection of dissimilar service requirements simultaneously. Its scope is sufficiently broad to include an examination of how to support a variety of service levels within a single IP network service platform. So, rather than describing the precise nature of a variety of multimedia applications and their associated encoding and signaling requirements, this study looks at the general network services required to support a multiservice environment and, more particularly, at the ways an IP service network can be tuned to respond to such service profiles.

How This Book Is Organized

This book is structured to progress logically from mechanisms and tools to architectures and solutions, building an understanding of service management environments by starting with the base network building blocks and moving through to complex service architectures.

Chapter 1: Internet Performance and Quality of Service. In this chapter, you will learn the various definitions of performance management and quality of service, as well as what they are perceived to provide. You will come to understand how QoS needs have evolved, and discover what service performance mechanisms exist, how they are being used. Finally, you will recognize why this topic is important today and for the immediate and more distant future.

Chapter 2: The Performance Toolkit. Any complex system is the sum of a number of simple components, and performance management systems are no exception. This chapter details the components of the network, with particular focus on the performance and service implications of each. The first component we describe is the Internet Protocol itself, comprising IP, UDP, and TCP, and the performance characteristics of each of these protocols Next is an examination of the physical elements of the network, followed by a look at routers and switches, the functions that these network elements perform, and those design aspects that have a direct bearing on service performance. You will also examine the Multi-Protocol Label Switching Protocol (MPLS) and its potential in relation to delivering managed service. The chapter then moves on to transmission systems and service performance, the carrier's service portfolio of synchronously clocked digital circuits, and their service characteristics. Packet- and cell-switched service offerings are discussed next, including X.25, Frame Relay, and ATM, along with a description of the service profiles offered by these systems, and their relationship to managed IP performance outcomes. The intent of this chapter is to build a basic set of per-element performance tools that can be assembled to form service delivery architectures.

Chapter 3: Performance Tuning Techniques. This chapter provides an in-depth examination of the actions of an IP router, which includes how each of the router's actions can be configured to produce various service performance outcomes. You will learn about admission classification actions that enable a router to map the traffic to a set of defined service classes. You will then learn about various queuing disciplines and the service outcomes that are achievable within each. Of course, if the queue is full queuing disciplines cannot help, and the router has no choice but to discard a packet. The chapter describes various packet discard behaviors and their impact on service performance, and reviews a number of early packet discard techniques that are used to generate particular service outcomes. The chapter also explores the traffic profile and admission control mechanisms that allow a router to manage the traffic load admitted into the network. This examination includes a look at QoS routing mechanisms, a variant of conventional routing protocols that present a view of a network topology filtered by some service classification. We will also examine packet fragmentation issues and header and payload compression techniques.

Chapter 4: Quality of Service Architectures. Having assembled a collection of basic performance tools, the next step is to assemble them within a coherent architecture. In this chapter, you will learn the various approaches to a service performance architecture. The first is the original IP Type of Service architecture where the host application uses a field in the IP header to inform the network of the desired service profile on a packet-by-packet basis. This examination of architectures continues with the Integrated Services architecture, in which the host systems make a resource reservation along the path, specifying a traffic profile. Traffic within the profile is performance-managed according to the reserved service profile. This per-application reservation architecture is then compared to the Differentiated Services architecture, which uses a per-packet service marking system similar in style to the Type of Service approach. This section concludes with an examination of the service management architecture, including the mechanisms to implement a network operator's chosen admission policies for managed services.

Chapter 5: Performance Engineering. This chapter addresses the typical applications for managed service performance, and how these outcomes can be implemented using the available architectures. It examines at per-application services; real-time services, such as voice and video streams; premium virtual circuit services, such as those required in virtual private networks; and premium service profiles, which may be part of a service level agreement. The chapter also examines the role of service management and the related area of accounting for the use of managed services.

Chapter 6: What Is Quality of Service Anyway? The final chapter reviews all the possible methods of implementing quality of service, including their merits and weaknesses. It also examines the marketing and economic factors that play a role in the rationale for managed services, and the alternative approaches to supporting multiple service profiles. This chapter also offers a simple summary of actions that you can undertake within the network and within host systems to improve the network's service efficiency and service flexibility. The effort to create a managed service environment within the Internet is certainly not yet finished, and the conclusion describes those aspects that are not adequately addressed in the managed service story so far. It is highly probable that these open issues will form the basis for further refinements of the managed service environment.

What's the Objective for Service Management?

As mentioned at the start of this introduction, it is important to understand how a managed service network is defined. To some, it means introducing an element of predictability and consistency into the existing variability of best-effort network delivery systems. To others, it means obtaining higher transport efficiency from the network and attempting to increase the volume of data delivery while maintaining characteristically consistent behavior. To others, managed services are a means of differentiating classes of data service, offering network resources to higher-precedence service classes at the expense of lower-precedence classes. Managed services may also refer to attempting to match the allocation of network resources to the characteristics of specific data flows. Others may see this as a means to construct the platform of a multiservice Internet network. All these definitions fall within the generic area of managed services, quality of service, or performance management, as we understand it today.

It is important that the chosen QoS implementation function as expected; otherwise, all the theoretical knowledge in the world won't help you if your network is broken. It is no less important that the selected QoS implementation be cost-effective to operate, because an economically prohibitive QoS model may be dismissed in favor of an inferior, but cheaper approach. Also, sometimes brute force is better—buying more bandwidth and more switching capability can often be a more effective solution than attempting to implement a complex and convoluted QoS option.

Applying the theory of service management to the realities of production networks is at the heart of engineering practice. To that end, this book examines service management tools, their application, and their service outcomes, to provide some insights into the state of service management capabilities in today's Internet networks.

chapter one

Internet Performance and Quality of Service

The art of elegant compromise

Over the course of the last decade, many exciting and surprising technologies have become part of the evolution of the Internet. One of the most unpredictable outcomes of this evolution is the assumption that the Internet, as a service platform, can be deployed into not just one market segment, or take the form of a single-functional model, but that it can support every form of communications services. In short, the assumption is that every known communications service can, somehow, be shoe-horned into an Internet service environment. The expectation is that the high-volume use of a common Internet platform will drive communications service charges down, resulting in cost efficiencies for Internet-based competitive communications service providers. Behind this *ubiquitous service platform* model is the belief that no other basic platform technology will be able to compete in the marketplace against the benchmark levels of price performance established by the Internet.

This assumption encompasses not just a traditional model of data transfer, such as computer-to-computer file movement, or even the more general case of nonreal-time and reliable data transfer service, but also various forms of time- and performance-critical services, such as interactive voice, streaming video, various e-commerce trading models, and real-time remote data-manipulation service models. Thus, the Internet is being cast into a unified role, in which the concept of the Internet Protocol as the single switching technology for an entire carrier service portfolio is cited as the ultimate future of the Internet evolution. As for most other modern-day concepts, a T-shirt summarizes this expectation, stating it simply as "IP Everywhere."

There is now tremendous interest in placing diverse application types onto a common IP platform, which already provides highly efficient support for reliable data flows and nonreal time data streams. The challenge is to add to this collection a set of real-time media services that can support voice and video applications, while retaining the efficiencies of data carriage. Most of the issues raised while attempting to meet this challenge are economic, political, and social, rather than technical. The traditional voice, video, and print communications services have dominated the communications industry for many decades, and the pressures for change are often seen in business terms rather than technology issues.

The recent entrance of IP into this environment does threaten the longer-term stability of this sector. In political terms, this is not a smooth process of *convergence* (as many industry commentators have labeled it), but a very real struggle for dominance over an historically lucrative and powerful industry sector. IP technology establishes new cost-efficiency benchmarks for transmission services that weaken the longer-term financial viability of the current communications infrastructure, which uses traditional time division multiplexed (TDM) circuit-switched technologies. The IP service model uses a simple, stateless packet-switching network; in so doing, it places responsibility for service definition and management out to the periphery of the network. This transition from "smart network, dumb handsets" to "dumb network, smart PCs" exposes potentially massive shifts in the associated communications service economy.

Such are the major issues exposed by the shifting of IP into a role of a single, ubiquitous communications service platform. Some of the issues raised by this shift are technical in nature, and it is these issues that we put under close scrutiny here, including issues regarding supporting the multiservice IP network, and the associated areas of service performance and service differentiation capability.

Filling the role of *universal carrier* requires IP to have a collection of performance attributes that can be negotiated between the end-to-end application and the Internet network. The tools to create these attributes, and the methods for their deployment within the network to affect performance outcomes, are the topic of this book.

Answering the question of how to design and apply these tools to a network presents significant challenges. Performance engineering is not the domain of an automated network wizard, where a set of scripts can be used to configure the network's service objectives, after which the wizard automatically configures and deploys relevant performance management tools. Nor is performance engineering a precise science, whereby every engineering team will inevitably arrive at the same network design when addressing the same objectives. Rather, performance engineering can be described as the *art of elegant com-*

__W__here Did the Internet Come From? *The development of the Internet is a classic example of the path a technology takes from academic study through phases of applied research to industry adoption and, finally, to widespread public deployment.*

The original academic studies of this area took place during the 1960s, and became an officially designated research project at the end of that decade. The Internet Protocol was developed in the mid-1970s. Its initial design, originally termed the Kahn-Cerf protocol after its primary architects (later renamed TCP/IP for Transmission Control Protocol/Internet Protocol), took place between 1973 and 1974, culminating with publication, in May 1974, of the paper titled "A Protocol for Packet Network Interconnection" [Cerf 1974]. This paper embraced the model of diverse communications elements with differing characteristics and made no assumptions regarding the imposition of state within the network. This design was refined in the light of experience gained through experimental deployment as part of the Advanced Research Projects Agency (ARPA) Internet program.

The formal specification of the Internet Protocol was published in September 1981 as part of a United States Defense Advanced Research Projects Agency (US DARPA) Internet program. The protocol was described in three parts: the Internet Protocol (IP) [RFC791], the Internet Control Message Protocol (ICMP) [RFC792], and the Transmission Control Protocol (TCP) [RFC793].

This version of the Internet Protocol is in widespread use today, universally referred to as IP. There is, however, another version of the Internet Protocol. Concerns over the consumption of the address space and the scaleability of the protocol led to the development of a refined Internet Protocol in the 1990s. This protocol is commonly termed IP version 6 (IPv6) [RFC2460], in reference to the value of the protocol version field in the packet header. When compared to IPv6, the current widely-used version is termed IP version 4 (IPv4).

promise. Under this definition, the engineer is faced with a set of often conflicting requirements, and the task is to respond in a way that results in a simple and elegant network design, one that attempts to minimize the negative impacts of the conflicting requirements.

A good example of making design choices from among conflicting requirements is the selection of buffer sizes within the network's routers. To efficiently support remote interactive access, electronic commerce transactions, and a large proportion of Web transfers, the network should minimize packet loss, or avoid it altogether. Large internal buffers can assist in achieving this objective. In addition, support for high-performance, reliable TCP-based data requires internal network buffers at a size comparable to the delay-bandwidth product of the link each buffer drives. However, while such a large buffer requirement will provide efficient management of some forms of data flow, it will also raise the delay variation, or *jitter*, considerably. Jitter can be defined as distortion of the inherent timing of the data flow. A large jitter component should be avoided when transferring real-time voice and video data streams. Preservation of the timing of the original signal as it is sent through the network is more critical than any small-scale incidence of packet loss.

To date, a prevalent approach in many Internet service networks has been to apply the *principle of commonality* to the performance problem space. Where it is necessary to compromise between conflicting requirements, frequently, engineers create a service environment that is equally ill-suited to all requirements! Such an approach is termed *uniform best-effort* service. A best-effort service network uses a single-level service response, usually implemented within the network's routers through simple first-in-first-out (FIFO) queue management mechanisms with medium-sized buffers. This is, essentially, a compromise between the differing requirements of the network, and it generates a service response that is not geared to providing high-quality service support to any particular network application. The inevitable result is that a best-effort network offers poor response to all applications, and hence can also be called a uniformly poor-effort network!

For performance-tuned networks, a uniform best-effort service response is simply not good enough. If constructed upon such a compromised service model, the vision of a single IP switching layer as a ubiquitous service platform for the entire communications industry will not survive. The basic IP suite is capable of supporting a diverse set of communications applications, whereas the service IP network must be capable of generating different responses to different application requirements. For example, such a differentiated service environment should allow real-time traffic streams to be passed through the network with minimal delay and minimal jitter. At the same time, the differentiated service environment should also permit large-volume data to be transferred reliably, making extensive use of internal network buffers to maximize application throughput. This is a *multiservice* response, meaning the network is intended to respond in different ways to different applications (see Figure 1.1).

Figure 1.1 The multiservice network.

The requirements for a multiservice network platform is definitely not a new idea in the data communications environment. One rationale given to support the transition of the carrier networks from a switching regime based on Time Division Multiplexing (TDM) to that of Asynchronous Transfer Mode (ATM) switching is the capability of ATM to support a diverse set of service profiles at the same time. However, while some carrier networks saw the multiservices architecture as one with a solid ATM foundation, other carrier networks (most notably those within the Internet Service Provider sector) identified an opportunity to expand their IP network platform into the same multiservice environment.

There is now a clearly visible trend within the Internet service environment to construct IP-based switching systems that have an even broader service profile than that envisioned for ATM. For this highly flexible Internet service model to be achieved, the service model of the IP architecture must be refined in two important ways. First, it must manage the performance profile of the network to deliver consistent performance outcomes; and, second, it must encompass a broad spectrum of performance profiles. This effort has been termed *Quality of Service* (QoS), although the effort is more accurately described as one of *performance engineering*, whereby the network is attuned to deliver particular performance profiles for various classes of traffic.

What Is Quality of Service?

What's in a name? In this case, not very much, because QoS has been given wildly varying definitions. This is partly due to the ambiguity and vagueness of the words quality and service.

Quality of Service: The Elusive Elephant

When sorting through the various definitions of QoS, it might be helpful to recall the old story of the three blind men who on their journey happen upon an elephant. The first man touches the elephant's trunk and determines that he has stumbled upon a huge serpent. The second man touches one of the elephant's massive legs and determines that the object is a large tree. The third man touches one of the elephant's ears and determines that he has confronted a huge bird. The three men envision different things, because each examined only a small portion of the elephant.

Think of the elephant as the concept of QoS. Different people interpret QoS variably, because numerous and sometimes ambiguous QoS problems exist. In the attempt to position IP as a universal carriage service, various constraints are imposed upon the performance and quality of the IP architecture, and the result can lead to contradictory service requirements. Unfortunately, these contradictions usually are not evident, as people seem to have a natural tendency to adapt an ambiguous set of concepts to a single paradigm that encompasses their particular problems. By the same token, within QoS this ambiguity yields different possible solutions to various problems, which has opened a schism in the networking industry where the issue of QoS is concerned.

Another analogy often used when attempting to define QoS is to the generic 12-step program for habitual human vices: The first step on the road to recovery is acknowledging and defining the problem. Of course, in this case, the immediate problem is the lack of consensus on a clear definition of QoS. To examine the concept of QoS, we must first examine the two operative words: *quality* and *service*, both of which can be equally ambiguous. Next we will explore why this situation exists, and provide a laundry list of available definitions.

What Is Quality?

The term quality can encompass numerous attributes in many domains. Within the realm of performance networks, quality is often used to describe the process of delivering data in a manner superior to normal expectations. The characteristics sought in such elevated expectations may include aspects of reduced probability of data loss, minimal (or no) network-induced delay or latency, consistent delay characteristics (also known as minimal jitter), and the ability to determine the most efficient use of network resources (such as using paths that have the shortest transmission distance between two end-points, or achieving maximum efficiency of circuit bandwidth). In this sense, quality is defined as *superior performance*.

In another sense quality can be considered synonymous with reliability and predictability, in which case quality refers to a constant level of loss, latency or jitter, or any other distinguishing property. Quality is also used in reference to particular characteristics of specific networking applications or protocols, and the level of constancy of those characteristics. In this context, quality is defined as reliability of the service in terms of constancy of the outcome.

We tend to use the term quality loosely, to define both concepts interchangeably, when in fact, in an engineering sense, they are very different concepts. The interpretation of *superior performance* can be translated as the capability of the network service to offer preemptive resource allocation; in contrast, reliability can be translated to mean the capability of the network to precisely deliver an allocated level of resource to a particular service profile.

What Is Service?

The term service also carries implicit ambiguity. Depending on how an organization or business is structured, service itself may have several meanings. People generally use service to describe the communications capabilities offered to the end users of any network, such as end-to-end communications or client/server applications. In such applications, services can cover a broad range of offerings, from electronic mail to desktop video and from Web browsing to chat rooms.

In multiprotocol networks, a service also may have several other definitions. In a Novell NetWare network, for example, each Service Advertisement

Protocol (SAP) advertisement is considered an individual service. In other cases, services may be categorized according to the various protocol suites, such as SNA, DECnet, AppleTalk, and so forth. In this fashion, you can bring a finer degree of granularity to the process of classifying services. It's not too difficult to imagine a more complex service-classification scheme in which you might identify services by protocol type (such as IPX, DECnet), then classify them further to a more granular degree within each protocol suite (such as SAP types within IPX).

Service Guarantees

Traditionally, network service providers have used a variety of methods to accommodate quality of the service they offer to their customers. A common method is through a *service guarantee*, normally expressed within the terms of a formal service contract. Network availability, for example, is one of the more prevalent and traditional measurements for a *service level agreement* (SLA) between a provider and a subscriber. Here, access to the network is the basic service, and failure to provide this service is a failure to comply with the contractual obligation. If the network is inaccessible, the quality aspect of the service is clearly questionable.

Internet service providers (ISPs) have experimented with more specific criteria for a service guarantee, to define additional aspects of service quality; these criteria include some forms of performance-related metrics, such as average packet loss or delay bounds for customer traffic. If a service provider delivers only 90 percent of your traffic over a period of six hours, for example, you might decide that the service quality violates or doesn't fulfill your contracted service guarantee. A service provider may specify an SLA that stipulates an average packet latency between the client and a defined set of locations or one that stipulates a specified percentage of packets will be passed through the network within a given maximum delay, bounding both loss and delay within the agreement. Another form of service guarantee places constraints on both the client and the provider. As long as the client limits the amount of traffic passed to the provider to fit within an agreed traffic load profile, the provider will deliver the traffic within an associated service level. The guarantee is intended to provide some level of assurance in the overall quality of the service offered by the provider.

Understandably, a segment of the networking community abhors the term service guarantee, because its use can be indiscriminate, ambiguous, misleading, and even a contradiction of terms. In particular, this may be true in an environment where virtually all traffic is packet-based, and structured packet loss may not be to the detriment of an application; indeed, as in the case of TCP transfers, packet loss may be used as a signal to dynamically determine optimal data flow rates. Offering a guaranteed service of some sort implies not only that no loss will occur beyond the limits stipulated in the guarantee, but that the performance of the network will be consistent and predictable. In a world of packet-based networks—a world that spans many thousands of service providers internationally—offering such a guarantee takes on the dimensions of a major engineering challenge. The service performance tools typically available to the network architect do not have intrinsic capabilities to enforce a particular service guarantee; consequently, the network's design rules have to be sufficiently conservative to accommodate the anticipated load profile while remaining within the general parameters of the associated network's service level guarantee. As you can imagine, there is constant tension between the demand for rigorous service level guarantees and the realistic expectation of adaptive Internet Protocol performance.

Quality of Service versus Classes of Service

The marriage of the terms quality and service has produced a fairly straightforward pair of definitions: a measurement of how transparently the network behaves in response to imposed traffic, and a structure intended to define the characteristics and properties of specific services. But this definition leaves a lot of latitude for creative interpretation, which is reflected in the confusion evident in the networking industry and in the hyperbole of promotional material. One common thread does run through most of the definitions of quality of service: the ability to differentiate between traffic or service types so that the network can treat one or more classes of traffic differently than other types. (The mechanisms to differentiate between traffic and service classes are discussed in more detail throughout this book.)

It is important to distinguish between what is appropriately called *differentiated classes of service* (CoS) and the more ambiguous *Quality of Service*. Although some may argue that these terms are synonymous, in practice, they have subtle distinctions. QoS has a broad connotation that encompasses many aspects of overall performance and service characteristics within the network. CoS implies that services can be categorized into separate classes, which can, in turn, be managed individually. This differentiation is the operative concept in CoS.

Implementing Performance Quality

Using the definition of QoS as a mechanism to set expectations of service performance across the network, QoS certainly encompasses CoS approaches. In environments where a spectrum of outcomes is available to the service application, the application must negotiate with the network on the determined service class for the application. Consequently, regardless of the mechanism used, the network requires the ability to assign traffic flows to one of a number of service classes, each of which offers a form of a predictable service outcome.

In other words, in these types of multiservice environments the critical outcome is not whether some traffic can be given preferential treatment over other types of traffic, but whether the characteristics of each network service class remain predictable. These behavioral characteristics rely on managing several aspects of the network's performance, including end-to-end response time (also known as round-trip time, RTT), latency, queuing delay, available bandwidth, and other criteria. Some of these characteristics may be more predictable than others, depending on the application, the type of traffic, the queuing and buffering characteristics of the network devices, and the architectural design of the network.

Latency as a Performance Factor

Of these network characteristics, latency is perhaps the most critical in performance management. Two forms of latency exist: real and induced. *Real latency* is considered to be the physical, binding characteristics of the transport media (electronic signaling and clocked speed) and the RTT of data as it travels between two points, as bound by the speed of propagation of electromagnetic radiation. This lower bound on delay is referred to as the *speed of light problem*, because changing the speed at which light travels is generally considered to be impossible. It represents the ultimate boundary on how much real latency is actually present in a network and how quickly data can be transmitted across any arbitrary distance.

Induced latency is introduced to the network in several ways:

♦ Packet reassembly delay within the network switching devices
♦ Processing delay
♦ Queuing delay within the network switches

Of these, the most significant is queuing delay. Queuing delay is not well-ordered in large networks, and its variation over time can be described simply

*The Speed of Light The speed of light in a vacuum, or the physical sciences constant **c** is probably the most researched constant in all of science. According to electromagnetics theory, its value, when measured in a vacuum, should not depend on the wavelength of the radiation. According to Einstein's prediction about the speed of propagation of light within the general theory of relativity, the measured speed of light does not depend on the observer's frame of reference; the speed of light in a vacuum is a universal constant.*

Estimates of the value of c have been undergoing refinement since 1638, when Galileo's estimate of "If not instantaneous, it is extraordinarily rapid" was published in Two New Sciences. *The currently accepted value is 299,792.458 kilometers per second.*

The speed of light in glass or fiber-optic cable is significantly slower, at approximately 194,865 kilometers per second.

The speed of propagation of electrical charge through a conductor is a related value; it, too, has been the subject of intense experimentation. Perhaps the most bizarre experiment was conducted in Paris, in April 1746, by Jean-Antoine Nollet. Using a snaking line of some 200 monks, connected by a mile-long iron wire, Nollet observed their reactions when he administered a powerful electric shock through the wire. The simultaneous screams of the monks demonstrated that, as far as Nollet could tell, voltage was transmitted through a conductor instantaneously. Further experimentation has managed to refine this estimate, and the current value of the speed of voltage propagation in copper is 224,844 kilometers per second, slightly faster than the speed of light through fiber-optic cable.

as highly chaotic. The resulting induced latency is the major source of the uncertainty in protocol-level estimates of RTT.

Why is an accurate estimate of RTT important for protocol performance? Maximized throughput efficiency is the most obvious answer. For a packet protocol to use a network efficiently, the steady-state objective of the protocol is to inject a new data packet into the network at exactly the same moment a packet in the same flow is removed from the network. This action simulates the function of a synchronized data clock in a transmission circuit, where 1 bit enters the circuit at precisely the same time as another bit leaves the circuit. This is shown in Figure 1.2. If data is injected into the network more rapidly than it is removed, then the network will ultimately overload, and

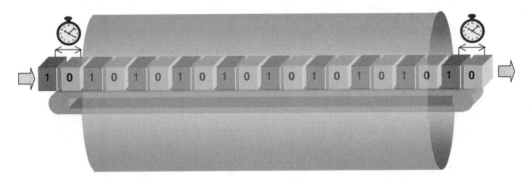

Figure 1.2 Data clocking.

throughput efficiency will plummet. If data is injected at a lower rate than it is being removed, then the network will become idle, again reducing throughput efficiency. The RTT estimate is used to govern TCP's time-out and retransmission algorithms and the dynamic data rate management algorithms.

There is a third latency type that can be introduced at this point: *remembered latency*. Craig Partridge makes a couple of interesting observations in his book, *Gigabit Networking*, concerning latency in networks and its relationship to human perception [Partridge 1994]. He refers to this phenomenon as the "human-in-the-loop" principle; that is, humans can absorb large amounts of information and are sensitive to delays in the delivery and presentation of the information. This principle describes some fascinating situations. For example, users of a network service have a tendency to remember more clearly the occasional failure of the network rather than its overall success in delivering information in a consistent and reliable manner. This perception leaves users with the impression that the quality of service is poor, when the overall quality of the service actually is quite good. More generally, humans note behavior that is outside the general pattern. Therefore, a burst of high-quality performance will also be noted, and, once experienced, the elevated service level will become the new expectation. ("Once you've traveled first class, economy class is very hard to handle" is how service quality engineers describe it.) As you can see, predictable service quality is of critical importance in any QoS implementation, because human perception is the determining factor in the success or failure of a specific implementation.

This characteristic of remembered latency has been particularly troublesome in the search for acceptable QoS mechanisms. Achieving a baseline of acceptable performance, while eliminating occasional variations, is a daunting challenge because of the variety of issues that can affect the response time in a network. Furthermore, the Internet product marketing community seems to expect that some sort of QoS technology will evolve to provide highly stable predictable end-to-end behavior in the network, regardless of the architectural framework of the network, the construction and design of the underlying infrastructure, the interconnection model, and the routing system. Certainly, perfection is a worthy goal, but it's unachievable, and network reality is a little more humdrum. There is no substitute for the basic investment in prudent network design and adherence to the fundamental principles of network engineering.

QoS and Fairness

Many heated discussions have centered on providing a "fairness mechanism" in the network. In this context, fairness is a resource management strategy that at the very least prohibits any one user from consuming all available network resources in the presence of other concurrent resource requests. In other words, fairness is intended to prevent network hogs from taking over so much of the network that other users receive no service at all. In general, though, fairness extends beyond this base level of control. Fairness may attempt to enforce *equal sharing* of the network. Of course, equal sharing is another term open to many interpretations. Where there are a number of unequal demands, sharing fairness may offer an equal response to all demands, a response that penalizes the large demands at a greater proportionate level than smaller demands. Alternatively, sharing fairness may offer the same proportionate response to all demands, allocating a larger share of the resource to larger demands. Let's examine the common *equal sharing* fairness strategies.

One strategy for implementing sharing fairness of resource allocation is to deploy a fixed level of resource to each client. Under such a scheme, no single client can consume all available resources, nor can any client deny resources to any other client (see Figure 1.3). Such a statically allocated resource scheme is entirely fair, in that every active client receives the same resource allocation from the network. Its disadvantage is that it is also very inefficient in terms of overall network utilization, as this kind of resource management structure has a high idle overhead. To implement such a fixed resource allocation strategy, the network itself is effectively segmented into resource units, and each active client is allocated a resource unit from the idle pool. For this scheme to operate effectively, the network manager must ensure that the total capacity of the network is large enough so that there are available idle slots at all times, even

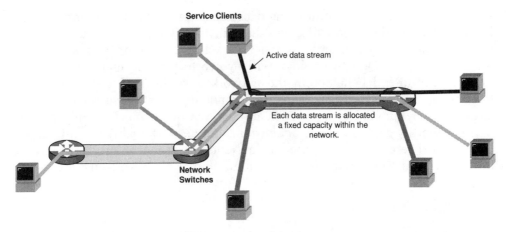

Figure 1.3 Fixed resource allocation for fairness.

during peak usage. When client demand is characterized by strong peak load clustering, it's likely that the network will have to be provisioned to a total capacity level well in excess of average demand. The result is likely to be low average network utilization levels and low operating efficiency.

The direction chosen by the TCP protocol is one of *dynamic sharing* fairness, where the objective is to ensure that the entire resource is fairly shared between all currently active clients. In this environment, the objective is for the actual amount of resource available to any single client to vary in inverse proportion to the number of active clients; any excess is shared equally across the other clients. Here the intent is to operate the network at full capacity all the time, with the total available resources deployed to service clients at all times. Such a resource management scheme is highly efficient in terms of minimizing idle time network overheads, and it is fair, in that every active client can receive the same share of the available resource as all other clients.

In an ideal simple network, there is a direct relationship between the number of active application flows and the total amount of available bandwidth per flow, which can be expressed as:

Available Bandwidth / Number of Flows = Notional Bandwidth Per Flow

In more complex environments, where at any time a number of active flows are operating between a variety of network ingress and egress locations, the resulting resource allocation pattern is very similar to the fluid dynamics problem of establishing an equilibrium state within a set of interacting fluid flows. The outcome is one of approximate fairness, rather than absolute fairness, across all active flows.

This fair resource allocation is normally the outcome of a passive network configuration interacting with active end-host systems. Within the network there exists a simple non-differentiating queuing discipline, operating within each router and interacting with the TCP end-to-end flow control algorithm operating on the sending and receiving hosts. It is the actions of the end systems dynamically adjusting their sending rates to match their view of current network load that implements this dynamically adjusting fair sharing algorithm.

Of course, a system that creates a dynamic equilibrium requires some time to stabilize. The underlying assumption is that each self-adjusting traffic flow is large enough so that it has sufficient time to insert itself into the active flow set, then interact with other active flows to establish a new point of dynamic equilibrium. This view is modified by the trend toward shorter TCP transactions in the Internet, brought about by the small transactions of the World Wide Web. Such small TCP transactions (or *mice*, as they are commonly called) contribute additional noise into the service environment. These transactions are often completed before they manage to equilibrate with the other active TCP sessions. The result is that the longer-held TCP sessions will react in various semi-chaotic ways to this additional load. Some TCP sessions may close down their sending rate and not recover, while others may be able to continue to operate at sustained high transfer rates.

The result of these factors is that while such a mechanism is intended to be fair, it is also unpredictable. Depending on the amount of traffic within the network from other concurrent active applications, the network performance of a particular application will vary considerably. Where the performance profile calls for some form of bounded predictability, fairness is not an adequate service management method.

What is needed for *differentiated* quality of service is *unfairness*; for example, the ability to provide premium treatment for one class of traffic and best effort for another class of traffic. This *unfairness*, if you will, allows the premium traffic to receive a greater level of the network's resources and to achieve a greater level of network performance than that available to the best effort traffic. For this to be achieved, the network must be changed from a simple, passive, uniform system to one that contains active elements that alter the flow characteristics for certain classes of traffic.

We now can begin to formulate a straightforward definition of QoS: any mechanism that provides distinction of traffic types, which can be classified and administered differently throughout the network. This is exactly what is meant by differentiated classes of service (CoS). Should we simply refrain from using the terms *Quality of Service* or *Differential Performance* in favor of *Differentiated Classes of Service*? Perhaps it is a more accurate description of the field; and in this book, we will examine several methods of differentiating traffic classes. That said, common use sets a powerful precedent, and old habits die hard. At this point in time, *Quality of Service* is the accepted term used to describe this broad area of class of service and performance management.

The Current Situation: QoS without the Hype

So what is QoS all about? There are a very small number of basic tools available to modify the network's service response to traffic. To achieve QoS and to create a network that will respond with differentiated performance profiles, the engineering task is to assemble these basic tools in various orders and with various triggers. However, in all cases, the elements of QoS engineering are the same. The basic set of actions used to create different performance behaviors in a packet-switched network are:

Filter. The packet header can be inspected to determine its protocol, source and destination addresses, type of service field, and, if applicable, the application port addresses and the state of the flow of which the packet is a component. This filter information can be used to selectively trigger one or more of the following switch actions.

Drop. The packet is silently discarded by the switch.

Delay. The packet is delayed, commonly by waiting in a queue within the switch.

Forward. The switch makes a forwarding decision as to the next hop switch. This decision can be based on a number of path attributes, and it may have implicit performance implications.

Fragment. Each transmission link may have an associated maximal packet size. When a large packet is forwarded over such a link, the packet may need to be fragmented into a number of segments to fit within the link's characteristics. Fragmentation may also be used to alter performance characteristics of a link. For example, low speed links may be configured with a smaller maximal packet size, in order to avoid head-of-line blocking conditions.

These five basic actions can be combined to form more complex network performance outcomes, some of which are briefly described here:

Simple precedence. Using the precedence bits in the IP header to vary the queuing priority, and hence alter the delay according to the precedence level, up to eight (0 through 7) classes of traffic, can be supported (see Figure 1.4).

Shaped admission control. A network can control its ingress ports by physically limiting the link speed through clocking a circuit at a specific data rate. An alternative approach is to use a token bucket filter to throttle incoming traffic, thereby controlling the drop action of the switch. A token represents the capability to admit a certain amount of data. Tokens are generated into a token bucket at the determined admission rate. When tokens are in the bucket, packets are admitted into the network. If no tokens are in the bucket, arriving packets are discarded (see Figure 1.5).

Differential congestion management. A congestion-management scheme that provides preferential treatment to certain classes of traffic in times of congestion. In such a case, packets belonging to a lower-precedence class have a higher probability of drop when the buffer occupancy passes a certain threshold, so that additional buffer space is only avail-

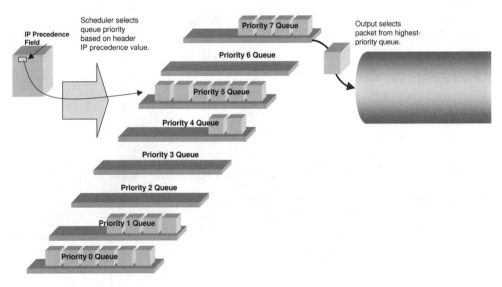

Figure 1.4   Precedence structures.

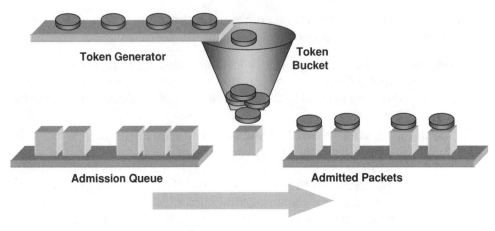

Figure 1.5 Admission control.

able for higher-precedence traffic. This can be combined with weighted queuing algorithms that give greater preference to the higher-precedence traffic. The result of these two measures is one of lower drop probability, larger buffer space, and lower average queue delay for higher-precedence traffic, eventuating in greater network resource allocation to such traffic (see Figure 1.6).

A maxim in the network engineering world says there are no perfect solutions, just different levels of pain. There is much that is true in this statement; it boils down to an exercise of finding the lowest pain threshold of a particular solution.

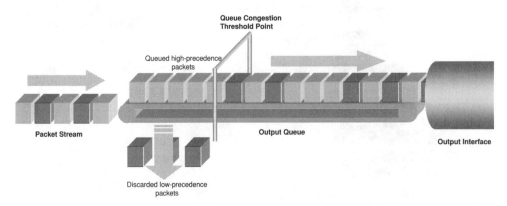

Figure 1.6 Congestion management.

Early QoS Experience

In earlier days of IP networking, the QoS was not an issue in deployed networks; getting packets to their destination was the first and foremost concern—although getting packets to their destinations successfully might be considered a primitive binary form of QoS—the traffic was either transmitted and received successfully, or it wasn't. It turns out that part of the underlying mechanisms of the TCP/IP suite evolved to make the most efficient use of this paradigm. In fact, TCP has matured to behave somewhat gracefully in the face of packet loss, by shrinking its transmission window when packet loss is detected. This congestion avoidance mechanism was pioneered by Van Jacobson at Lawrence Berkeley Laboratory in 1988 [Jacobson 1988].

In a heterogeneous networking environment, a significant and potentially complex addition to the IP protocol is the introduction of a network layer signal of underlying congestion. The modifications to the TCP flow control algorithm circumvented this explicit network layer signaling, and instead used packet loss itself as this congestion indicator; loss of a packet indicated that the network was unable to sustain the service quality at the particular transmission rate. The sender's response to packet loss was to reduce the sending rate, adjusting the rate of data entering the network to ensure that the service quality was sustained.

The original IP design placed a Type of Service (TOS) field in the IP header to allow networks to offer various service profiles. However, no real deployment of either TOS-enabled network services or TOS-enabled host systems or applications was deployed in the initial research phases of the Internet, though some successful experimentation was undertaken by the U.S. National Science Foundation-funded NSFnet, where TOS-based priority queuing was enabled in the routers. In this experiment, traffic generated to support the operation of the routing protocol used elevated TOS priority. The result was that the in-band data exchange required by the routing protocols was not unduly compromised by user-generated traffic congestion events within the network. The overall outcome of this experiment was more robust routing protocol operation, ensuring that any congestion condition was not exacerbated by route oscillation. This effort, however, was not extended to include differential service offerings from the NSFnet to its clients.

---

*T*he NSFnet was a remarkably fertile proving ground for many Internet performance technologies. For more background on the NSFnet, see www.cise.nsf.gov/ncri/nsfnet.html and www.isoc.org/internet-history/.

One reason for lack of further activity in this area of service quality and congestion was that the introduction of these technologies to the network was not seen as critical at the time. Prior to the transition of the NSFnet from the supervision of the National Science Foundation in the early 1990s to the commercial Internet service providers that now make up the national Internet backbones in the United States, congestion management and differentiation of services was not nearly as critical an issue as it is now. The principal interest was simply keeping the traffic flowing, the links up, and the routing system stable.

Not much has changed in that regard in the Internet of the late 1990s, other than that the same basic network integrity problems have worsened significantly. ISPs still primarily focus on solving the problems of basic operational integrity and stability, compounded by issues of more complex interaction policies. Only very recently has interest in QoS gained momentum within the ISP community. A cynic might say that ISPs are only interested in deploying QoS to achieve competitive market differentiation and as a means of generating additional revenue.

Why Is QoS Compelling?

In the commercial Internet environment, QoS can be a mechanism to provide more distinguished service than those offered by competing ISPs. Many clients believe, with some justification, that all ISPs offer the same basic service, namely best-effort Internet packet transmission, and that they are operating in a commodity market where price alone is the major factor used to distinguish among ISP service offerings. Armed with QoS, the ISP has this additional market differentiator, in the form of an agreed service level offered to its clients, which can stand up as a value-added offering over and above the basic commodity of best-effort services.

Many ISPs regard QoS as a means to an end: greater market coverage across various different market segments. An undifferentiated best-effort service offers a uniform service quality to all its clients, and in so doing places itself in a particular market niche of quality and price. Differentiated service networks enable the operator to offer a number of discrete service offerings while continuing to implement the same base service platform, freeing the ISP to offer services to a broader range of market segments. For this reason, QoS is viewed as a valuable service, as well as an additional source of revenue. If the mechanisms exist to provide differentiated CoS levels, so that one customer's traffic can be treated differently from another customer's, it's certainly possible to develop different marketing and tariffing models on which to base these services. The level of granularity of differentiated service can be finer than that of all the traffic associated with a customer, and can extend to applications or even individual instances of an application. This is the basis for a service provider offering premium-priced elevated service levels to quality-critical applications, such as voice and video, while offering other applications a lower-cost, best-effort service. Such an application-based service profile is another instance of broadening the service market base while operating a single-service platform.

By the same logic, there are equally compelling reasons to provide differentiated QoS within corporate, academic, and research networks. Although the competitive and economic incentives might not apply identically in these types of networks, providing different levels of QoS still may be desirable. Imagine a university campus whose network administrator would like to give preferential treatment to professors conducting research on computers that span the diameter of the campus network. It certainly would be ideal to give this type of traffic a higher priority when contending for network resources, while giving secondary resources to other types of best-effort traffic such as student e-mail.

Perhaps the greatest driver for QoS today is the lure of voice-over-IP (VoIP). Telephony has matured for over a century in an environment that historically has been dominated by monopoly service providers. In many markets, this monopoly position allowed the price of telephony to rise well above the basic cost of provision of the service, making it a highly lucrative activity. Deregulation of the voice markets—allowing the entry of competitive telephone service providers—did not substantially alter this financial picture, and the voice market has remained a major revenue source for communications enterprises around the world.

However, deregulation of the communications sector in many countries also made it possible for the ISP sector to enter the voice market. Many ISPs recognize the financial opportunities in providing voice services in addition to Internet data services. If the ISP network's performance can be managed so that it can pass voice data streams across the network with a quality comparable to the telephone service, the ISP can enter this market with a competitive service offering. For ISPs, the lure of voice is the potential to resell their installed network capacity as voice-over-IP for a significantly higher financial return than that provided by simple best-effort Internet services. Of course, the critical assumption here is that the QoS aspects of supporting voice data flows inside the IP network can be managed such that the quality of the voice conversation using an IP network is no different from using a conventional telephone service.

Performance and Over-Engineering

One tried-and-true method of providing ample network resources to every user and subscriber is to *over-engineer* the network. This implies that there will never be more aggregate traffic demand than the network can accommodate; thus, there will be no resulting congestion. The objective of over-engineering is to provision the capacity of the interior of the network such that it can readily manage the traffic levels imposed from receiving traffic at the network boundary.

The general rule of thumb with this method is to ensure that at any location the provisioned internal capacity is greater than the sum of capacity of all the network's ingress services. For example, where a network access point is provisioned with 60 ISDN B channels for customer access, the sum of the ingress capacity for these access ports is 4Mbps. Over-engineering would call for an egress capacity in excess of 4Mbps for this access point. A very simple network design that illustrates this principle is shown in Figure 1.7.

Of course, in larger networks, it's not practical to believe that this can always be accomplished; but, believe it or not, this is how many Internet networks have traditionally been engineered, and indeed such over-engineering is the hallmark of many carrier networks as well. Over-engineering is not, however, always economically feasible. Coupled with usage-insensitive pricing strategies, over-engineering would appear to contradict one of the fundamental rules of resource management economics: The demand for goods

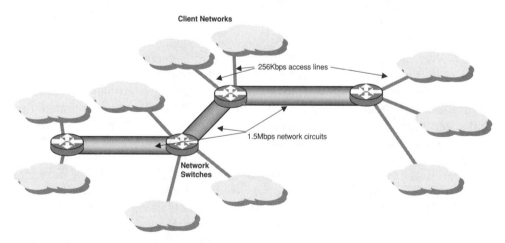

Figure 1.7 Over-engineered network design.

supplied without marginal cost will always exceed the capability to supply. But, some say, there can be a strong balance between demand and supply for Internet carriage capacity, and over-engineering coupled with both usage-sensitive and usage-insensitive pricing methods may be a feasible engineering approach in certain situations [Odlyzko 1998].

In networks such as the Internet, demand is highly variable, and over-engineering margins can prove difficult to calculate. A sudden change in traffic volume or flow patterns can dramatically affect the performance of the overall network and erode any engineering margins. Also, the behavior of TCP as an adaptive flow-management protocol can undermine such margins. As additional network capacity is installed, TCP sessions will expand into the available network capacity. In this case, over-engineering is not simply a case of multiplexing constant rate sessions by examining the time and duration patterns of the sessions, but of factoring in adaptive rate behavior within the long-held TCP sessions. If the network is being over-engineered so that the network experiences no congestion load whatsoever, this will lead to the assumption of relatively high margins of over-supply across the entire network. The result will be one of high levels of idle network capacity, which in turn leads to inefficient network utilization. Such inefficiencies impact the network's overall cost efficiency. If network's unit costs of packet carriage are a critical factor for the network operator—and this is the case in many situations—then over-engineering is a luxury, unrealistic in today's competitive commercial network services market.

Over-engineering itself is a somewhat fuzzy term, as data traffic itself does not conform to well-defined demand models. Viewed at a fine enough level of detail, traffic on a network resembles a fractal pattern of spikes, and these spikes fill all available capacity [Leland 1993]. From that perspective, there will always be some level of resource contention within any shared data network. This observation tends to negate the basic proposition that it is possible to over-engineer to the extent that there is absolutely no level of resource contention within the shared network infrastructure. The response to the theory that it is technically and economically feasible to install sufficient network infrastructure to preclude resource contention is that in an environment of a very large number of adaptive data systems, demand will always exceed capacity at some time. Where demand exceeds capacity, some form of demand management is necessary: Either the end-systems work adaptively to establish a point of equilibrium of demand (the best-effort response) or there is an imposed resource management policy that selectively delivers resources to particular network uses (a managed QoS response).

Why Hasn't QoS Been Deployed?

The most straightforward answer to this question is that, in general, ISPs have been dealing with more critical operational challenges. Simply sustaining growth of the network to match the growth in demand is perhaps the common primary focus of all ISPs today. Most of the larger ISPs are simply peddling as fast as they can just to meet the growing demand for bandwidth. In the past, they have not been concerned with looking at ways to introduce QoS differentiation in a stable fashion. There are two reasons for this: First, to date, QoS tools have been somewhat coarse and primitive, and their implementation on high-speed links has often had a negative impact on packet-forwarding performance. The complex queue-manipulation and packet-reordering algorithms that implement queuing-based QoS differentiation have, until recently, been a liability at higher-link speeds. Second, reliable traffic load management tools have not been available to ensure that one class of traffic actually attains a higher performance outcome over other classes of traffic, nor are there adequate management tools to measure the outcomes. If providers cannot adequately demonstrate to their customers that differentiated traffic is being treated with a quantifiable priority, and if the customers cannot independently verify these outcomes from such differentiation, the QoS implementation is effectively valueless as a commercial offering.

The accounting and billing issues raised by QoS are also a significant blocking factor to its deployment. Many Internet service operations use very simple and uniform accounting systems that reduce the service tariff to factors of time and access capacity. In such accounting systems there is no metering of the precise nature of the use of the network; individual packets are passed through the network without metering of packet payload volume, distance traversed, propagation delay, imposed jitter, or level of queuing response. QoS implies some level of preemptive reservation and use of network resources, and with that preemption comes an associated tariff premium. The client base expects to pay a tariff premium for just the additional resource demands of traffic passed through the network at a premium service level. For the service provider, such an accounting and billing system can quickly become highly complex. In the extreme, each switch generates a steady stream of reporting records that detail QoS actions at the packet level; an accounting server attempts to relate a set of such records to a packet's path through a network, then map the accounting of each packet's service to an individual client.

The complexity of these issues also impacts on interprovider interconnection structures, where the addition of premium traffic across an interconnection requires reconsideration of the financial basis of interconnection. If a network service provider accepts a premium service packet from an interconnected network, it will be looking for some form of financial compensation from the interconnected party if it is to honor the premium service request.

There are other reasons why, historically, QoS has not been deployed in corporate and campus networks. Within the campus network, some level of over-engineering has been a sustainable design approach in numerous cases. In such environments (private wiring systems), bandwidth is relatively cheap, so it's sometimes is a trivial matter to increase the bandwidth between network segments when the network spans only several floors in the same building or several buildings on the same campus. For these locally contained private networks, vendors have already created solutions for increasing the switching capacity at the local switching hubs. In environments of low-delay high and very high bandwidth, traffic-engineering tools simply may not be the right answer, as it is often preferable to simply throw more bandwidth at local network congestion problems.

This does not imply a universal condemnation of QoS by the service provider industry. The solutions to providing superior performance to a subset of network clients is either to deploy QoS mechanisms within the network or over-engineer the entire network to offer access at the same high-performance levels to all clients. Within a well-defined scope of deployment, over-engineering can be a cost-effective alternative to QoS structures. But this is an increasingly artificial environment. As noted, in price-sensitive competitive Internet service markets, over-engineering is an economically prohibitive luxury. For as long as growth of the network dominates most service provider networks, QoS deployment will be a secondary option to simply scaling the network at a speed equal to the growth in demand. It is reasonable to predict that, at some stage, the market will be saturated and growth in demand will taper off. In such an environment, the search for cost efficiencies and market differentiators will probably turn industry attention to QoS deployment as the basic performance-tuning mechanism. (Market growth trends and QoS deployment are graphed in Figure 1.8.) That said, even in a high growth environment there are still some important factors to consider in relation to QoS.

Scaling Pressures and Performance

The QoS problem is evident when trying to efficiently use wide area links, which traditionally have a much lower bandwidth and much higher cost of access than the one-time capital cost of much of the campus infrastructure.

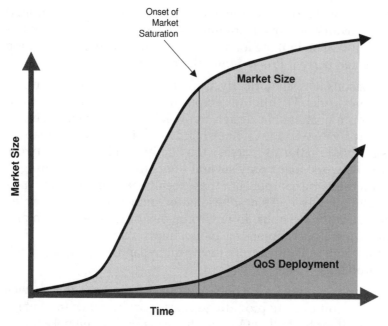

Figure 1.8 Market growth and QoS deployment.

This can be translated to a unit cost, such as a cost per byte, where the total cost of operation of the network is divided by the sustainable traffic volume. In terms of such unit costs, the wide area links have significantly higher unit costs compared to local network infrastructure. This higher unit cost then imposes a higher value on traffic carried across wide area networks (WANs), which may in turn lead to differential performance requirements being placed on the network. In a corporate network with geographically diverse locations connected by a wide area network, it may be imperative to give some mission-critical traffic higher priority than other types of application traffic.

Several years ago, it was not uncommon for the average local area network (LAN) to consist of a 10Mbps (megabits per second) Ethernet segment, or perhaps even several of them, interconnected by a router or shared hub. During the same time frame, it was also not uncommon for the average WAN link to consist of a 56Kbps circuit that connected geographically diverse locations. Given the disparity in bandwidth, only a very small amount of LAN traffic could be accommodated on the WAN. If the WAN link was consistently congested and the performance proved to be unacceptable, an organization would most likely get a faster link. Unfortunately, it was also not uncommon for this to occur on a regular basis. If the organization obtained a T1 circuit (approximately 1.5Mbps), application traffic increased to consume this link as well, and the process would be repeated.

This paradigm of aggressive consumption of available WAN resources follows a corollary to Moore's Law:

> As you increase the capacity of any system to accommodate user demand, user demand will increase to consume system capacity.

More simply, this is also known as a "vicious cycle."

Moore's Law In 1965, Intel co-founder Gordon Moore predicted that transistor density on microprocessors would double every two years. Moore was preparing a speech when he made this memorable observation. When he started to graph data to illustrate the growth in memory-chip performance, he realized a striking trend. Each new chip contained roughly twice as much capacity as its predecessor, and each chip was released within 18 to 24 months of the preceding chip. If this trend continued, he reasoned, computing power would rise exponentially over relatively brief periods. Moore's observation, now known as Moore's Law, described a trend that has continued and that is still remarkably accurate. In fact, it is the basis for many network planners' performance forecasts.

Demand for Fiber

Recent years have witnessed large Internet service providers become caught in the following cycle: They ramp up their infrastructures to add bandwidth, and customer traffic consumes it; they add more bandwidth, and traffic consumes that, too. On and on. Two technology factors have been fueling this cycle for some years. First, there has been an ample pool of available bandwidth. The telephone industry was the original consumer of carriage bandwidth, and telephony is generally engineered using significant margins of over-engineering. It is this excess carriage capacity from the telephony environment that has been deployed to fuel the rapid growth of the Internet. Second, a number of technology refinements have been affecting the carriage capacity of fiber itself, with the introduction of optical amplifiers and *wave division multiplexing* (WDM) techniques. These refinements have resulted in almost a thousandfold increase of the bandwidth of a fiber cable, from some hundreds of megabits per second to speeds of up to 600 gigabits per second.

Interestingly, fiber capacity itself is not a near-term bottleneck. Fiber-optic cable has three 200-nanometer-wide transmission windows at wavelengths of 850, 1300, and 1500 nanometers, and the sum of these cable-transmission windows could, in theory, be used to transmit between 50 and 75 terabits per second [Partridge 1994]. The electronics at either end of the fiber can be considered as a capacity bottleneck. Currently deployed transmission technology uses 2.5Gbps (gigabits per second) transmission systems on each fiber-optic cable, with some recent cable systems now supporting a 10Gbps channel. The deployment of wave division multiplexing (WDM) has lifted the total per-fiber transmission rates to between 100Gbps and 600Gbps per cable.

This situation very much depends on the availability of fiber in specific geographic areas. Some areas of the planet, and, more specifically, some local exchange carriers within certain areas, seem to be better prepared to deal with the growing demand for fiber. In some areas of North America, for example, there may be a six-month lead time for an exorbitantly expensive DS3/T3 (45Mbps) circuit, whereas in another area, an OC-48c (2.4Gb) fiber loop may be relatively inexpensive and can be delivered in a matter of days.

Traffic Engineering

Bandwidth may be provided either by increasing the data clocking rate of a circuit, or by increasing the number of available circuits in the network. Internet networks are constructed using stateless destination-based address datagram switching systems. This switching paradigm scales very effectively when the available bandwidth is augmented through increases in the data clock rate. In these situations, the underlying switching paradigm is effectively unaltered. When network capacity is provided through the provisioning of additional discrete circuits, scaling issues must be addressed within the design of the network.

The point is that IP networks require careful engineering if they are to scale adequately. If faster pipes are not readily available, network operators can do only one of two things: oversubscribe the existing capacity or run parallel links to compensate for the bandwidth demand. After a network begins to insert parallel internal paths in response to a lack of availability of faster links, efficiency and stability issues begin to creep into the engineering design. In a connection-based network, where each application connection can be forged over a different path of underlying circuits, parallel circuits can be readily integrated into the network in response to scaling pressures. In a packet-based datagram network, the tools to manage performance over parallel circuits involve a combination of switch architectures, routing protocol configuration, and even the introduction of connection-oriented elements with technologies such as Multi-Protocol Label Switching (MPLS) [Davie 1998]. This aspect of performance engineering, directed primarily at achieving an efficiently loaded network, is commonly termed *traffic engineering*.

Availability of Delivery Devices

There is a disparity in the availability of devices that can handle these higher speeds. A couple of years ago, it was common for a high-speed circuit to consist of a T3 point-to-point circuit between two routers in the Internet backbone. As bandwidth demands grew, a paradigm shift began to occur, whereby ATM was the only technology that could deliver OC3 (155Mbps) speeds, and router platforms could not accommodate these speeds. After router vendors began to supply OC3 ATM interfaces for their platforms, the need for bandwidth increased yet again to OC12 (622Mbps). This shift could seesaw for the foreseeable future or another paradigm shift could occur to level the playing field—it is difficult to predict (see Figure 1.9).

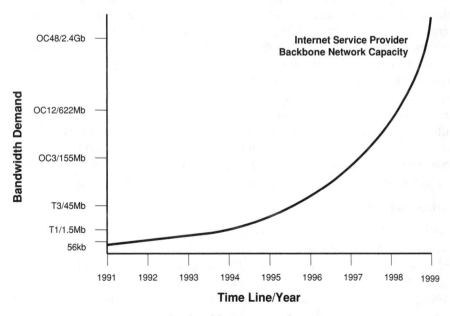

Figure 1.9 Service provider backbone growth curve.

Changing Traffic Patterns

It is also difficult to gauge how traffic patterns will change over time. Paradigm shifts happen unexpectedly. For example, until recently, network administrators engineered WAN networks under the 80/20 rule (80 percent of the network traffic stays local, within the campus LAN; only 20 percent of local traffic is destined for other points beyond the local WAN gateway). Given this rule of thumb, network administrators usually could be reasonably safe designing a network if they kept an ear to the ground and an eye on WAN-link utilization. But remember, paradigm shifts happen without warning. The Internet "gold rush" of the mid-1990s revealed that the 80/20 rule no longer held true; a larger percentage of traffic was destined for points beyond the LAN, largely as a result of the overwhelming uptake of the distance-insensitive point-and-click model used by Web browsers. Some parts of the world are now reporting a complete reversal of traffic dispersion along a 25/75 pattern, whereby 75 percent of all delivered traffic not only originates nonlocally, but is imported from an international source.

If you combine this paradigm shift of traffic with the fact that LAN media speeds have increased a hundredfold in recent years, and that a further ten-fold increase in these speeds is right around the corner, the result is more pressure on the WAN links to operate at speeds in excess of the LAN links—a complete reversal of more traditional network design guidelines. The capability of the long-distance carriage systems to cost-effectively deliver on such requirements is questionable. Although WAN link speeds have also increased by a hundred over the same period, with an even greater decrease in cost, long-distance carriage circuits have not uniformly undergone the same degree of change. Long-distance circuits of the appropriate capacity are sometimes economically prohibitive and may simply be unavailable in certain locations. Obviously, a WAN link will be congested if the aggregate LAN traffic destined for points beyond the gateway is greater than the WAN bandwidth capacity. It becomes increasingly apparent that the LAN-to-WAN aggregation points have become a performance bottleneck and that disparity in LAN and WAN link speeds still exists.

Congestion Indicators

Other methods of connecting to network services also have changed dramatically. Prior to the phenomenal growth and interest in the Internet in the early to mid-1990s, providing dial-up services was generally a method to accommodate off-site staff members who needed to connect to network services. There was no need for a huge commodity dial-up infrastructure. However, this changed dramatically as the general public discovered the Internet, the number of ISPs offering dial-up access to the Internet skyrocketed. Of course, some interesting problems were introduced as ISPs tried to capitalize on the opportunity to make money and connect more subscribers to the Internet. But when the subscribers they had connected vastly outnumbered what they had provisioned for network capacity, the ISPs soon had to face customers complaining bitterly about network performance. Many network operators learned the hard way that you can't put 10 pounds of traffic in a 5-pound pipe.

The development and deployment of the World Wide Web fueled the fire. As more content sites on the Internet began to appear, people everywhere clamored to get connected. This trend has continued to accelerate considerably over the last few years, to the point at which services once unimaginable now are offered as a matter of course—indeed are expected—on the Web. The original Web browser, Mosaic, which was developed at the National Center for Supercomputing Applications (NCSA), was the initial implementation of the "killer app," which has since driven the hunger for network bandwidth over the past few years.

Corporate networks have also been affected by this trend, but with a slightly different twist. Telecommuting—employees dialing into the corporate network while working from home or at other off-site locations—became very popular during this period; it, too, continues to gain favor and practice. Corporate networks also reaped the value of the Web by deploying *intranets*—internal Web servers that enable easy employee access to company information, documents, policies, and other corporate assets.

The magic performance bullet that network administrators now are seeking has been targeted to this gray area called QoS, in hopes of efficiently using network resources, especially where the demand for resources could overwhelm the bandwidth supply. Congestion-control mechanisms are increasingly important in heavily used or over-subscribed networks, and QoS is being placed in the role of mediator, between demand and supply.

QoS Characteristics

Engineering a network for performance involves using a number of tools and techniques. Ideally, the available tools would be compatible with each other, so that any combination would function in a predictable fashion, with the outcome calculable as the sum of each tool's actions. Unfortunately, the real-world engineer's toolkit for performance includes, metaphorically speaking, both chainsaws and the more delicate instruments of microsurgery. Needless to say, the result of using such a diversity of tools and techniques in various combinations can be simply disastrous!

Thus, before you can deliver QoS, you must first define the methods for differentiating traffic into certain classes. You must then provide mechanisms for determining the way traffic classes will be handled, thereby defining service outcomes for each traffic class. Achieving such outcomes is where preferential queuing and congestion-control mechanisms can play a role.

To understand the process of differentiated services, consider the airline industry's method of categorizing passenger seating. The manifest for a typical commercial passenger aircraft lists those passengers who have purchased "premium" seating in the first-class cabin and those who have purchased "coach" tickets—lower-priced, no-frills seating. The aircraft has only a certain number of seats—a finite bandwidth, if you will—and the differentiator is the quality of seating and service.

Similarly, network administrators are looking for mechanisms to identify specific traffic types that can be treated as first class and as coach. Of course, this is a simplistic analogy, but for the most part, it is accurate. And, additional service classes may exist as well, such as second class, third class, and so forth. Over-booking the aircraft results in passengers being "bumped" and having to take another flight; this is analogous to over-subscribing a portion of the network. In terms of networking, this means that an element's traffic is dropped when congestion is introduced.

Quality of service encompasses all these factors: differentiation of CoS, admission control, non-FIFO and preferential queuing, and congestion management. It also may encompass other characteristics, such as deterministic path selection based on lowest latency, jitter control, and other end-to-end path measurements.

Differentiated Classes of Service

Whether they realize it or not, when most people refer to quality of service, they're actually referring to differentiated classes of service, coupled with perhaps a few additional mechanisms that institute traffic policing, admission control, and administration. Differentiation is the operative word here, because before you can provide a higher quality of service to a particular customer, application, or protocol, you must classify the traffic into classes, then determine how to handle the various traffic classes as traffic moves throughout the network. This brings up several important concepts.

When differentiation is performed, it invokes some unique criteria to put incoming traffic into classes. Each packet is distinguished by the classification mechanisms at the network ingress point, as well as farther along in the network topology. The differentiation can be done using a variety of methods, but some of the more common consist of identifying and classifying traffic by:

Protocol. Network and transport protocols such as IP, TCP, UDP, IPX, and so on.

Source protocol port. Application-specific protocols such as Telnet, IPX SAPs, and so on, dependent on their source host address.

Destination protocol port. Application-specific protocols such as Telnet, IPX SAPs, and so on, dependent on their destination host address.

Source host address. Protocol-specific host address indicating the originator of the traffic.

Destination host address. Protocol-specific host address indicating the destination of the traffic.

Source device interface. Interface at which the traffic entered a particular device, otherwise known as an *ingress interface*.

Flow. A combination of the source and destination host address, as well as the source and destination port.

Service Mark. In this case, the Type of Service (TOS) field of the packet header is the identifying field.

Differentiation usually is performed as traffic enters the network, ensuring that the traffic is subject to performance management mechanisms that will force it to conform with whatever user-defined policy is desired. This entry-point enforcement is called *active policing* (see Figure 1.10). It is normally

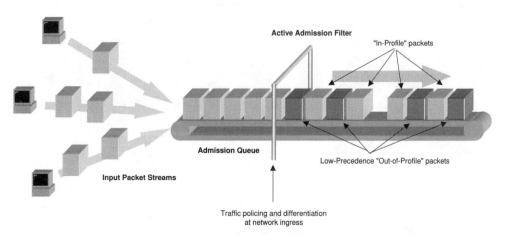

Figure 1.10 Differentiation and active policing.

performed at the network entry points to ensure that traffic presented to the network conforms to the network's performance policies. Enforcement of another sort is also conducted at intermediate nodes along the transit path; however, instead of resource-intensive enforcement, such as various admission-control schemes, it is a process of administering what already has been classified, so that it can be queued, forwarded, or otherwise handled accordingly. This is a less resource-intensive practice.

The alternative approach to active differentiation is known as *passive admission,* during which the traffic is accepted as-is from a downstream subscriber and carried across the network (see Figure 1.11). The distinction is made between doing initial differentiation through the use of selective policing and allowing differentiation to be done outside the administrative domain of your network. In other words, you would allow downstream connections outside your administrative domain to define their own traffic service policies and simply let this traffic enter the service network without further modification.

But there are certain risks inherent in providing a passive-admission service. For example, downstream subscribers may try to mark their traffic at a higher priority than they are entitled, and subsequently consume more network resources than was intended in the performance management framework. The marking of the traffic may take different forms; it may be marked as RSVP reservation requests or by setting IP precedence to indicate relative queuing priority; or it may be a Frame Relay DE (discard-eligible) bit or an ATM CLP (cell loss priority) bit set to indicate the precedence of the traffic in

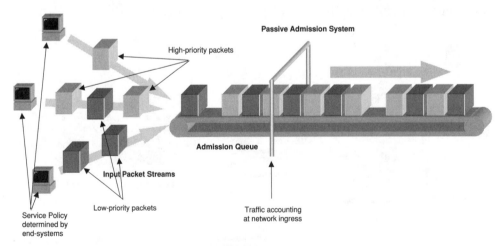

Figure 1.11 Differentiation and passive admission.

a congestion situation. The level difficulty in policing traffic at the edges of a network depends on how and at which layer of the protocol stack it is done.

Based on the preceding criteria, a network edge device can take several courses of action after the traffic identification and classification has been accomplished. One of the simplest is to queue the various classes of traffic differently to provide diverse servicing classes. Several other choices are available, including selectively forwarding specific classes of traffic along different paths in the network. You could do this by using traditional (or nontraditional) packet-forwarding schemes or by mapping traffic classes to specific Layer 2 (Frame Relay or ATM) switched paths in the network cloud. A variation of mapping traffic classes to multiple Layer 2 switched paths is to also provide different traffic-shaping schemes or congestion thresholds for each virtual circuit in the end-to-end path. Probably, there are more ways to provide differentiated service through the network than are outlined here, but the important aspect of this discussion is that the concept of identification, classification, and marking of the traffic is fundamental to providing differentiated CoS support. Without these basic building blocks, chances are that any effort to provide any level of QoS will not result in the desired behavior in the network.

Layer 2 Switching Integration

Another chink in the armor of QoS deployment momentum is the integration of Layer 2 switching into new portions of the network topology (see Figure 1.12). With the introduction of Frame Relay and ATM switching in the Internet carriage environment, isolating network faults and providing adequate network management have become more difficult. Layer 2 carriage switching is unique in a world accustomed to point-to-point circuits, and sometimes it is considered a mixed blessing. Switched data systems, such as Frame Relay and ATM, are treated as a godsend by some, because they provide the flexibility of provisioning virtual circuits (VCs) between any two endpoints in the switched network, regardless of the number of Layer 2 switches in the end-to-end path. Also, both Frame Relay and ATM help reduce the amount of money that must be spent on a monthly basis for local loop circuit charges, because many VCs can be provisioned on a single physical circuit. In the past, this would have required several point-to-point circuits to achieve the same connectivity requirements (consider a one-to-one relationship between a point-to-point circuit and a virtual circuit) and, subsequently, would have incurred additional monthly circuit charges.

One of the more unfortunate aspects of both ATM and Frame Relay is that although both provide a great deal of flexibility, they also make it very easy to

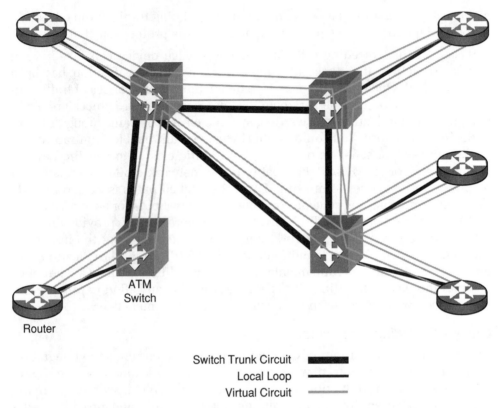

Figure 1.12 Hybrid switching networks.

construct networks with poor service performance. By implementing virtual circuits, the network designer can configure logical circuits that may bear no relationship to the underlying physical topology of the network. The consequence is that a chosen virtual circuit topology may appear to be well designed in terms of interconnecting major traffic points with minimal hop counts, but, in fact, the actual traffic loads imposed on the physical network may be highly inefficient, with packets performing double, or even triple, traversals of the same physical cable when moving through a sequence of virtual circuits. In very large networks, it is unwise to configure all Layer 3 devices (routers) only one virtual circuit hop away from one another (a fully meshed virtual circuit logical topology). Depending on the routing protocol being used, as the network grows larger, this may introduce instability into the routing system, because of the excessive computational overhead required to maintain a large number of adjacencies. These are instances of the so-called *more rope* principle, whereby the increased number and flexibility of features can lead to feature abuse, rather than improve the quality of the service. Simplicity of design features often leads to equally simple and highly robust service delivery systems.

Problems of this nature are due to the *ships in the night* paradigm that occurs when devices in the lower layers of the Open Systems Interconnection (OSI) reference model stack (in this case, the link layer) also provide buffering and queuing, and in some cases, routing intelligence (for example, PNNI). This can cause performance and stability problems as the IP network layer also performs buffering and queuing and supports IP level routing. Where there are a number of separate traffic management systems operating at different protocol levels, their interaction often introduces chaotic elements into the overall traffic flow pattern. The performance problems arise due to the end system's protocol stack attempting to sense the prevailing network load and adjust its behavior to match this sensed load. The interaction of the multiple traffic management systems tend to confuse the end system's load sensing algorithms by creating a level of introduced chaotic fluctuation in the network's traffic flows.

Performance Management

One of the major activities required to deliver QoS comes under the broad heading of performance management. Managing the performance of a network is one of the greater challenges of network management. It is necessary to define performance measurements that have some relevance and bearing to the network and its services, then match such service-oriented measurements to the data collected from the various managed network elements.

The generic tasks within this area of management fall into four areas:

- Data collection
- Analysis of data for current performance levels and trends
- Setting performance thresholds
- Network planning and engineering

The objective is to help the network manager reduce levels of network congestion and service degradation, and provide a consistent base level of performance to all users. By implementing performance management utilities, the network operator can monitor the service performance of the network's transmission, switching, and service delivery platforms. Performance data also can use statistical analysis to project longer-term trends for the capacity planning process.

Data collection. The primary source of data for performance management is polling network components for current performance measurements. Typically, this involves polling the network's routers using the Simple Network Management Protocol (SNMP), gathering data on link occupancy and average queue length at frequent intervals.

Data analysis. This activity requires a thorough understanding of the system being examined, in order to undertake a meaningful analysis of the data. Internet networks present some challenges in this area. Unlike a telephony network, in which each call takes a fixed amount of resources from the network, an Internet network has a variable traffic load from each end-to-end transaction, where the per-transaction load can adapt to the total network load. The majority of traffic, in many cases, up to 90 percent of the total traffic, is carried within the TCP. The self-adjusting mode of operation of TCP will confound simple traffic load analysis, so that more sophisticated demand and traffic flow models will be required to undertake this analysis.

At what point is a network link operating at maximum efficiency? The answer is not 100 percent utilization, but it is not far from this. In an ideal world, the buffer would absorb most of the speed variations that arise from the slow start of new sessions and congestion avoidance in the steady state. The buffer queue depth oscillates quite markedly between a full queue with packet overflow and an empty queue. Regardless of the number of concurrent TCP sessions across the circuit, a single-circuit network can have an average utilization of close to 100 percent when there is any form of network activity. Of course, real networks do not reproduce this theoretical efficiency of loading. A number of factors tend to reduce this point of attainable efficiency, including diversity of paths, a mix of UDP and TCP traffic, and a mix of long- and short-duration TCP flows.

Setting performance thresholds. To manage performance levels, it is necessary to add a further step in the data collection and analysis process: the addition of threshold levels on the analyzed data to indicate bounds of acceptable performance for the network. For example, a performance threshold may be phrased in terms of *link peak load,* such that a link is considered saturated when load exceeds 85 percent of available capacity for more than 10 percent of the time within a week-long analysis window. A threshold also may be phrased in terms of *overall average load*, which will be more helpful in the business cost analysis. A threshold may be that a link is saturated when the average weekly line occupancy in the direction of the heaviest traffic load exceeds 55 percent of total capacity. However, if a slightly degraded peak performance were considered acceptable, a higher-performance threshold might be used as a mechanism to decrease the average unit cost of traffic passed through the circuit. The reason that unit cost decreases with more traffic is that the cost of the circuit is assumed to be fixed for a certain capacity. The more traffic that can be passed across this circuit for a fixed price, the lower the unit cost of data transmission for a fixed unit of data. Likewise, lower thresholds also can be applied to produce an outcome of superior performance characteristics, at a price of a higher unit cost of traffic. Further analysis of the situation can provide a more generic relationship between overall performance quality and the unit cost of data transmission, indicated in a general fashion in Figure 1.13.

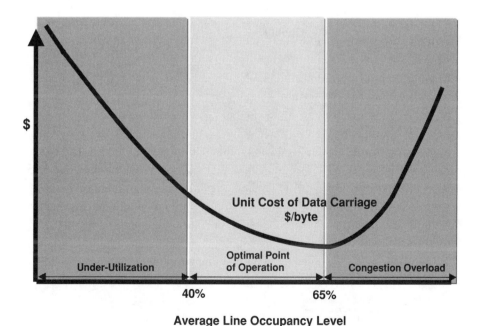

Figure 1.13 Relationship between line load and unit cost.

The real question is who should set the general performance threshold for an ISP network? The answer is that within the bounds of technically acceptable thresholds, the decision is a marketing or service management matter, not an engineering problem. The chosen performance target has a direct bearing on transmission costs, which in turn has a direct bearing on the price position of the ISP service product. In the ISP marketplace, a range of price and performance positioning is visible. In general, higher pricing is associated with higher service performance, while lower-priced services will operate within a lower performance range, as shown in Figure 1.14. Exactly where each ISP positions itself within the market is a business, not a technical, decision. When a range of acceptable performance target values has been determined as potential outcome of network engineering, the decision to use a particular set of performance target values becomes a managerial decision.

Network planning and engineering. Certainly, it's good to know when a network is operating with poor levels of service performance, but, obviously, it's better to use this information to do something to fix the situation. However, this is a reactive mode of operation, where the network operator monitors the performance indicators and deploys addi-

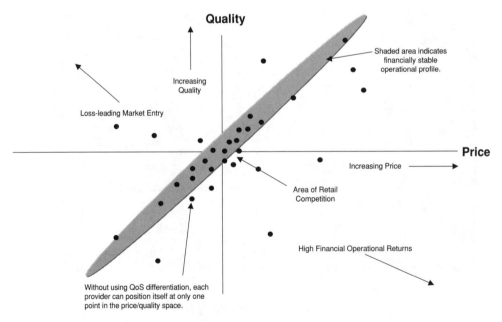

Figure 1.14 Price performance relationship.

tional network capacity when the indicators fall below some threshold level. In such cases, the service performance provided by the network may cycle dramatically between good and poor. The better response is to use the data to predict when service performance levels are likely to be at risk, and deploy additional capacity in advance of that time. This activity involves projecting demand levels and service performance, identifying critical dates and locations where capacity will be placed under stress, and formulating responses to such forecast events. The planned responses may involve engineering changes to the routing system to alter traffic flows within the network, or it may involve the deployment of additional capacity to address forecast loads. The overall strategic goal of this activity is to ensure that the network performance remains relatively stable at the target service levels, while accommodating various levels of growth and change in the network demand characteristics.

Critical to an ISP operation in performance management is that the ISP is forced to operate at a single point in the price/performance space. This is not an optimal position within a competitive market, in that competitors can enter the market at different price performance points. The typical competitive response to such situations has been to reduce revenue levels and offer high-performance services to all market segments. While such a strategy does protect market share, it also seriously erodes the value of the operation, forcing the service provider into areas of low, if any, return on operational expenditure.

The preferred position for providers is where they can offer services at a number of distinct price performance points, so that they can also offer a performance- and price-differentiated product to each distinct market segment, while maintaining the value or revenue potential of a particular performance point. The promise offered by QoS is the ability to effectively segment a single-service network into a number of discrete performance levels, and individually price access at each level (see Figure 1.15). Whether this is a realistic expectation remains unclear, and many of the issues underlying the operation of such a multilevel managed performance network are discussed in detail in later chapters.

Insurmountable Complexity?

One point of view is that most of the methods of implementing a QoS scheme are too complex, and that until the issues surrounding QoS are simplified, a

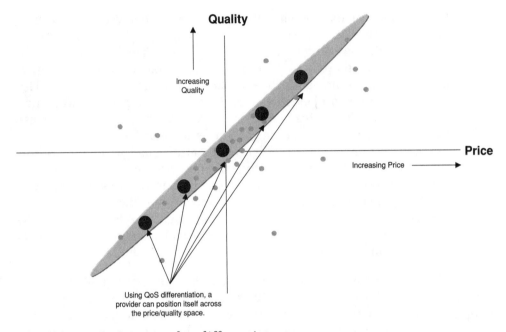

Figure 1.15 QoS as a market differentiator.

majority of network engineering professionals won't even consider deployment of the more complicated mechanisms currently available.

Without question, there is considerable tension between the objectives of
various players, including the Internet Engineering Task Force (IETF) (the
Internet standards body), the vendor community, the Internet service provider
industry sector, the corporate network services sector, and, of course, the end
consumer of network-based services.

A number of the concerns expressed by network engineers are valid in
terms of the technical complexity, stability, and scalability of the proposed
technical standards. However, the complexity of older protocols and other
technical documentation pales in comparison to that of the documents
recently produced in the area of the RSVP protocol and Integrated Services
technology. Therefore, it is understandable why there are those who consider
some of this technology to be overwhelming and abstract in its relevance to
the real world. Within the IETF community, this complexity led to the consideration of simpler mechanisms; and, specifically, the QoS area now has
technical specifications for service delivery from both the Integrated Services

The IETF *The IETF is an international group of network engineers, protocol designers, researchers, and other technical geeks. One view of the IETF's role is as developer of the technologies and protocols that keep the Internet running. Importantly, the IETF is the body that standardizes Internet technologies and protocols, using a methodology that focuses on producing functioning and relevant standards for the Internet. An important distinction of the IETF Standards Process is the omission of any form of formal voting, and a strong emphasis on achieving consensus in defining the most suitable technology. This has been encapsulated by the often quoted phrase "rough consensus and running code." The majority of the work produced by the various working groups (and individuals) within the IETF is documented in Internet Drafts (I-Ds). After the I-Ds are refined in final documents, they generally are advanced and published as Requests for Comments (RFCs). Advancement to an Internet Standard requires further steps, involving the demonstration that the proposed technology can be used to construct robust interoperable implementations. Standardization also requires some form of demonstration that the technology is relevant and is used in real networks. Virtually all the core networking protocols used in the Internet are published as standards-track RFCs. For more information on the IETF, see www.ietf.org.*

Working Group and the Differentiated Services Working Group. QoS is also not being introduced into an eagerly receptive environment. The need for differentiated services within Internet service networks has been questioned; even within the groups working in this area there is a body of opinion that simpler methods are available to achieve similar performance management outcomes.

On the other hand, there is a good reason why these topics are complex: The technology itself is complex. These service performance management technologies are difficult to envision, difficult to develop, difficult to implement, and difficult to manage. And just as with any new technology, it will take some time to evolve and mature. As our understanding of this technology matures, some level of clarity will emerge, to separate those aspects of this space that are crucial from those that are simply insignificant.

Quality of service is an ambiguous term with multiple interpretations, many of which have emerged in response to the uniqueness of the problems facing the networking community, which in turn is faced with managing infra-structure and providing services within its organizations. QoS generally is used to describe a network providing preferential treatment to certain types of traffic. But such descriptions are too vague to adequately describe the precise mechanisms used to provide these types of services or even what it is these services actually provide. By contrast, it may be more appropriate to use the term differentiated classes of service (CoS). At least CoS implies that traffic types can be distributed into separate classes, each of which can be treated differently as it travels through the network. Of course, several methods are available for delivering differentiated CoS—and as many opinions as to which approach is best, which provides predictable behavior, which achieves the desired goal, or which is more elegant than another.

The first issue a network administrator must tackle, however, is to under-stand the problem to be solved. This is paramount in determining the appro-priate solution. There are almost as many reasons for providing and integrating performance management support into the network as there are methods of doing so, among them, capacity management and value-added services. On the other hand, understanding the benefits and liabilities of each approach is equally important. Implementing a particular technology to achieve service differentiation without completely understanding the limitations of that implementation can yield unpredictable and perhaps disastrous results.

References

[Cerf 1974] *A Protocol for Packet Network Interconnection,* V.G. Cerf and R.F. Kahn, *IEEE Transactions on Communications*, Col. COM-22, no. 5, 1974.

The original paper describing the TCP/IP protocol.

[Davie 1998] *Switching in IP Networks,* B. Davie, B. Doolan, and Y. Rekhter, Morgan Kaufmann Publishers, 1998.

A book describing the background to IP switching technologies, as well as the various IP switching technologies currently available. It describes Toshiba's Cell Switching Router, Ipsilon's IP Switching, Cisco's Tag Switching, and IBM's Aggregate Route-Based IP Switching, as well as the IETF standardization activity of Multiprotocol Label Switching.

[Jacobson 1988] *Congestion Avoidance and Control*, V. Jacobson. *Computer Communication Review*, vol. 18, no. 4, August 1988.

A landmark paper on end-system flow control mechanisms for TCP, intended to avoid network congestion collapse.

[Leland 1993] *On the Self-Similar Nature of Ethernet Traffic*, W. Leland, W. Willinger, M. Taqqu, and D. Wilson. ACM SIGComm '93, San Francisco, 1993. A copy of this paper is available online at www.sobco.com/e.132/files/sigcomm93.ps.

It is one of a number of studies on the nature of data traffic. The paper shows a self-similar traffic pattern that has a strong fractal nature, where bursting is visible over any chosen timebase. This traffic pattern has serious implications for the design, control, and analysis of high-speed data networks.

[Odlyzko 1998] *The Economics of the Internet: Utility, Utilization, Pricing, and Quality of Service*, A. M. Odlyzko. www.research.att.com/~amo, 1998.

A paper arguing the point of view that over-provisioning data networks is a viable and economically sustainable response to the demands for service quality within data networks, and that such a response is technically and economically superior to implementing QoS responses within the network.

[Partridge 1994] *Gigabit Networking*, C. Partridge, Addison-Wesley, 1994.

This book surveys the issues and available technologies for very high-speed data networks.

RFCs

Request for Comments documents (RFCs) are documents published by the RFC Editor. They are available online at www.rfc.editor.org.

[RFC791] *Internet Protocol, DARPA Internet Program Protocol Specification*, J. Postel, ed., RFC791, Standard RFC, September 1981.

The specification of version 4 of the Internet Protocol.

[RFC792] *Internet Control Message Protocol, DARPA Internet Program Protocol Specification*, J. Postel, ed., RFC792, Standard RFC, September 1981.

The specification of the Internet Control Message Protocol component of version 4 of the Internet Protocol.

[RFC793] *Transmission Control Protocol, DARPA Internet Program Protocol Specification*, J. Postel, ed., RFC793, Standard RFC, September 1981.

The specification of the Transmission Control Protocol component of version 4 of the Internet Protocol.

[RFC2460] *Internet Protocol, Version 6 (IPv6) Specification,* S. Deering, R. Hinden, RFC2460, Draft Standard RFC, December 1998.

The specification of version 6 of the Internet Protocol.

chapter two

The Performance Toolkit

*Complex systems are usually constructed from a small set
of very simple actions.*

Performance engineering steps are similar to using children's building blocks. The engineering task involves taking a small set of very basic performance building blocks and locking them together in various ways to produce larger-scale complex outcomes. The performance of networks, and the capability for a network manager to impose various performance behaviors within the network, relies on the complex manipulation of a relatively small set of tools. In this chapter, we will look at these basic tools and assemble a toolkit of basic performance characteristics.

In many engineering disciplines, complex behaviors are constructed using a simple and limited set of components, requiring an equally simple set of actions. The same holds for Internet performance engineering, where very subtle and powerful network behaviors are the result of the application of a set of basic actions. This chapter addresses these basic building blocks of performance engineering by examining each of the components of a network from the perspective of performance management.

We will examine the three basic components of an Internet network, taking a top-down approach, starting at the protocol and drilling down through the routers to the underlying transmission systems. First up is the Internet Protocol (IP) itself, both IP versions 4 and 6. Our objective is to expose what is contained within each protocol, and describe the detailed components that have a direct bearing on performance. In a similar vein, we will look at the function of a router, and describe its four basic actions in terms of impact on

Acknowledgment is made to *The ISP Survival Guide*, by Geoff Huston, published by John Wiley & Sons, 1999, for material relating to the description of router functions and transmission systems.

network performance. Last, we will turn to transmission systems, looking at the various systems in use today and their potential performance role. In each case, we will highlight those aspects of the technology that have a direct bearing on the performance of individual network applications and those aspects that allow active differentiation of network performance.

Performance Terminology

To define the actions of various network components and their overall impact on performance, it is necessary to define commonly used terminology. But note, this section is necessarily a brief overview, as this is an area that abounds in terminology, with associated detailed definitions; it is beyond the scope of this chapter to treat terminology comprehensively. For a more thorough look at the framework that underlies the definitions of performance metrics, refer to RFC2330.

The following are the major terms, along with an informal definition of each.

Delay (or Latency). A measurement of time, specifically, the period of time it takes a data packet to traverse from one point in the network to another. Delay generally refers to the time required for a packet to be passed from the sender to the receiver, in which case the measurement is that of *one-way delay* [ID-ippm-delay]. This is a relatively simple metric in concept, but quite challenging to measure, requiring precise clock synchronization of both the sender and receiver. Delay is an attribute of most network components, and is the cumulative result of propagation delay in the transmission systems, and packet assembly, scheduling, switching, output buffering delay in the switching systems within the network path. Reliable data transfer protocols use the concept of a *round trip* to support their operation, where the receiver sends a signal stream back to the sender, indicating which parts of the data have been successfully received. The cumulative delay of a signal loop, where the sender sends a block of data then awaits the corresponding acknowledgment of successful reception, is referred to as the *round trip time* (RTT). Many performance attributes are affected by delay; indeed, one school of thought attributes all performance characteristics to the basic element of delay. In any case, delay is a critical attribute of a system, and one that is carefully managed in all network components.

Jitter. The variation of delay over a period of time. Within packet-switched networks that use variable-sized packets are a number of network components that operate with variable delay. Packet assembly delay will vary for different-sized packets; the varying load on the switch may induce different scheduling delay times; and competition for the output interface will cause buffer delay. Transmission systems may also exhibit variable propagation delay, potentially in the case of switched transmission systems, such as Frame Relay, X.25, or ATM. There are a number of ways to measure jitter, including the maximal variance of delay measurements, the deviation from the average delay, or as the first-order derivative of the delay curve. This latter attribute is referred to as *instantaneous packet delay variation* (IPDV) [ID-ippm-ipdv]. Each measurement describes an aspect of delay variance that is relevant to performance and quality management.

Loss. An attribute of packet networks commonly expressed in a metric as a probability value. A network is not a constant load-bearing system; the traffic passed into the network will vary over time, and will at times exceed the capacity of some component of the network. At such times, the network component must shed some proportion of the load. It does so by discarding packets. Packet loss is typically expressed as a probability value, indicating the average probability that an individual packet will be discarded by the network. As we will see when we examine the behavior of the TCP protocol, some small level of packet loss within a network is to be anticipated as normal network operational behavior. As always, however, too much of a good thing can be a disaster; and excessive loss is symptomatic of a critical problem, because if the offered load is consistently greater than the network's capacity, the average loss level will rise dramatically. The probability of the loss of a single packet is referred to as *one-way packet loss*, and refers to the loss probability for a packet traveling in a single direction through the network [ID-ippm-loss]. In terms of reliable transfer protocol behavior, the round-trip loss probability is also of relevance to performance, as the loss of an acknowledgment signal is, in general terms, equivalent in impact to loss of a data packet.

Bandwidth. An attribute generally used to refer to clocked transmission systems. The term bandwidth refers to the data-clocking rate of the system, typically expressed in bits per second (bps). This is a physical level attribute of the transmission system. An associated transmission attribute is the *bit error rate* (ber), typically expressed as a fraction, indicating the average ratio of errored bits to correctly transmitted bits.

Throughput. A more generic term used to describe the capability of a system to transfer data. The most useful form of this metric is a measurement of the reliable data transfer rate, which includes the performance of the transmission systems, the switching systems, the host systems, as well as the protocol itself. The instance of data throughput that includes consideration of the protocol as well as the state of the network is also termed *bulk transfer capacity* [ID-ippm-btc].

Efficiency. A generic term that refers to a measure of productivity. In terms of network performance, efficiency refers to the level of overhead required to support a data transfer. This can be a volumetric measurement, calculating the amount of protocol and transmission level data that is wrapped around the original payload data, or a throughput measurement, comparing the available bandwidth for a data flow to the achieved throughput of data. The general trade-off that occurs in performance engineering is one of efficiency versus robustness. Additional flow state information passed between sender and receiver allows the sender to rapidly recover from various error conditions; but in doing so, an efficiency penalty is imposed in carrying this additional signaling information within the protocol.

With this brief terminology review complete, we are now in a position to examine the Internet Protocol, highlighting those aspects that have a direct bearing on performance and quality.

The Internet Protocol and Performance

The performance of a network application is dependant on three major factors: the state of the systems supporting the application, the state of the network, and the protocol used to support the application's data transfer. In this chapter we will review the architecture of the Internet Protocol (IP).

The Internet Protocol

The overall architectural philosophy of IP can best be expressed by observing that it combines the use of adaptive end systems and a stateless network switching paradigm; or, to phrase it more succinctly, IP assumes smart hosts and a dumb switching network. This generic description includes the design concept of a distributed communications network that does not attempt to impose a restricted or fixed data flow model upon the network's application set. Instead, the IP protocol defines a simple and well-defined network switching task and allows individual implementations to undertake this task as efficiently as possible. More complex application transaction models can be constructed above and beyond this simple network model by adding functionality to the end-to-end, host-based protocol, rather than by attempting to add functionality into the network model itself.

The intended behavior of the host-based protocol is one that can adapt its behavior to the perceived state of the network at each point in time, using intelligence derived from the state of the end-to-end traffic flow to derive the current network state. In this way, the host protocol stacks can continually modify their flow behavior to match the dynamic capabilities of the network. This model of adapting to prevailing conditions is like a road traffic system: When traffic on the road is light, individual cars travel faster; and as the road starts to fill with traffic, individual cars reduce their speeds. This host-controlled adaptive model of the distributed network allows every host to behave cooperatively in an effort to optimize the transmission-throughput efficiency of the network. This model results in an optimal fair share of transmission capacity being allocated to each active host. But no network-based traffic control exists within the Internet. Instead, the network is managed by the dynamic behavior of the host systems on the periphery of the network.

This architectural view of network performance, one described as an adaptive interaction between the host and the network, was first presented in a paper titled *End-to-End Arguments in Systems Design*, by Saltzer, Reed, and Clark, [Saltzer 1984]. This philosophy, often termed the *end-to-end* architecture, was championed by Clark during his tenure on the Internet Architecture Board (IAB) [Clark 1988]. The end-to-end architecture advocated that decisions regarding the interaction between the host and the network should, as far as is possible, be delegated to the host, rather than attempt to supplant or duplicate host functionality within the network. The implication of this architecture is that functions such as connection establishment and termination, traffic flow control, error detection, and retransmission of lost packets, should be host functions rather than network operations. To achieve this, hosts should be able to insert control signals into the data stream, and listen for corresponding signals from the remote host both to support integrity of the data transfer and to ensure that optimal data flow rates are maintained.

The intention with IP is to support an end-to-end communications architecture capable of spanning a diverse sequence of individual networks [RFC1958]. Given a heterogeneous collection of network media, including various LANs and WANs, and the requirement to transparently communicate end-to-end across various sequences of such media, it is necessary to support *gateways* (or *routers*, as they are normally called) between each network segment. These gateways strip off the network-dependent encapsulation, then apply a forwarding decision to the packet, based on the packet's IP destination address. This action generates a network-to-network hop, so that the packet must be encapsulated once more for the next network transit. This generic function is indicated in Figure 2.1. The function of end-to-end sequence control and data transfer reliability is designed into the end-to-end control algorithm, as is flow control. The overall transmission design criteria are to make an absolutely minimal set of assumptions regarding the capability of each component network in the end-to-end path.

The basic assumption is that the network undertakes best-effort delivery of data packets, without any associated performance guarantee. This lack of any guarantee implies that there is no implicit indication from the network to the sender or receiver when a packet is delayed or discarded. Likewise, there is no network-generated signal directed to the sender to indicate successful delivery to the receiver. It is within these parameters that any network component may elect to silently discard the packet as a response to local congestion. The router model is also simplified by the deliberate omission of *state* in the router design. With the omission of state there is no concept of a virtual circuit, and

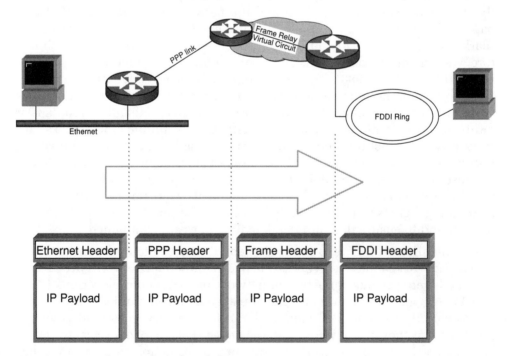

Figure 2.1 End-to-end packet delivery.

the resultant use of locally significant circuit labels for controlling switching, as is found in connection-oriented packet switched protocols. A stateless system uses only a global addressing level, where each packet contains the globally significant address of the packet's destination. Each router requires sufficient knowledge of the global address state in order to undertake a consistent local forwarding decision. In addition, routers perform neither flow control nor retransmission functions, as the stateless mode of operation assumes that once a packet is forwarded, no further action is required from the router in relation to the packet.

Internet architecture uses four core protocols: IP, UDP, TCP, and ICMP. Their relationship is shown in Figure 2.2. In terms of a fully functional protocol stack, a number of functional elements are missing here, namely the lower data link layer protocols and the upper-layer application interface protocols. This omission was deliberate. The relationship of these component protocols into the general Open Systems Interconnection (OSI) networking model is shown in Figure 2.3.

IP

"The internet protocol is specifically limited in scope to provide the functions necessary to deliver a package of bits (an Internet datagram) from a source to a destination over an interconnected system of networks. There are no mechanisms to augment end-to-end data reliability, flow control, sequencing, or other services commonly found in host-to-host protocols." [RFC791].

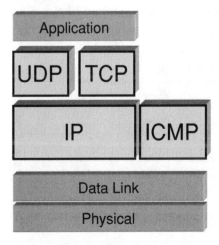

Figure 2.2 Internet layering.

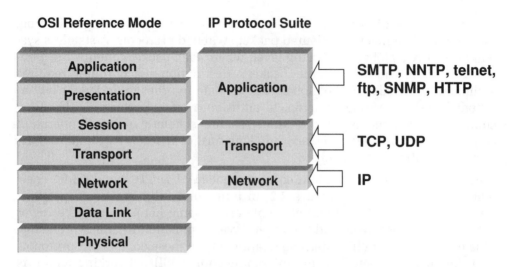

Figure 2.3 The OSI Reference model and the Internet Protocol suite.

IP is an encapsulation method that surrounds its payload, using a header structure prefixed to the payload (see Figure 2.4). It is a datagram-based protocol, in which individual packets are passed through the IP network on a best-effort basis. No guarantee is made that individual packets will arrive at the destination, nor that a sequence of packets will arrive in the same order as they were sent. An IP packet includes a single level of addressing for the destination address, using an address that has global significance. In the delivery of a datagram, all routers along the path process each datagram individually, with local forwarding decisions based on this unique destination address. The

Version	IHL	Type of Service	Total Length	
Identification			Flags	Fragment Offset
Time To Live		Protocol	Header Checksum	
Source Address				
Destination Address				
Options				Padding

Figure 2.4 An IP datagram.

router is not required to maintain a record of how previous packets were processed nor to maintain a completely fixed view of the path of any given destination. In this way, IP makes minimal demands on the underlying network structure in terms of its functionality.

There are a number of fields in the IP packet header that are relevant to performance and quality of service, namely Type of Service (TOS), Fragmentation Control, Protocol Identifier, and the Source and Destination address fields.

Type of Service

The TOS field is an 8-bit field intended to enable the packet to specify the form of service it is requesting from the network. The original semantics for this field were specified in RFC791, and were intended to indicate the level of service requested from the network. This field can be broken down into two components, a precedence value and a service type, as shown in Figure 2.5.

The 3-bit precedence subfield is used to indicate the queuing function within routers. The IP specification defines eight values and their suggested use, as shown in Figure 2.6. The implication of the RFC791 specification is that the precedence field defines an ordered set of service levels, where a higher precedence value indicates an elevated claim for network resources. The specification does not precisely describe the characteristics of the elevated precedence. The precedence value can be interpreted as an inverse of a drop preference value or as a queue delay avoidance preference.

The three service type bits were intended to be used as a single-level indicator of service selection. The objective of the field was to request that the packet be passed through the network with minimal delay, maximal through-

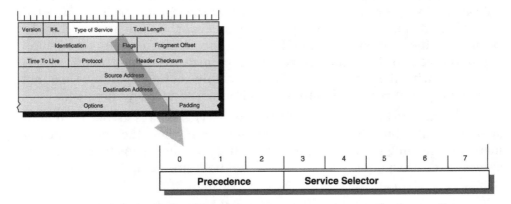

Figure 2.5 IP Type of Service field selector.

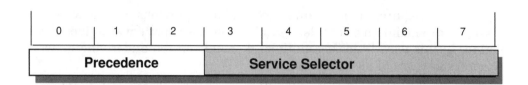

Precedence Field:

111 - Network Control
110 - Internetwork Control
101 - Critical functions
100 - Flash Override
011 - Flash
010 - Immediate
001 - Priority
000 - Routine

Figure 2.6 IP Type of Service precedence field.

put, or maximal transmission reliability. The specification noted that requesting elevated performance within one of these parameters might adversely affect performance in another parameter; it also noted that setting all three bits simultaneously was not an anticipated occurrence.

The specification did not define how such elevated service response could be achieved for each parameter. A network could use this service field as a path metric selector, in conjunction with a routing protocol that supported multiple route metrics. Multiple path selection capability in a routing protocol implies the maintenance of multiple routing topologies, with each topology optimized so that it minimizes (or maximizes) the path metric that is associated with the service type. However, this routing technique has not been deployed in large-scale production Internet service environments.

Another way to think about the IP Type of Service field is to treat each indicator as a single-bit selector of a switching option, where the delay indicator maps to a queuing precedence, the reliability indicator maps to a drop preference, and throughput is a combination of both precedence and drop preference.

The original specification of the TOS field was subsequently refined in RFC1349; it is shown in Figure 2.7. The changes in this revision included the definition of the service selectors as a 4-bit field. Three values were retained, matching the single-bit selectors as defined in RFC791; a fourth value was added to indicate minimization of monetary cost. While the 4-bit field has 16 possible values, RFC1349 defines only 5 values, as shown in Figure 2.7. The result is that only one of the four service types can be selected by any packet. The specification also indicates that these selectors should trigger some form of premium response from the network. Its intention is that the selection of TOS indicators in the packet header should not result in a lower quality of service than that available if the TOS field were clear (all zeros). The specification also stated that the TOS field in the IP datagram was primarily intended to act as a path selector, where the selection of a service type would imply the router's use of a route topology based on optimizing the associated path metric. The specification does note that an alternative approach to supporting the TOS field setting may involve a modification to the default queuing behavior within the router.

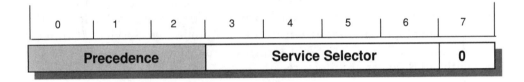

Service Selector Field:

    Bits 3–6:

        1 0 0 0   Minimize delay
        0 1 0 0   Maximize throughput
        0 0 1 0   Maximize reliability
        0 0 0 1   Minimize monetary cost
        0 0 0 0   Normal best-effort service

    bit 7:

    0   Reserved for future use

Figure 2.7 IP Type of Service selector field (RFC1349).

The default configuration of most operational routers is to ignore the service selector field when switching a packet. As a result, this field, and the concept of providing differentiated service, is simply ignored over much of the Internet today. Although the use of this field may appear to be a relatively simple issue, it has proved not to be so.

To date, the common deployment of these values and their implicit intention of supporting service differentiation have gained little acceptance. The effort of RFC791 and RFC1349, and the related Internet host and router requirements documents RFC1122, RFC1123, and RFC1812 was intended to define a standard interpretation for this service selector field. In practice, this has not happened, and this field continues to have no effective networkwide significance. As a result, these 8 bits of the IP header can be regarded as significant only within the context of a local network; they have no end-to-end significance.

If we accept that that the service selector field will not have a uniform interpretation, then the field can be defined as a per-ISP local service selector, used to trigger various locally defined performance responses. We can expand with this approach by standardizing a number of performance behaviors, as an alternative to attempting to rigidly define both the specific bit settings and their interpretation. This task of standardizing performance behaviors is the subject of developmental effort within the Differentiated Services (DiffServ) Working Group of the IETF. We'll return to this field when we look at the Differentiated Services architecture.

TOTAL PACKET LENGTH AND FRAGMENTATION CONTROL

Total Packet Length is a 16-bit field that specifies the total length of the packet, including both the header and the payload. The value is in units of octets, and the field size of 16 bits allows the maximum IP packet size to be 65,535 octets.

Remember that when the IP protocol was designed in the mid-1970s, transmission speeds of 56Kbps were considered state of the art. At this speed, a maximally sized 64Kbyte packet would take slightly more than nine seconds to enter the circuit. This maximum packet size value probably was intended to address potential high-speed networks of the future, as it certainly was an over-engineered value at the time. At 100Mbps, a maximally sized packet would take five milliseconds to enter the circuit, which is a far more reasonable time span. It is an interesting observation that the designers of the IP protocol envisaged the potential of thousandfold increases in network capacity within the lifetime of the protocol.

All hosts and routers must pass packets up to 576 octets in length without needing to invoke fragmentation. Packets of a larger size may be fragmented by the host or the router.

The Packet Identification, Fragmentation Flags, and Fragment Offset fields comprise a 32-bit header segment used to control packet fragmentation. Within the design of the IP protocol, every network has a maximum packet size, or maximum transmission unit (MTU), and these sizes vary. The MTU is an outcome of the network's design and is a product of the network's bandwidth, maximal diameter, and desired imposed jitter. Because the path selection is dynamic, the sender is not fully aware of the sequence of MTUs supported on the end-to-end sequence of networks. To reduce the risk of sending oversized packets, the IP specification allows a router to fragment an oversized packet into smaller units that match the MTU of the next network hop.

Fragmentation may occur more than once within an end-to-end transit path, and an already fragmented packet may be further fragmented without change to the fragmentation functionality. To accommodate this, and to allow some degree of transit efficiency, no reassembly is attempted within the network. Fragments may take different paths through the network, and the various fragments may not necessarily reassemble at any subsequent intermediate location within the network. This leads to the fragmentation condition that fragmentation of an IP packet is irreversible within the network. Once fragmented, a packet is reassembled only at the final destination.

Three fields are used in fragmentation control: Packet Identifier, Fragmentation Flags, and Fragmentation Offset Value.

Packet Identifier. A 16-bit value used to identify all the fragments of a packet, allowing the destination host to perform packet reassembly. Note that the packet identifier value cannot be reused while fragments of a previous incarnation of this identifier value remain within the network. For low- to medium-speed networks, this constraint is not a problem, but at gigabit speeds, the wraparound limitation may prove to be a significant limitation. When using maximum-sized packets, fragmentation identifier wraparound occurs every 32 gigabits. When using the more common 576-byte packet size, the fragmentation identifier wraparound is every 256 megabits. As traffic flow speeds increase, minimum packet sizes will need to increase, while at the same time fragmentation capability may have to be dropped as an available option for IP routers, because the combination of the Packet Identifier field and packet size acts as a limit on the amount of data held in transit within the network.

Fragmentation Flags. The three-bit Flag field has the first bit flag reserved. The second bit flag is the Don't Fragment flag. When a router attempts to fragment an IP packet with this flag set, no fragmentation occurs, and an ICMP error message is sent back to the sender to inform of the delivery error, and the packet is discarded. The third bit flag is the More Fragments flag. When a packet is fragmented, all packets except the final fragment have the More Fragments field set. The fragmentation algorithm operates such that only the final fragment of the original IP packet has this field clear (set to zero). Even when a fragment is further fragmented, this rule remains in force.

Fragmentation Offset Value. This 13-bit value counts the offset of the start of this fragment from the start of the original packet. The unit used by this counter is *octawords*, implying that fragmentation must align to 64-bit boundaries.

The fields altered by fragmentation are shown in Figure 2.8, where, a 1300-byte IP packet has been fragmented into two 532-byte packets and one 276-byte packet. The IP packet length has been altered to reflect the fragment size,

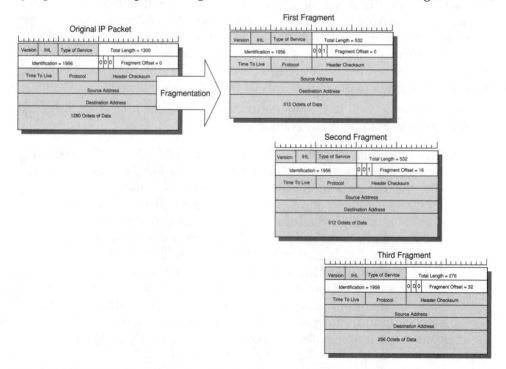

Figure 2.8 Fragmentation of an IP packet.

and the Fragmentation Offset Value field has been set to 0, 16, and 32 respectively. The final fragment has the More Fragments flag cleared to show that it is the final fragment of the original packet.

Fragmentation is not completely transparent in terms of performance. If one fragment is lost within the network, the entire packet is discarded by the receiver, because retransmitting just the lost fragment is not a viable option for the protocol (the sender has no knowledge that the packet was fragmented, and the router that performed the fragmentation maintains no copy of the original packet). This retransmission of a full packet upon the loss of a single fragment has serious performance implications. Rather than rely on IP packet fragmentation to adapt to various per-hop maximum packet sizes, TCP can use a path MTU discovery algorithm [RFC1191] to allow the TCP session to proceed without the use of fragmentation.

PACKET CLASSIFICATION: PROTOCOL IDENTIFIER, SOURCE, AND DESTINATION ADDRESSES

These fields can be used by a router profile filter to identify a sequence of packets that form part of a traffic flow between two hosts, or for traffic originating from or destined to a particular host. Such information may be used to trigger an administratively assigned performance filter.

The Protocol Identifier field is used to identify the protocol carried within the Internet packet. Currently assigned values are maintained by the Internet Assigned Numbers Authority (IANA) (available at www.iana.org).

The Source and Destination fields are 32-bit values that are the addresses of the sender and intended recipient of the IP packet. One of the more remarkable aspects of the original design is the use of such large fields for the address, and the intention to use globally significant address values, rather than locally significant addresses. Remember that when using low-speed transmission lines, every bit of protocol overhead impacts the data transmission efficiency. Other protocols in use at the time IP was designed used 8-bit address values; or, as a bold initiative, included 16-bit values, ascribing them local corporate or campus network significance. The adoption of an address space encompassing 4 billion addressable entities, when the original requirement of connectivity appeared to be in the hundreds or low thousands, was either an act of inspired genius or outlandish optimism. Either guess could be correct.

IP OPTIONS

Base IP packet headers are 20 octets long and contain no other fields. However, the protocol design does allow for the addition of a variable number of options to the IP packet. These options are appended to the base IP header and precede any IP data payload. In general, IP options are rarely used these days, and, with two exceptions, the set of defined IP options are effectively obsolete due to this infrequent use. The loose source-routing and strict source-routing options are used in many operational networks as a valuable debugging tool; they allow the sender of the packet to override the normal forwarding path selected by the network. However, a large number of client networks disable these options at the boundary to the larger Internet, due to the security risks such source-routed packets can pose.

Normally, IP options are implemented within a significantly slower processing path within the router, not when application performance is a critical requirement.

ICMP

Within a datagram protocol architecture, failure to deliver a datagram, or *packet loss*, is not in itself an error condition. Some losses are the result of local congestion, where a part of the network is overloaded and packets are discarded. Some level of packet loss is normal in IP networks, to the extent the upper-level transport protocol relies on packet loss as a signal of local congestion. TCP, as we will see in a following section, uses packet loss as a rate control mechanism. Some losses are the result of errors in the packet header, where the packet is undeliverable by the network under any circumstances. In certain circumstances, the IP architecture makes some provision to notify the sender of this error condition, although not every loss event triggers a notification. The Internet Control Message Protocol (ICMP) is used for loss and error notification.

ICMP is layered above IP and uses IP as its network layer protocol. The packet format is shown in Figure 2.9. If ICMP messages are generated in response to certain IP packets that encounter an error condition, and if ICMP packets are themselves IP packets, what is to stop a recursive loop of ICMP messages from saturating the network? ICMP messages are not generated in specific circumstances. They are not generated as a result of processing an ICMP error message, or in response to an IP datagram addressed to or from a broadcast, or multicast address, or in response to a noninitial IP fragment. In terms of Internet performance, one type of ICMP error message merits further consideration, the ICMP Source Quench message.

ICMP Source Quench messages may be generated by a router when a packet is discarded due to local buffer congestion. The packet is directed to the source address of the discarded packet, and is intended to signal the sender to reduce its sending rate. As a flow-control method, ICMP Source Quench has garnered neither wide acclaim nor widespread acceptance. The reason is because this method of flow management is very indirect and often ineffectual, as it

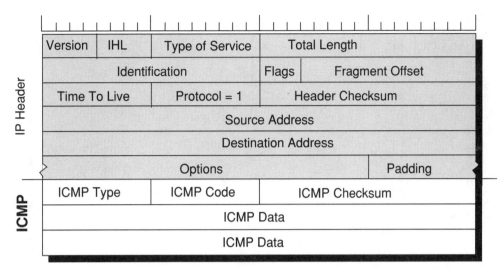

Figure 2.9 ICMP packet format.

increases the amount of traffic at the point in time when traffic already has reached overload conditions. In addition, nothing ensures that the ICMP Source Quench will be sent to the host system that is sending at the highest data rate; nor is any undertaking provided that the sending system will respond to the ICMP Source Quench message. For this reason, use of Source Quench is not encouraged in the Router Requirements specification [RFC1812].

There are mechanisms that offer greater control over congestion events than the ICMP Source Quench message. Queue management practices, coupled with end-to-end flow control, are very effective methods of ensuring that the router does not remain in a state of chronic buffer space exhaustion.

UDP

UDP is an application protocol layered above IP that offers end-to-end functionality between a pair of host systems. The format of a UDP packet is shown in Figure 2.10.

The end-to-end address used by UDP includes *source* and *destination ports* that are 16-bit values. The 96-bit, 4-tuple of:

```
(source address, source port, destination address, destination port)
```

is used by the destination system as an index to identify the correct local process to receive the datagram.

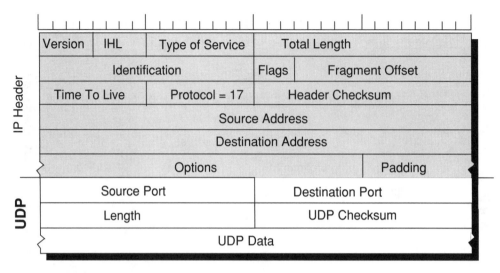

Figure 2.10 UDP packet format.

UDP is effectively a null protocol. It offers port addressing and header checksums, but no other functionality other than that provided by IP itself. Within the network, a sequence of UDP packets may be reordered or lost. No guarantee is made for the reliability of data transfer or for the preservation of order on an end-to-end basis; nor are there any control mechanisms relating to flow control and performance.

UDP has been used to support the Trivial File Transfer Protocol (TFTP), Domain Name System (DNS) query protocols across the wide area network, and network file systems across local networks. UDP also has been used to support authentication applications in which a remote authentication server is queried by local agents. In this context, the UDP attributes are: efficiency in minimal protocol transmission overhead, and no overhead in establishing, maintaining, and then terminating a synchronized connection state. Where the probability of packet loss is relatively low, and the data to be transferred is minimal (in that it fits within a single IP packet), then UDP is a highly efficient protocol. The efficiency is gained through the elimination of the additional data exchange required to support a synchronized state. UDP is often used for real-time streaming applications, such as voice or video transport protocols, and the Network Time Protocol (NTP), where the ability to time packet transmission against an external clock source is of paramount importance to the application.

UDP is a nonadaptive protocol, in that it will not react to local congestion within the network. UDP can exacerbate local congestion conditions by continuing to send into a congestion point. For networks where UDP is a small percentage of total traffic volumes, this may not be a serious performance issue. But where the proportion of nonadaptive UDP traffic becomes significant, such as more than 50 percent of the total traffic volume, the consequence is that the network will not recover quickly from local congestion conditions. Additional traffic control mechanisms, such as admission filters, may need to be deployed where UDP is a very significant proportion of total network load.

TCP

TCP is the embodiment of reliable end-to-end transmission functionality within the overall Internet architecture [RFC793]. All the functionality required to take a simple base of datagram delivery and impose upon this an end-to-end signaling model that implements reliability, sequencing, flow control, and streaming is embedded within TCP. TCP is a bilateral duplex protocol; it allows both parties to send and receive data within the context of the single TCP connection. Rather than impose a state within the network to support the connection, TCP uses synchronized state between the two end points. Much of the protocol design ensures that each local state transition is communicated to, and acknowledged by, the remote party. TCP is also a rate-adaptive protocol, in that the rate of data transfer is intended to adapt to the load conditions within the network and to the available capacity along the data path. There is no predetermined TCP data transfer rate; if the network has available capacity, a TCP sender will attempt to inject more data into the network to take up the available space. Conversely, if there is local congestion, a TCP sender will reduce its sending rate to allow the network to recover.

Like UDP, TCP provides a communication channel between processes on each host system. The channel is reliable, serialized, and full-duplex. The stream of octets passed to the TCP driver at one end of the connection will be transmitted across the network so that the stream is presented to the remote process as the same sequence of octets, in the same order as that generated by the sender. However, TCP is a true streaming protocol, and application-level network operations are not transparent. Some protocols explicitly encapsulate each application transaction; for every *write*, there must be a matching *read*. In this manner, the application-derived segmentation of the data stream into a logical record structure is preserved across the network. TCP explicitly does not preserve such an implicit structure imposed on the data stream, so that there is no explicit pairing between write and read operations within the network protocol. For example, a TCP application may write three data blocks in sequence into the network connection, which may be collected by the remote reader in a single read operation. This is called a *streaming protocol*. The size of the data blocks (segments) used in a TCP session is negotiated at the start of the session. The sender attempts to use the largest segment size it can for the data transfer, within the constraints of the receiver's maximum segment size and the configured sender's maximum segment size.

The only stream formatting permitted within TCP is the concept of *urgent data,* whereby the sender can mark the end of a data segment that the application wants to bring to the attention of the receiver. The TCP segment that carries the final byte of the urgent data segment can mark this data point; the TCP receiving process has the responsibility to pass this mark to the receiving application.

TCP allows data to be passed in both directions simultaneously. The TCP connection is identified by the hosts at both ends by a 96-bit, 4-tuple of:

```
(source address, source port, destination address, destination port)
```

in which the port identifiers are 16-bit values. Note that the same source port number can be associated with a number of distinct connections, because a connection includes the local and remote address fields, rather than only the port identifiers.

The abstraction used here is that of a *socket*, which acts as the interface between the process and the network. The socket is bound to a TCP port address. TCP uses the mechanism of a synchronized state between two sockets. The process expects to direct the socket to *open* a connection to the remote socket, and following a success indication from the open operation, to *send* and *receive* data blocks and query the *status* of the socket. When the transaction is complete, the process must be able to *close* the connection or, in the event of an abnormal condition, to *abort* the connection (see Figure 2.11).

The TCP header structure (shown in Figure 2.12) uses a pair of 16-bit source and destination *port addresses*. The next field is a 32-bit sending Sequence Number, which identifies the sequence number of the first data octet in this packet. The sequence number is not an absolute value; the selection of an initial sequence value is critical, as the initial sequence prevents the misinterpretation of delayed data from an old connection being incorrectly interpreted as being valid within a current connection. The sequence field is necessary to ensure that arriving packets can be reordered in the sender's original order. This field is also used within the flow control structure, as it allows each side of the connection to estimate the current round-trip time across the network.

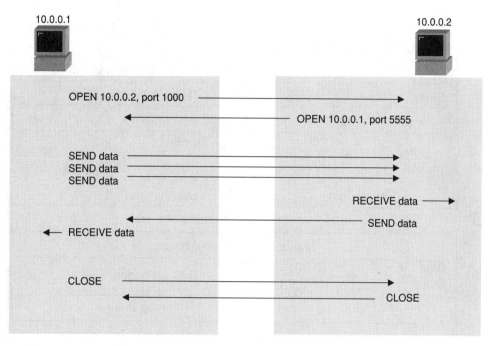

Figure 2.11 TCP control operations.

The next field is the Acknowledgment Number, used to inform the remote end of the data that has been successfully received. The acknowledgment sequence number is actually one greater than that of the last octet correctly received at the local end of the connection. The Data Offset field indicates the number of four-octet words within the TCP header. Six single *bit flags* are used to indicate various conditions. URG is used to indicate whether the *urgent pointer* is valid. ACK is used to indicate whether the Acknowledgment field is valid. PSH is set when the sender wants the remote application to push this data to the remote application. RST is used to reset the connection. SYN is used within the connection startup phase, and FIN is used to close the connection in an orderly fashion. The Window field is a 16-bit count of available buffer space. It is added to the acknowledgment number to indicate the highest sequence number the receiver can accept. As with UDP, the TCP checksum is applied to a synthesized header that includes the source and destination addresses from the IP datagram. The final field in the TCP header is the Urgent Pointer, which, when added to the sequence number, indicates the sequence number of the final octet of urgent data.

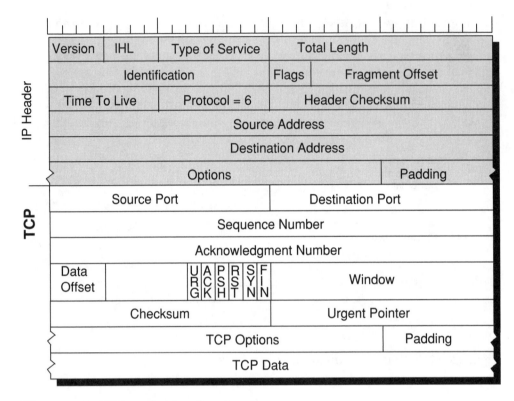

Figure 2.12 TCP packet header structure.

A number of options can be carried in a TCP header. Those relevant to TCP performance are:

Maximum Receive Segment Size option. Used when the connection is being opened. This option is intended to inform the remote end of the maximum segment size, measured in octets, that the sender is willing to receive on the TCP connection. This option is used only in the initial SYN packet (the initial packet exchange that opens a TCP connection). It sets both the maximum receive segment size and the maximum size of the advertised TCP window, passed to the remote end of the connection. In a robust implementation of TCP, this option should be used with MTU discovery to establish a segment size that can be passed across the connection without fragmentation, an essential attribute of a high-performance data flow.

Window Scale option. Intended to address the issue of the maximum window size in the face of paths that exhibit a high-delay bandwidth product. This option allows the window size advertisement to be right-shifted by the amount specified (in binary arithmetic, a right-shift corresponds to a multiplication by 2). Without this option, the maximum window size that can be advertised is 65,535 bytes (the maxim value obtainable in a 16-bitwide field). The limit of TCP transfer speed is effectively one window size in transit between the sender and the receiver. For high-speed, long-delay networks, this performance limitation is a significant factor, as it limits the transfer rate to at most 65,535 bytes per round-trip interval. If network capacity is available, preferably the sender could use the network resource. Use of the window scale option allows the TCP sender to effectively adapt to high-bandwidth delay network paths, by allowing more data to be held in flight. The maximum window size with this option is 2^{30} bytes. This option is negotiated at the start of the TCP connection, and can be sent in a packet only with the SYN flag. Note that while an MTU discovery process allows optimal setting of the Maximum Receive Segment Size option, no corresponding bandwidth delay product discovery allows the reliable automated setting of the Window Scale option.

SACK Permitted option and SACK option. Alter the acknowledgment behavior of TCP. SACK is an acronym for selective acknowledgment. SACK Permitted is offered to the remote end during TCP setup as an option to an opening SYN packet. The SACK option permits selective acknowledgment of permitted data. The default TCP acknowledgment behavior is to acknowledge the highest sequence number of in-order bytes. This default behavior is prone to cause unnecessary retransmission of data, which can exacerbate a congestion condition that may have been the cause of the original packet loss. The SACK option allows the receiver to modify the acknowledgment field to describe noncontinuous blocks of received data, so that the sender can retransmit only what is missing at the receiver's end.

Any robust high-performance implementation of TCP should negotiate these parameters at the start of the TCP session, ensuring the following: that the session is using the largest possible IP packet size that can be carried without fragmentation, that the window sizes used in the transfer are adequate for the bandwidth-delay product of the network path, and that selective acknowledgment can be used for rapid recovery from line-error conditions or from short periods of marginally degraded network performance.

TCP Operation

The setup of TCP requires a three-way handshake, ensuring that both sides of the connection have an unambiguous understanding of the remote side's sequence number. The operation of the connection is as follows:

1. The local system sends the remote end an initial sequence number to the remote port, using a SYN packet.
2. The remote system responds with an ACK of the initial sequence number and the remote end's initial sequence number in a response SYN packet.
3. The local end responds with an ACK of this remote sequence number.
4. The connection is opened.

The operation of this algorithm is shown in Figure 2.13. The performance implication of this protocol exchange is that it takes one and a half round-trip times (RTTs) for the two systems to synchronize state.

After the connection has been established, the TCP protocol manages the reliable exchange of data between the two systems. Just as IP provides no specification of how routing and forwarding is undertaken within the specification of the IP packet and the semantics of the IP header fields, the TCP packet specification provides no predetermined method of data-flow control and reliable data-transfer management. The algorithms that determine the various retransmission timers have been redefined many times, and the refinement of flow-control TCP algorithms remains an active area of IETF activity.

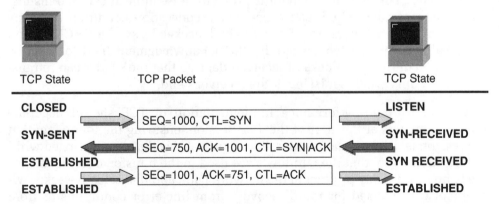

Figure 2.13 TCP three-way handshake.

The general principle of flow control is based on the management of the advertised window size and the management of retransmission time-outs.

The majority of TCP protocol stacks in use on the Internet today do not implement state-of-the-art flow control mechanisms, and they suffer performance degradation as a result. Tuning a TCP protocol stack for optimal performance over a very low-delay, high-bandwidth LAN requires different settings to obtain optimal performance over a dial-up Internet connection, which in turn is different for the requirements of a high-speed wide area network. Although TCP attempts to discover the delay bandwidth product of the connection, and to optimize its flow rates within the estimated parameters of the network paths, some estimates will not be accurate, and the corresponding efforts by TCP to optimize behavior will not be completely successful.

Another critical aspect is that TCP is an adaptive flow-control protocol. TCP uses a basic flow-control algorithm of increasing the data-flow rate until the network signals that some form of saturation level has been reached (normally indicated by data loss). When the sender receives an indication of data loss, the TCP flow rate is reduced; once reliable transmission has been reestablished, the flow rate slowly increases again. If no reliable flow is reestablished, the flow rate backs further off to an initial probe of a single packet, and the entire adaptive flow control process starts again.

This process has a number of results relevant to service quality. First, TCP behaves *adaptively*, rather than *predictively*. The flow-control algorithms are intended to increase the data-flow rate to fill all available capacity, but they are also intended to quickly back off if the available capacity changes due to interaction with other traffic, or if a dynamic change occurs in the end-to-end network path. A single TCP flow across an otherwise idle network attempts to fill the network path with data, optimizing the flow rate within the available network capacity. If a second TCP flow opens up across the same path, the two flow-control algorithms will interact so that both flows will stabilize to use approximately half of the available capacity per flow (see Figure 2.14). The objective of the TCP algorithms is to adapt so that the network is fully used whenever one or more data flows are present. In design, tension always exists between the efficiency of network use and the enforcement of predictable session performance. With TCP, you give up predictable throughput but gain a highly utilized, efficient network.

T/TCP

For network applications that generate small transactions, the application designer is faced with a dilemma. The application may be able to use UDP, in

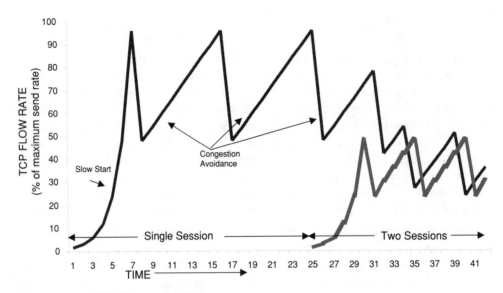

Figure 2.14 TCP flow adaptation.

which case the sender must send the query and await the response (see Figure 2.15). This operation is highly efficient, as the total elapsed time for the client is a single RTT. However, this speed is gained at the cost of reliability. A missing response is ambiguous, in that it is impossible to tell whether the query was lost or the response was lost in transit. If multiple queries are generated, it is not necessarily true that they will arrive at the remote server in the same order as they were generated.

Alternatively, the application can use TCP, which will ensure reliability of the transaction. As noted previously, TCP uses a three-way handshake to complete the opening of the connection, and uses acknowledged FIN signals for each side to close its end of the connection after it has completed sending data (see Figure 2.16). The sending TCP driver will retransmit the query until the sender receives an acknowledgment that the query has arrived at the remote server. Similarly, the remote server will retransmit the response until the server receives an indication that the response has been successfully delivered. The cost of this reliability is application efficiency, as the minimum time to conduct the TCP transaction for the client is two RTT intervals.

TCP for Transactions (commonly referred to as T/TCP [RFC1644]) attempts to improve the performance of small transactions while preserving the reliability of TCP. The T/TCP places the query data and the closing FIN in the initial SYN packet. If the server accepts this format, the server responds with a

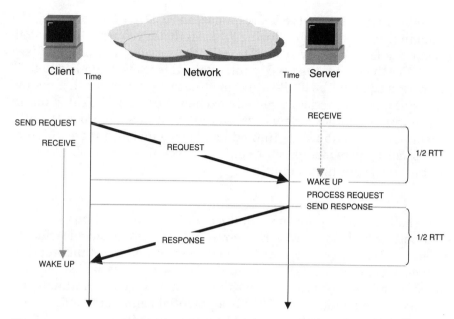

Figure 2.15 Simple UDP transaction.

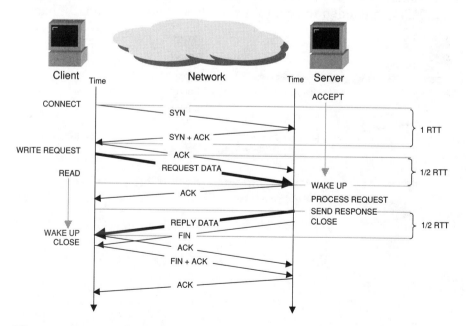

Figure 2.16 Simple TCP transaction.

single packet, which contains its SYN response, an ACK of the query data, the server's data in response, and the closing FIN. All that is required to complete the transaction is for the query system to ACK the server's data and FIN (see Figure 2.17). For the client, the time to undertake this transaction is one RTT interval, a period equal to the UDP-supported transaction, while still allowing for the two systems to negotiate a reliable exchange of data. T/TCP is not in common use in the Internet today, because though it does improve the efficiency of simple transactions, the limited handshake makes it more vulnerable from a security perspective, and concerns over this vulnerability have been a prohibitive factor in its adoption.

RTP

Part of the challenge of supporting a multiservice communications environment is supporting both adaptive variable-rate data transfer protocols (such as file transfers, or Web page downloads) and constant-rate real-time applications (such as telephony or interactive video). TCP with its variable transmission rate and its retransmission algorithm is not a good candidate for supporting real-time applications. UDP is a potential candidate, but requires an additional layer of protocol to support timing information.

This additional layer of protocol is termed the Real-Time Protocol (RTP) [RFC1889]. RTP is intended to be a thin shim layer between the UDP packet

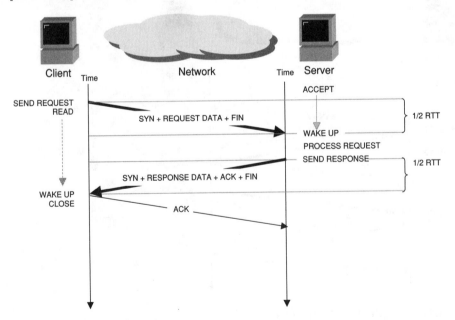

Figure 2.17 T/TCP transaction.

header and the payload. The information in the RTP protocol is intended to provide support for applications with real-time properties, such as streaming audio or video, allowing the application to undertake timing reconstruction, loss detection, security, and content identification. RTP is accompanied by a control protocol, the Real-Time Control Protocol (RTCP), which provides support for real-time conferencing of groups of any size within an internet. The RTCP functionality includes source identification, support for audio and video bridges, and multicast-to-unicast translators. The protocol also offer service-quality feedback from receivers to the multicast group, as well as support for synchronization of different media streams.

RTP is intended to be an application-visible end-to-end protocol, and RTP processing is envisaged to be part of the application, rather than part of the transport protocol stack. RTP itself provides a timestamp for each unit of data, as well as a synchronization source, to assist in the accurate playback of the data; it also identifies the type of payload and the identity of the sources.

The RTP protocol header is shown in Figure 2.18. The first 12 octets are present in every RTP packet, while the list of contributing source (CSRC) identifiers are present only when inserted by a mixer. The fields of RTP are:

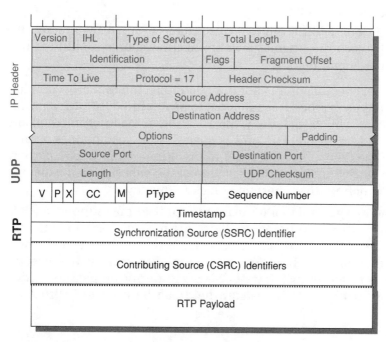

Figure 2.18 The RTP header.

Version. The protocol version, currently 2.

Padding. If set to 1, the payload contains additional trailing padding octets.

Extension. If set to 1, the RTP header is followed by one variable-length extension header.

CSRC Count. Contains the number of CSRC source identifier fields appended to the base RTP header.

Marker. May be defined by the payload profile.

Payload Type. Specifies the format of the RTP payload. It is used by the application to determine how to interpret the data.

Sequence Number. The source's packet sequence count, used by receivers to detect lost and out-of-sequence packets.

Timestamp. Represents the sampling time of the first octet in the payload. The Timestamp field is measured against a random time base, with the requirement that it is based against a linear clock. The clock frequency is not specified within the RTP protocol itself; it is part of the payload-specific set of definitions.

Synchronization Source. Identifies the synchronization source associated with this data stream.

Contributing Sources. If a number of sources is joined by a mixer, the mixer will place the original synchronization source identifiers into this variable-length field.

The RTP protocol specification is not intended to be functionally complete for all application types; rather, it is intended to be modified through a profile specification, each of which includes a specification of the interpretation and use of the header's Marker bit. The profile specification also defines a value for the Payload Type field. Any additional information required for a particular application is carried in the payload section of the packet. Finally, the protocol specification defines the timestamp clock frequency.

RTP is intended as a general tool for real-time applications, supporting both unicast and multicast transport models. Three main aspects are added to the functionality by use of RTP. First, a receiver can use the Source Identifier field to identify an individual data flow. Second, the receiver can use the received sequence numbers to detect lost, out-of-order, and duplicate packets. Third, and perhaps most critical, the timestamp field can be used to instruct the receiver's playback application to generate the correct timing for playback of the reassembled data, reproducing the timing integrity of the original data.

The associated control protocol, RTCP, allows information to be exchanged between source and destination of the real-time data flow. The primary function of this information exchange is to provide feedback to the source on the quality of the data flow. Because an SSRC identifier may change as the application restarts, RTCP also carries a persistent source identifier that is session-independent. This allows a receiver to associate multiple data streams as coming from a single source. In a multicast environment, RTCP also conveys current session membership information to all participants. Note that neither RTP or RTCP undertake explicit QoS functions; it is left to other control mechanisms to provide such services to RTP sessions.

Of course, all this comes at a price, and the highest price is that of protocol overhead. A reexamination of Figure 2.18 shows that RTP adds 12 octets of header data to a packet. Together, the IP, UDP, and RTP sections of the packet account for some 40 octets of overhead; and to this must be added any layer 2 packet framing overhead and any encoding overhead used to covert the real-time analog signal to a real-time digital data stream. The challenge is to manage the efficiency and quality of the end-to-end data flow. One possibility is to increase the size of the payload by reducing the sampling rate to reduce the proportion of packet header to payload, but larger packets are prone to higher jitter levels, making interactive real-time applications cumbersome to operate. A reduced payload size, achieved by an increased sampling rate, will address some of these quality issues, but at the expense of a greater overhead of the packet protocol headers.

IPv6

IPv6, also known as IPng or IP Next Generation, version 6 of the IP protocol, is included here because it is now gaining increasing levels of industry attention [RFC2460]. A number of implementations of the protocol are now available, and the Internet Regional Registries are now undertaking allocation of IPv6 address prefixes.

The primary reason for the development of the IPv6 protocol was to introduce longer address fields into the Internet Protocol. The version 4 protocol uses 32-bit addresses, for a maximum of some 4 billion host addresses. This may sound like a large number, but almost no address deployment plan achieves 100 percent address utilization rates. A deployment plan that achieved 10 percent utilization (where one-tenth of the addresses are actually used to address host systems) would be considered an outstanding success in many communications environments [RFC1715]. Current version 4 address allocation practices aim for utilization rates of around 40 to 60 percent, which impose significant discipline on network administrators to produce an address deployment plan that has few margins for growth. Even with this very strict allocation policy, we cannot achieve an equilibrium between the ever increasing demand for version 4 addresses and the finite total address pool. Given the pervasive nature of Internet-based applications, and the volume market for consumer electronics, the exhaustion of the version 4 32-bit address space is a likely outcome.

IPv6 certainly attempts to address this basic problem. Increasing the size of the address fields in the protocol to 128 bits allows for almost any conceivable deployment scenario, even accounting for relatively low levels of deployment utilization. The IPv6 basic header is delineated in Figure 2.19. In addition to the longer address fields, IPv6 developers chose to rationalize the use of other header fields, partly to improve the potential for the protocol to support differentiated service responses from the network. So far, this potential has not proved to be a major selling point for IPv6; and of the reasons used to rationalize an industrywide migration to IPv6, integrated QoS is not one of the most compelling arguments. It is a common misperception that the IPv6 specification somehow includes a magic knob that, when turned, will provide QoS support. Although the IPv6 packet structure is significantly different from version 4, the basic functional operation of the protocol and its associated performance capabilities, are still quite similar; and there is no breakthrough in QoS support in IPv6.

IPv6 Service Performance Fields

While there is no single QoS knob in IPv6, two significant components of the IPv6 protocol packet header may assist in delivering differentiated service responses. The first is an 8-bit Traffic Class field in the IPv6 header, which is equivalent to the IP differentiated service bits in the IPv4 specification. The Traffic Class field can be used in a fashion similar to that described earlier for IPv4, in an effort to identify and discriminate traffic types based on contents of

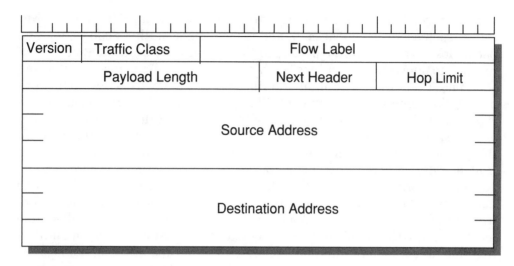

Figure 2.19 IP version 6 packet header.

this field. The second component is a Flow Label, which was added to enable the labeling of packets that belong to particular traffic flows for which the sender might request special handling, such as nondefault QoS or real-time traffic.

*Y*ou can find more information on IPv6 by visiting the IETF IPng Working Group Web page at www.ietf.org/html.charters/ipngwg-charter.html, or the IPv6 Web page hosted by Sun Microsystems, located at playground.sun.com/pub/ipng/html/ipng-main.html.
    You can find information related to the experimental deployment of IPv6 on the 6Bone Web site at www.6bone.net.

The IPv6 Traffic Class Field

The Traffic Class field has an interesting history. The original 1995 IP version 6 specification [RFC1883] identified this field as a 4-bit Priority field designed to allow the traffic source, or some other external agent, to identify the desired delivery priority of its packets (as compared to having the network impose a priority based on some administrative policy filter). The 4 bits in the Priority field were divided into two ranges: Values 0 through 7 specified the priority of traffic for which the source was providing congestion control (in other words, traffic, such as TCP, that backs off and responds gracefully to congestion and packet loss); values 8 through 15 specified the priority of traffic that did not respond to congestion situations, such as real-time traffic being sent at a constant rate. There was no implicit ordering between the two classes, so that priority 7, the highest congestion-controlled priority, could be implemented at a significantly higher priority than any of the priorities 8 through 15. The interesting aspect of this proposed prioritization scheme was to recognize within the protocol itself that the network might need to handle congestion-controlled traffic differently from uncontrolled traffic sources. There was no "default" priority for traffic within this specification, meaning that the source had to assign "base" traffic a priority of 0 for uncharacterized congestion-controlled traffic, and 8 for uncharacterized uncontrolled sources. In effect, the most significant bit of this field represented a "congestion control" identification bit, allowing an intermediate system the capability to manage its response in accordance with the congestion mechanism. For example, a queue could include Random Early Detection admission (see Chapter 3, "Performance Tuning Techniques," for a description of this mechanism) when the bit was clear, and use a shorter tail-drop queue when the bit was set. In itself, the interpretation of this priority field represented a major shift in thinking by protocol designers, as it forced the recognition that a single prioritization number space, and an associated single network service response mechanism, were simply inadequate for effective performance management of varied traffic types.

Of course, technology often changes in midstream, and IPv6 certainly is no exception. The current proposed standard document for IPv6 changes this specification. The 4-bit Priority field is replaced by an 8-bit Traffic Class field (shrinking the Flow Label from 24 bits to 20 bits in length). In accordance with the Differentiated Services architecture, the document makes no firm proposals as to the internal structure of this 8-bit field, nor does it associate any particular performance behaviors with any bit value. Instead, the document points to the IP version 4 proposals for differentiated service models, and notes that, "Detailed definitions of the syntax and semantics of all or some of the IPv6 Traffic Class bits, whether experimental or intended for eventual standardization, are to be provided in separate documents" [RFC2460]. The advantage of this approach is that it does allow the consistent use of the 8-bit IP version 4 Type of Service and the 8-bit IP version 6 Traffic Class fields, if so desired. Unfortunately, the somewhat imaginative use of a congestion control indicator has been dropped from this updated standard.

The current proposed semantics for the Traffic Class field describe it as a "Differentiated Services Field," and this is meant as a common standard for both IP version 4 and IP version 6 [RFC2474]. Interestingly, the architecture for Differentiated Services is more limited in scope than the previous priority settings; and, as noted in the Differentiated Services architecture, "This architecture only provides service differentiation in one direction of traffic flow, and is therefore asymmetric. Development of a complementary symmetric architecture is a topic of current research but is outside the scope of this document" [RFC2475]. The proposed IPv6 Traffic Class field provides the capability of requesting some form of differentiated packet-level response on a per-hop basis from the network, but the semantics of the various code points within the field are not tightly coupled to the protocol definition. It is anticipated that network administrators would want to signal relative queuing priority, delay sensitivity, and discard preference within this field; to achieve this, the network may use locally defined code points within this field. When traffic enters the network, it would be classified according to some type admission filter, and the field set according to the classification process. The subsequent service response from the network would be determined by this field setting. We will return to the Differentiated Services architecture when we explore a range of service delivery architectures in Chapter 4, "Quality of Service Architectures."

The IPv6 Flow Label

The Flow Label in the IPv6 header is a 20-bit value designed to uniquely identify any traffic flow so that intermediate nodes can use this label to identify flows for special handling, for example, with a resource reservation protocol, possibly RSVPv6 or a similar protocol. For this reason, the Flow Label is of interest when examining protocol-specific hooks that can be used to create differentiated service outcomes. At this stage, the specification notes that the Flow Label is an experimental aspect of the protocol and may be subject to change in the light of deployment experience. The Flow Label is assigned to a flow by the flow's source node or the flow originator. The IPv6 specification, RFC2460, requires that hosts or routers that do not support the functions of the Flow Label field set the field to 0 when originating a packet, pass the field on unchanged when forwarding a packet, and ignore the field when receiving a packet. It also specifies that flow labels must be chosen randomly and uniformly, ranging from hexadecimal 0x000001 to 0xFFFFF. Additionally, all packets belonging to the same flow must be sent with the same source address, destination address, and Flow Label. Interestingly, the requirement for consistency of priority within a flow was dropped with the change to the protocol specification, in favor of the use of a Traffic Class field; and there is no corresponding protocol-mandated requirement for consistency of Traffic Class within a flow. The implication of the constrained fields is that the true Flow Identification field is the vector of *(source address, destination address, flow label)*. This is not altogether a unique introduction to the IP protocol, given that an equally valid flow identification within IPv4 is the vector of *(source address, destination address, source port, destination port)*.

Currently, aside from the suggestion mentioned in the protocol specification, there are still no clearly articulated methods for using the Flow Label in the IPv6 header, apart from with a resource reservation protocol such as RSVP. However, at least one recommendation has been published that provides a generalized proposal for using the Flow Label in the IPv6 header. RFC1809 suggests that Flow Labels may be used in individual sessions, during which the flow originator may not have specified a Flow Label; and an upstream service provider may want to impose a Flow Label on specific flows as they enter their network, in an effort to differentiate these flows from others, possibly for preferential treatment. Currently, it is unclear what intrinsic value the Flow Label actually provides. Commonly devised methods are available to build flow state in routers without the need for a special label in each packet, using the IPv4 vector of end-point IP addresses and end-point port addresses.

IPv6 QoS Observations

Dissension continues in the networking community over the need for IPv6; and when closely examining the opposing viewpoints, you might be hard-pressed to find a compelling reason to pervasively deploy IPv6, at least in the short term. In fact, most of the enhancements designed into the IPv6 specification have been retrofitted for IPv4. The utility gained by migrating to IPv6 in an effort to benefit from the dynamic neighbor-discovery and address-autoconfiguration features, for example, might be negated by the availability and subsequent wide-scale deployment of Dynamic Host Configuration Protocol (DHCP) for IPv4 [RFC2131].

In a similar vein, the IPv6 Traffic Class field does not offer substantial improvement over the utility of the IP Precedence field in the IPv4 header. Indeed, the current thinking within the Differentiated Services architecture community is to treat the IPv4 Type of Service field and IPv6 Traffic Class field identically, so that there is no effective difference whatsoever in the capabilities of defining per-hop behaviors within the two protocols.

The IPv6 Flow Label, on the other hand, may prove to be quite useful—although so far the only immediate recognizable use for the Flow Label is in conjunction with a resource reservation protocol, such as RSVP for IPv6. The Flow Label possibly could be used to associate a particular flow with a specific reservation. The presence of a Flow Label in data packets also may help in expediting traffic through intermediate nodes that have previously established path and reservation states for a particular set of flows. However, these options are speculative in nature and, aside from using the IPv6 Flow Label with RSVP, its benefit is not immediately determinable. It should also be noted that it is readily possible to construct a Flow Label from an IPv4 TCP or UDP packet flow by combining the source and destination addresses with the source and destination port addresses of the flow.

The inevitable conclusion from this examination is that IPv6 does not offer any substantial QoS capabilities above and beyond what already is achievable with IPv4. This is certainly a disappointing conclusion, in that hopes were high that IPv6 would offer the industry an augmented set of performance capabilities for IP, allowing the clients and providers to enter into service relationships that included meaningful service performance characteristics. Instead, we appear to have found one more instance of the more general cynical observation: So far we have yet to find a single truly useful IP version 6 feature that we cannot retrofit into IP version 4. As a side note, it would appear that the most convincing case for IP version 6 takes us back to the basics of the extended address space, allowing for another wave of Internet devices to form part of the network without resorting to various partitioning mechanisms as required by network address translation schemes.

Protocol Performance

In this section we will examine the transfer of data using the TCP protocol, focusing on the relationship between the protocol and performance. To begin, we must point out that such an examination is relevant only to TCP; it is not strictly pertinent to either UDP or to the underlying Internet Protocol. The reason for this deliberate exclusion is that both UDP and IP are simple datagram transmission models, with no inherent capability to support reliable sequenced data transfer. Accordingly, any application can generate UDP packets (or IP packets) and submit them to the network for transmission at any arbitrary rate. There is no inherent limitation imposed by the protocol on the sending rate, nor any acknowledgment or sequencing identification that places a limit on the number of packets that can be in flight within the network during any time interval. From this perspective, the definitions of UDP and IP impose no performance limitations on the mode of unreliable data transfer supported by these protocols. But while there is no protocol limitation, there is also no protocol support for performance with UDP and IP. The limiting performance factor for such uncontrolled protocols is network congestion, and the outcomes of such congestion are degraded quality through escalating delay and loss within the network.

The picture changes substantially where the capability of the protocol is extended to support efficient and reliable data transfer, as is the case with TCP, and a number of performance aspects are defined by the protocol.

TCP is generally used within two distinct application areas: short-delay, short data packets (sent on demand, to support interactive applications such as telnet, or rlogin), and large packet data streams (supporting reliable volume data transfers, such as mail transfers, Web page transfers, and FTP). Different protocol mechanisms come into play to support interactive applications, as distinct from short- and long-held volume transactions. In this section, we will briefly examine these mechanisms and their performance limitations.

Interactive TCP

Interactive protocols are typically directed at supporting single-character interactions, where a keystroke is passed reliably from the client to the server, after which the server provides an echo of the character. The protocol interaction to support this is indicated in Figure 2.20. TCP makes some small improvement in this exchange through the use of *piggybacking*, where an ACK is carried in the same packet as data, and *delayed acknowledgment*, where an ACK is delayed up to 200ms before sending, to give the sending application the opportunity to generate data that the ACK can piggyback. The resultant protocol exchange is indicated in Figure 2.21. Even with this improvement, however, TCP still imposes significant overhead for interactive performance, because a minimum of 80 bytes of protocol overhead are added to the transmission of a single byte of data to the server and its corresponding echo.

For short-delay LANs, this protocol exchange offers acceptable performance. This three-way protocol exchange occurs within some 16ms on an Ethernet LAN, corresponding to an interactive rate of 60 characters per second [Stevens 1994]. Once the network delay is increased in a WAN, this is not necessarily an adequate approach. The operational result of this protocol in a WAN is an interlaced sequence of small packets and their corresponding ACKs, as indicated in Figure 2.22.

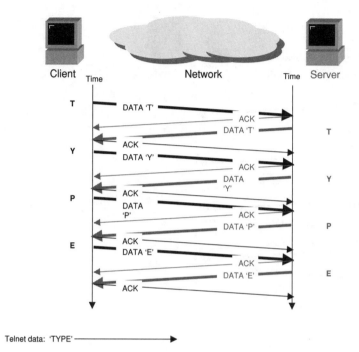

Figure 2.20 Protocol exchange for interactive applications.

The TCP mechanism to address this small packet congestion was described by John Nagle in RFC896. Commonly referred to as the *Nagle algorithm*, this mechanism inhibits a sender from transmitting any additional small segments while the TCP connection has outstanding unacknowledged small segments (see Figure 2.23). On a LAN, this modification to the algorithm has a negligible effect; in contrast, on a WAN, it has a dramatic effect in reducing the number of small packets in direct correlation to the network path's congestion level. The cost is an increase in the variability of interactive delay (or an increase in the jitter of the session), as well as an overall increase in average delay. Applications that are jitter-sensitive typically disable this control algorithm.

TCP is not a highly efficient protocol for the transmission of interactive traffic, as seen with the support of character-based remote terminal sessions. The carriage efficiency of the protocol across a LAN is 2 bytes of payload and 120 bytes of protocol overhead. Across a WAN, the Nagle algorithm may improve this carriage efficiency slightly by increasing the number of bytes of payload for each three-way transaction, although it will do so at the expense of increased session jitter. For character-based interactive traffic, normally this additional jitter is insignificant.

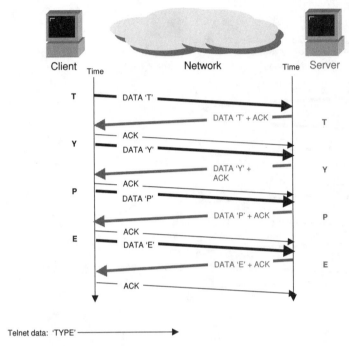

Figure 2.21 TCP interactive exchange.

TCP Volume Transfer

The other major TCP application is of bulk data transfer, such as that generated by the FTP application, or by Usenet news transfers. The protocol objective here is to maximize the efficiency of the data transfer, implying that the protocol mechanism should endeavor to locate a point of dynamic equilibrium, where the sending data rate is maximized and the packet loss is minimized. Simply increasing the sending rate will run the risk of generating a congestion condition within the network or of exacerbating an existing congestion condition. The outcome will be increased packet loss levels, which in turn will force the TCP protocol to retransmit the lost data, resulting in losing overall data transfer efficiency. Attempting to minimize packet loss rates implies that the sender must reduce the sending rate of data into the network, so as not to create congestion conditions along the path to the receiver. Such an action will, in all probability, leave the network with idle capacity, making inefficient use of available network resources.

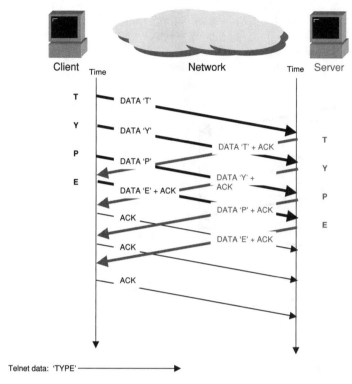

Figure 2.22 Small packet flow.

Accordingly, this functional objective implies a need to establish the threshold of packet loss; maximal efficiency is achieved where the rate of data packets entering the network is set to just under the point of the onset of packet loss. Of course, the network will not provide any advance indication of where such a point of threshold exists, nor will this point remain stable over time. Other sessions may be active on elements of this TCP path, and as they open, adjust their sending rates, and close down, the point of maximal efficiency for the active TCP session will vary. To achieve this dynamic setting of optimal carriage efficiency, TCP uses a feedback control loop to dynamically discover, and then maintain, maximal efficiency of the data transfer. The mechanism modifies the TCP window, adding another window control point; the manipulation of this window control point forms TCP's congestion control response [RFC2581].

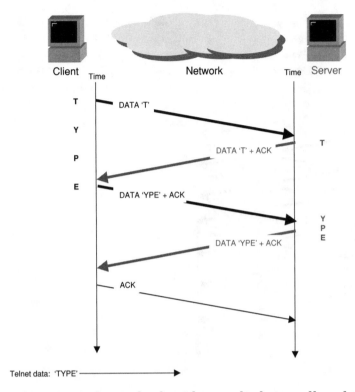

Figure 2.23 The Nagle algorithm applied to small packet flow.

TCP uses a sliding window protocol to support bulk data transfer. The receiver advertises to the sender the available buffer size at the receiver. The sender can transmit up to this amount of data, with the additional condition that the sender must retain a copy of the data until an acknowledgment is received. The send window is the minimum of the sender's buffer size and the advertised receiver window. Each time an ACK is received, the left side of the sender's window is advanced. The minimum of the sender's buffer and the advertised receiver's window is used to calculate a new right edge; and if this is greater than the already sent data, additional data can then be sent, up to the edge of the window (see Figure 2.24).

The size of TCP buffers in each host are a critical limitation to performance in WANs, as the sender can have at most one transfer buffer of data transferred per round trip, where the transfer buffer size is the minimum of the sender's

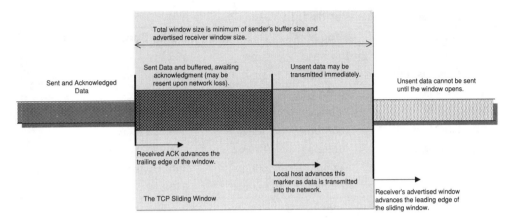

Total window size is minimum of sender's buffer size and advertised receiver window size.

Sent and Acknowledged Data

Sent Data and buffered, awaiting acknowledgment (may be resent upon network loss).

Unsent data may be transmitted immediately.

Unsent data cannot be sent until the window opens.

Received ACK advances the trailing edge of the window.

Local host advances this marker as data is transmitted into the network.

The TCP Sliding Window

Receiver's advertised window advances the leading edge of the sliding window.

Figure 2.24 TCP window.

buffer and the receiver's advertised window. For example, with a transfer buffer of 4096 bytes, and a transmission path with an RTT of 600ms, the maximal transmission rate of a single TCP session is throttled to no more than 48Kbps, regardless of the bandwidth of the path. To achieve maximal efficiency for the transfer, the server must be able to fill the path from the sender to the receiver with data. As the sender will have an equivalent amount of data awaiting reception of an ACK signal, both the sender's buffer and the receiver's advertised window should be no smaller than the *delay-bandwidth product* of the network path.

Window size >= bandwidth (bytes/sec) x round-trip-time (secs)

Table 2.1 provides some typical values of the delay-bandwidth product for a number of network paths. The 16-bit field within the TCP header can transmit values up to 65,535 values, imposing an upper limit on the available window size, thereby imposing an upper limit on protocol performance.

This limit can be modified by the use of a Window Scale option, described in RFC1323, effectively increasing the size of the window to a 30-bit field, but transmitting only the most significant 16 bits of the value. This allows the sender and receiver to use buffer sizes that can operate efficiently at speeds that encompass most of the current network transmission technologies.

The operation of the window is a critical component of TCP performance for volume transfer. The mechanics of the protocol involve an additional over-

Table 2.1 Typical Delay-Bandwidth Values

Network	Bandwidth	RTT Delay	Delay-Bandwidth Product
Ethernet	10Mbps	1ms	125 bytes
100Mbps Fast Ethernet	100Mbps	1ms	12 bytes
56K modem	56Kbps	20ms	169 bytes
WAN	1.5Mbps	40ms	7,720 bytes
ISP Core Network	155Mbps	50ms	968,750 bytes
Undersea Cable	45Mbps	180ms	1.006,560 bytes
Satellite	64Kbps	660ms	5,820 bytes
Satellite	8Mbps	660ms	675,840 bytes

riding modifier of the sender's window, the *congestion window*, referred to as *cwnd*. The objective of the window management algorithm is to start transmitting at a rate that has a very low probability of packet loss, then to increase the rate (by increasing the *cwnd* size) until the sender receives an indication (through the detection of packet loss) that the rate has exceeded the network's available capacity. The sender then immediately reduces its sending rate to a value anticipated to be below the loss threshold (by reducing the value of *cwnd*), and again commences another increase of the sending rate. The goal is to continually modify the sending rate such that the sending rate oscillates around the true value of available network capacity. This oscillation of the sending rate enables a dynamic adjustment that automatically senses any increase or decrease in available capacity through the lifetime of the data flow. An idealized view of the behavior of this algorithm is shown in Figure 2.25.

The intended outcome is that of a dynamically adjusting cooperative data flow, where a combination of such flows behaves fairly, in that each flow

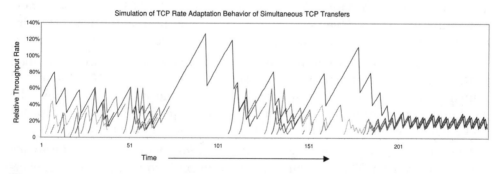

Figure 2.25 Flow rate sensing behavior.

obtains essentially a fair share of the network, and so that close to maximal use of available network resources is made. This flow control functionality is achieved through a combination of *cwnd* value management and packet loss and retransmission algorithms.

TCP Slow Start

The starting value of the *cwnd* window is set to that of the received Maximum Receive Segment Size option value, obtained during the SYN handshake. The sender's window is then set to the minimum of the three window size parameters: the sender's buffer size, the receiver's advertised window size, and the receiver's maximum segment size. Typically, this value is the final one, that of a single segment. An experimental extension to the TCP protocol allows this initial value to be as large as four times the maximum segment size [RFC2581]. The sender then enters a flow control mode termed *slow start*.

***S**low Start Value The choice of a single segment as the initial slow start value has been attributed to the characteristics of one of the first Ethernet network adapter cards. This interface card had a buffer that could store a single Ethernet packet. Starting a TCP session with any more than a single packet would overflow the device buffer. More recent review of this choice of a start value in RFC2414 has indicated that a value of up to four times the receiver's maximum segment size would allow for a more efficient startup of the data flow, without unduly exacerbating any prevailing network congestion condition.*

The choice of the initial segment size has a significant effect on the throughput of small TCP sessions. Many Internet networks experience an average TCP flow of some 17 segments, due to the predominance of Web page fetches within the overall traffic profile. With a initial segment size of 1, which doubles every round-trip time (RTT), 17 segments will be transferred in 5 RTT intervals, while an initial start of 4 segments will transfer all 17 segments in 3 RTT intervals, assuming no other performance limitations and no packet loss within the network.

The sender then awaits the corresponding ACK from this initial segment. When the ACK is received, the sender increases its window by increasing the value of *cwnd* by the value of one segment. This allows the sender to transmit two segments; at that point, the congestion window has been exhausted, and the sender must await the corresponding ACKs for these transmissions. This algorithm continues by increasing the value of *cwnd* (and, correspondingly, opening the size of the congestion window) by one segment for every ACK received.

The effect of this algorithm is that the data rate of the sender doubles every RTT. Obviously, this cannot be sustained indefinitely. Either the value of *cwnd* will exceed the advertised receive window or the sender's window, or the capacity of the network will be exceeded, in which case packets will be lost. We will look at packet loss in the next section.

One final aspect to highlight concerns the interaction of the slow start algorithm with high-capacity long-delay networks, the so-called Long Fat Networks, or LFNs (pronounced "elephants"). The behavior of the slow start algorithm is to send a single packet, await an ACK, then send two packets, and await the corresponding ACKs, and so on. The activity tends to cluster at each epoch of the round-trip time, with a quiet period that follows after the available window of data has been transmitted. The received ACKs arrive back at the sender with an inter-ACK spacing that is equivalent to the data rate of the bottleneck point on the network path. During slow start, the sender transmits at a rate equal to twice this bottleneck rate. The rate adaptation function that must occur within the network takes place in the router at the entrance to the bottleneck point. The sender's packets arrive at this router at twice the rate of egress from the router, and the router stores the overflow within its internal buffer. Once this buffer overflows, packets will be dropped, and the slow start phase is over. The important conclusion is that the sender will stop increasing its data rate when there is buffer exhaustion, which may not be the same as the true available data rate. If the router has a buffer capacity considerably less than the delay-bandwidth product of the egress circuit, the two values are certainly not the same. In this case, the TCP slow start algorithm will finish with a sending rate that is well below the actual available capacity. The efficient operation of TCP is critically reliant on adequately large buffers within the network's routers.

Packet Loss

How does a TCP sender know that it is sending at a rate greater than the network can sustain? The obvious answer is that data packets are dropped by the network. In this case, TCP has to undertake a number of functions:

- ◆ The packet loss has to be detected by the sender.
- ◆ The missing data has to be retransmitted.
- ◆ The sending data rate should be adjusted to reduce the probability of further packet loss.

There are two ways that TCP can detect packet loss. First, if a single packet is lost within a sequence of packets, the successful delivery packets following the lost packet will cause the receiver to generate a *duplicate ACK*. The reception of these duplicate ACKs is a signal of packet loss. Second, if a packet is lost at the end of a sequence of sent packets, there are no following packets to generate duplicate ACKs. In this case, there are no corresponding ACKs for this packet, and the sender's retransmit time will expire and signal packet loss.

A single duplicate ACK is not a reliable signal of packet loss. When a TCP receiver gets a data packet with an out-of-order TCP sequence value, the receiver must generate an immediate ACK of the highest in-order data byte received. This will be a duplicate of an earlier transmitted ACK. Where a single packet is lost from a sequence of packets, all subsequent packets will generate a duplicate ACK packet. On the other hand, where a packet is rerouted with an additional incremental delay, the reordering of the packet stream at the receiver's end will generate a small number of duplicate ACKs, followed by an ACK of the entire data sequence, once the errant packet is received. The sender uses the reception of three duplicate ACK packets as a signal of packet loss.

The duplicate ACK condition causes TCP to cease slow start mode, if it was currently active, and enter *congestion avoidance* mode. In addition, because there is still a sequence of ACKs arriving at the sender, the network is continuing to pass timing signals to the sender. In this case, the sending rate is halved, (by halving the current value of cwnd*)*, and this new window size is stored as the threshold of congestion avoidance. This value is commonly referred to as *ssthresh,* and this response is termed *fast retransmit*. A more detailed description of this algorithm is contained in [Stevens 1994].

The other signal of packet loss is a complete cessation of any ACK packets arriving to the sender. The sender cannot wait indefinitely for a delayed ACK, but must make the assumption at some point in time that the next unacknowledged data segment must be retransmitted. This is managed by the sender maintaining a *retransmission timer*. The maintenance of this timer has performance and efficiency implications. If the timer triggers too early, the sender will push duplicate data into the network unnecessarily. If the timer triggers too slowly, the sender will remain idle for too long, unnecessarily slowing down the flow of data. The TCP sender uses a timer to measure the elapsed time between sending a data segment and receiving the corresponding acknowledgment. Individual measurements of this time interval will exhibit significant variance, and most implementations of TCP use a smoothing function when updating the flow's retransmission timer with each measurement. The commonly used algorithm is described in [Jacobson 1988], modified so that the retransmission timer is set to the smoothed round-trip time value, plus four times a smoothed mean deviation factor [Jacobson 1990].

When the retransmission timer expires, the actions are similar to that of three duplicate ACK packets, in that the sender must reduce its sending rate in response to congestion. The threshold value, ssthresh, is set to half of the current value of cwnd, as in the duplicate ACK case. However, the sender cannot make any valid assumptions about the current state of the network, given that no useful information has been provided to the sender for more than one RTT time interval. In this time-out case, the sender closes the congestion window back to one segment, and restarts the flow in slow start mode by sending a single segment. The difference from the initial slow start is that, in this case, the ssthresh value is set so that the sender will probe the congestion area more slowly using a linear sending rate increase once the congestion window reaches the remembered ssthresh value.

Congestion Avoidance

Congestion avoidance is a more tentative probing of the network to discover the point of threshold of packet loss. Where slow start uses an exponential increase in the sending rate to find a first-level approximation of the loss threshold, congestion avoidance uses a linear growth function.

When the value of *cwnd* is greater than ssthresh, the sender increments the value of *cwnd* by 1/*cwnd*, plus one-eighth of the segment size, in response to each received ACK. This ensures that the congestion window opens by one segment within each RTT time interval.

The congestion window continues to open in this fashion until packet loss occurs. If the packet loss is isolated to a single packet within a sequence, the resultant duplicate ACKs will trigger the sender to halve the sending rate and continue a linear growth of the congestion window from this new point. Figure 2.26 illustrates this behavior of cwnd in an idealized configuration, along with the corresponding data flow rates. The overall characteristics of the TCP algorithm are: an initial relatively fast scan of the network capacity to establish the approximate bounds of maximal efficiency (slow start), followed by a cyclic mode of adaptive behavior that reacts quickly to congestion, and then slowly increases the sending rate when the area of maximal transfer efficiency is reached.

Packet loss, as signaled by the absence of any ACK packets, causes the sender to recommence slow start mode, following a time-out interval. The corresponding data flow rates are indicated in Figure 2.27. The inefficiency of this mode of performance is caused by the complete cessation of any flow sig-

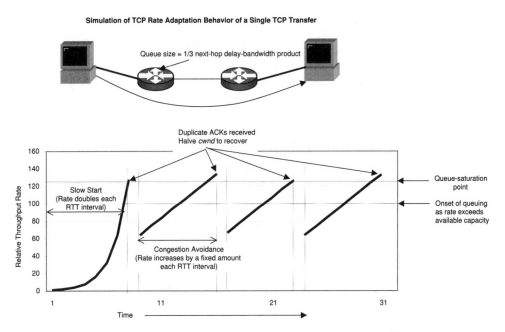

Figure 2.26 TCP flow behavior with slow start and congestion avoidance.

naling from the receiver to the sender. In the absence of any information, the sender can only assume that the network is heavily congested, and so must restart its probing of the network's capacity with an initial congestion window of a single segment. This leads to the performance observation that any form of packet drop management that tends to discard the trailing end of a sequence of data packets may cause significant TCP performance degradation, as such drop behavior forces the TCP session to continually time out and restart the flow from a single segment again.

Routers and Performance

In the previous sections, we examined Internet performance from the perspective of the interaction of the two protocol engines at either end of the connection. From the perspective of the operation of the protocol, the configuration that maximizes performance of the protocol is one where the two systems are privately connected in a back-to-back configuration. In such a configuration, there is no intervening network to add an element of distortion to the signals

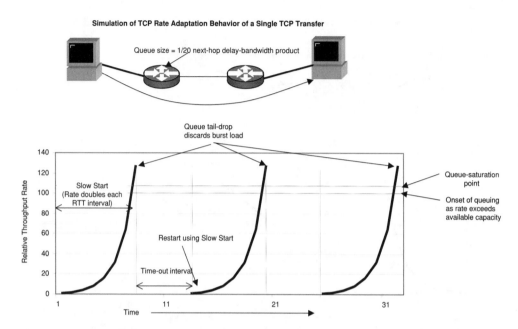

Figure 2.27 TCP flow behavior with time-outs.

passing between the two systems; the resultant performance of the interaction is a product of the performance of the systems themselves and the capabilities of the communications protocol.

For clients of a networked environment, the reality is some distance from this isolated idealized model. In between two communicating systems lies a network, comprising a set of switching elements and transmission segments, and these elements share the network's resources across a number of simultaneous network transactions. This environment adds a level of distortion to the end-to-end interaction defined by the communications protocol.

In this section we will examine the atomic functions of the IP switching elements, or routers, and examine how these basic functions interact with the end-to-end protocol.

What Is a Router?

In their most basic form, routers are packet-switching devices intended to interconnect two networks. That said, there are a set of network interconnection devices that can also perform this basic function in a wide variety of ways, so some elaboration is necessary to distinguish a router from these other devices.

The simplest network interconnection device is a *repeater*. A repeater's task is to reproduce data from one network to the other, working at the physical level of bits—1s and 0s. A repeater is in many ways analogous in function to a signal regenerator in transmission systems, faithfully reproducing the original transmission signal received from one network onto the other, preferably with minimal real-time delay and with no deviation from the original signal, either in value or in data rate clocking. In terms of describing its impact on the flow of data, a repeater can be regarded as a constant delay unit.

Where a repeater's function is unconditional forwarding, a *bridge* adds a basic level of network topology knowledge to the task. A bridge is a data link device, operating at the data link frame and address level of the protocol stack. In its normal mode of operation, a data link frame is forwarded by a bridge from one network to the other only if the bridge is aware that the destination of the frame is located on the other network, and that the task of forwarding the frame is necessary to ensure successful frame delivery. Bridges commonly assemble the entire packet from the input network before passing the packet to the appropriate output network. In terms of describing its impact on the flow of data, a bridge is a source of both delay and jitter, as the delay component includes a variable packet assembly delay and a variable delay in the output queue. A bridge can also operate with more than two interfaces, so that the bridge can also act as a *switch*, using the data link addresses to determine to which output the packet should be switched.

Routers undertake interconnection of two networks at the level of the network protocol. They are active protocol-specific elements that participate in a shared topology discovery algorithm. Protocol packets are addressed to the data-link level address of the router. The router strips off this data-link layer encapsulation and inspects and, potentially, modifies the network-level packet header fields. The protocol-specific information in the header (usually the destination address of the packet), is used as the basis of a forwarding decision to determine the correct next-hop network and next-hop router on that network. The packet then is encapsulated in the data-link information as required by the next-hop network and forwarded to that network. The sequence of processing operations internal to the router to achieve this functionality is shown in Figure 2.28.

Although a router can be constructed as a software process operating on a general-purpose computing platform, high-performance routers are constructed as

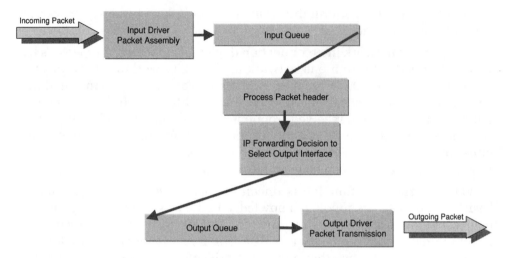

Figure 2.28 The steps involved in routing an IP packet.

specialized pieces of equipment. Their architecture can vary from simple platforms that use a very conventional single processor and shared memory hardware configuration (with a simple routing software kernel driving a set of interface devices) to a parallel-processing, high-capacity platform with multiple special-purpose processors driving particular components of the routing function. The drive to these specialized architectures is prompted by the objective to not only increase the switching capacity of the devices into gigabits per second transmission systems, but to increase the flexibility of control of the switching function, thereby allowing the units to become active elements in applying administratively defined performance policies into the network.

Router Functions

Routers undertake three major tasks within a network: packet forwarding between attached network interfaces, participation in the operation of routing protocols, and the operation of local network management tasks. We will examine each of these functions in turn to provide some insight on the internal operation of a router.

Packet Handling

We use a very generic description of packet handling here; it is not intended to include all possible variations of packet handling within a router. Although the set of tasks described here are common to all routers, the manner in which these tasks are implemented within various router hardware and software platforms varies considerably. These functions make reference to the processing steps indicated in Figure 2.28.

The network interfaces on routers assemble packets explicitly addressed to the router. The data-link level address of the packet must reference the router interface's data-link level address, the local network's broadcast address, or, if configured, one of the local multicast address groups. After the entire packet has been assembled by the interface driver, the data-link encapsulation is stripped off the packet and the protocol-specific payload is forwarded to the relevant protocol-forwarding process. This process may be a shared resource, so the packet is placed in an input queue while awaiting the scheduling of the processing resource.

The protocol process first decides whether the packet is addressed to this router. The verification must involve the protocol addresses for all connected interfaces. If this is the case, the packet will be processed as a host would process a packet addressed to it, stripping off the protocol transport encapsulation and forwarding the payload to the appropriate application process. Otherwise, the packet must be forwarded.

Forwarding involves determining the next hop by looking up (referencing) the destination address in a locally maintained forwarding table. If the IP packet has strict or loose source-routing options enabled, the destination is drawn from the option field of the packet. Otherwise, the IP header destination address is used in the lookup algorithm. The forwarding table contains entries for network address blocks, rather than a comprehensive list of 32-bit host addresses; therefore, the algorithm used locates the entry in the table corresponding to the best match within the table. Here, best is defined as most specific or longest match (which refers to the number of bits used in the match of the unmasked bits of the destination address to the corresponding unmasked bits of the table entry's network address). The more bits used in the match, the longer, or more specific, the match.

If no match is found in the table, no possible forwarding decision exists. In this case, under specific circumstances, the router generates an ICMP error response to the packet, addresses the ICMP packet to the original packet's source address, discards the original packet, and forwards the ICMP packet back to the sender. In many cases, routers use a *default route* entry, which is the match of last resort, so that the router does not need to operate with a comprehensive set of entries in its forwarding table. If the packet does not match an explicit entry in the forwarding table, the default route, if present, is used to make a forwarding decision. If the packet is forwarded, the packet header's TTL field is decremented by 1. If this action brings the TTL field to 0, under specific circumstances, an ICMP error response is generated, per the preceding procedure, and the packet is discarded. If loose or strict source routing options were used, the option-specific pointers are updated as necessary. At this stage, the packet is ready for forwarding, and the interface selected by the table match is selected as the next hop interface.

The next hop network's maximum transmission unit size is compared against the size of the packet. If the packet is too large, and the packet's header field permits fragmentation, the packet will be fragmented to match the maximum transmission unit size. Again, an ICMP error message may be generated if fragmentation is required and if the packet header does not permit such an action.

The packet is then queued for the next hop network interface, using the output queue associated with that interface. If the queue is full, the packet may be discarded. Such a discard operation may emit an ICMP Source Quench message to the sender, indicating that a local congestion condition has resulted in a packet discard. If the sender is using TCP, this message may be interpreted as a signal to reduce the congestion window and slow down the transmission rate. The utility of this action is a matter of some debate, and some routers elect not to generate ICMP Source Quench messages or do so under relatively strict rate control measures. The packet is encapsulated (or segmented, as in the case of ATM) as per the data-link layer of the next hop, including generating the data-link layer address, and the packet is then transmitted to the next hop network.

This narrative of the packet-handling algorithm highlights the atomic actions that a router will perform on a packet:

Discard. The router will discard the packet if: the packet is addressed to a location that is not described in the forwarding tables of the router, the packet is too large for the forward path and the packet cannot be fragmented, the packet's TTL field has expired, or the output queue is full.

Forward. If the packet is not discarded, it will be forwarded along the path to the destination, using the next hop stored in the corresponding entry in the router's forwarding table.

Delay. All packets will take some time interval to traverse the router. This delay includes the actions of packet assembly from the input interface, scheduling delay, switching delay, and output queuing delay.

Fragment. A packet may be fragmented if the packet's size exceeds the forwarding path's MTU.

Routing Protocols

To maintain an accurate and consistent set of forwarding table entries, routers participate in the operation of routing protocols. Routing protocols are used to maintain (in each router) a coherent picture of the current topology of the network and a set of minimum cost paths to every reachable destination address. Two families of routing protocols are in use on the Internet: interior and exterior routing protocols. Interior routing protocols are designed for use by a single organization's network, and are intended to maintain an accurate view of the topology of the network. Exterior routing protocols are designed to connect different organizations' networks and to deliberately mask out the fine-level details of the interior topology of each organization's network.

An interior routing protocol is a distributed algorithm that operates across all the participating routers; the basic objective of the algorithm is one of traffic management distribution and synchronization, where the routers distribute local topology information across the entire set of participating routers and then ensure that each router makes a local next-hop decision that is consistent with the minimum cost path to reach a destination. Therefore, the collection of such local next-hop decisions avoids loops and is consistent with minimal-cost end-to-end paths. An interior routing protocol is used to ensure that all local traffic management decisions are based on a current and consistent picture of overall network topology. Figure 2.29 itemizes the costs of each link in an example network. The routing protocol will select the path that corresponds to the minimal path cost for node A to reach node B. The path cost is determined by the sum of the individual link costs that make up the end-to-end path.

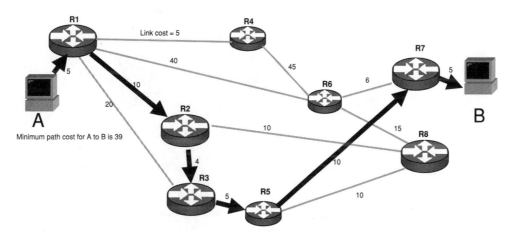

Figure 2.29 Path selection within a network.

The protocol must ensure that it has an accurate picture of the operational status of all local network interfaces. When the status changes, the router must inform the routing protocol so that any consequent adjustments to forwarding paths can be made. An interior routing protocol drives a topology-driven reachability state, where changes to the state are triggered by changes to the underlying network topology or by changes to the announcements of address reachability.

An exterior routing protocol is used as a peering protocol between two or more network domains. The aim of the exterior routing protocol is *not* to expose the dynamic state of the internal topology of one network domain to the other; indeed, it is the exact opposite. The functional objective of the exterior routing protocol is to allow two network domains to exchange information regarding destination addresses reachable from each network, without explicitly disclosing the internal paths used from the interconnection point to the destination's location. In other words, the exterior routing protocols also provide a means to describe third-party reachability, so that if network domain A is connected separately to network domains B and C, the exterior routing session between A and B can be used to inform B of destinations reachable in domain C, using domain A as a transit (see Figure 2.30).

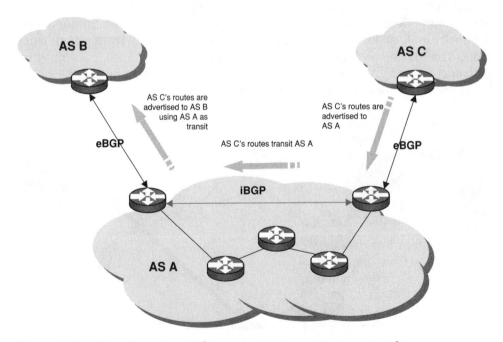

Figure 2.30 Transit routes within an exterior routing protocol.

Exterior domains need to be able to *tag* a collection of routes with the identity of the network domain where they originated. As these routes are promulgated from one network domain to the next, the routing protocol must record the sequence of domains through which the route has been promulgated. These network domains are called *autonomous systems* (ASes). A multidomain sequence is described as a sequence of these autonomous systems, and the sequence is termed an *AS path*. The role of the exterior routing protocol is to allow the network administrator to express external connection policies through the manipulation of the AS path entries within exterior routing protocols. The policy may explicitly permit or deny the routing advertisement of network addresses associated with an AS or AS path. The connection and transit policies of the local network and the network's neighbors are expressed as permitted AS path entries, which are accepted by the local network, and permitted AS path entries that are advertised to neighboring networks.

The common, and default, mode of operation of a router is to use a single forwarding table. This does not imply that an IP router is constrained to operate a single routing protocol; multiple routing protocols may be active concurrently. No matter what the number of concurrent routing processes, in this mode of operation, each router maintains a single IP forwarding table to make local switching decisions. Therefore, where multiple sources of routing information exist, including both static and dynamically learned routes, the router requires a set of precedence rules to determine in what order of precedence routing information is loaded into the forwarding table.

However, this mode of operation has performance implications, in that every packet to a given destination will be forwarded along the same path to that destination. The operation of the routing protocol generates a single view of the network topology, using a single metric of link performance, and attempts to optimize all traffic flows simultaneously, through a routing algorithm that attempts to uniformly minimize path metrics. If it is desired to allow a network to deliver differentiated levels of performance by utilizing different paths through the network, this single forwarding table and single-link metric model must be altered.

The alteration of the forwarding behavior to allow quality-based behavior modification can take a number of forms. It may be through the use of multiple forwarding tables, where the routing protocol must maintain a number of different metrics for a link and generate discrete network topologies that are a reflection of optimization for each metric. The alternative is that each router maintains a vector of local information relating to the performance attributes of each connected next-hop network. A particular flow sets up a sequence of per-flow forwarding entries in the sequence of routers between the source and the destination; the subsequent traffic flow travels along this sequence of remembered per-flow states. We will examine quality-based routing systems in further detail in Chapter 3, "Performance Tuning Techniques."

Network Management

Routers also can undertake a number of management functions to assist in network management. The router maintains a set of counters that indicate the following performance attributes:

♦ Router performance
♦ Router resource use (such as queue depth, queue drop counters, CPU load, and router memory use)
♦ Link use (transmitter and receiver counts of octets and packets and collision counters for shared access media)
♦ Link condition (error counters)

These functions are monitored by the router and aggregated into variables that are polled via a Simple Network Management Protocol (SNMP) network management-compliant utility. In addition to polling, SNMP traps can be used by the router to generate reports of real-time events, such as link status changes or connection events. SNMP traps are transmitted via UDP, and are prone to packet loss, so that a reliable management system uses both traps and polling to monitor the operational state of a network.

End-to-end integrity of network links is also part of the router management function. The router relies on two indicators of interface integrity:

♦ The control lines of the physical interface, which provide an indication of carrier status and clock integrity.
♦ The use of a lightweight end-to-end integrity monitoring function.

For point-to-point circuits, routers normally use a lightweight link-monitoring protocol, or *keepalive protocol,* to ensure: that the circuit remains up in both directions, that the circuit has not shifted to a loopback condition, and that the remote router is operational. This is accomplished by a simple counter reflection protocol undertaken at the link layer. Each router maintains a per-link counter variable and a timer. The counter is passed to the remote router, which responds with a reflection of this counter and sends its counter value. The local router increments its counter and sends this new value to the remote end, together with a reflection of the remote counter. If the router does not see a reflection within a set interval, the link is reset and marked as unavailable.

Additional Router Functions

Although the functions of packet forwarding, router protocol operation, and network management form the core router functions, routers have been considerably refined to adapt to increasing scale and broader utility of operation.

Routers can apply a set of classification controls (or filters) to modify the basic action of the forwarding process. Filters can specify a number of match criteria, which are applied to packets. These criteria can take the form of particular destination addresses, a range of addresses, a set of source addresses, or a combination of the two. The match criteria also can include specification of the protocol within the IP payload, whether it be TCP, UDP, or ICMP, and the port addresses used by TCP and UDP. For example, to match mail traffic to the address 218.130.1.1, a filter rule could be specified as in Figure 2.31.

The consequent actions of the router following a filter match could be to discard the packet (as may occur within a firewall function), to forward the packet to a particular interface regardless of the forwarding table entry (policy-based forwarding), or to change the packet's relative priority in the forwarding operation (quality of service policies). Essentially, the set of actions available with filters allow the network manager to alter the default router behavior, to ensure that traffic conforms to a set of local policy constraints, using a set of administrative tools and related criteria of applicability.

In addition to these classification functions, the router's queuing and discard functions can be manipulated to produce various performance outcomes.

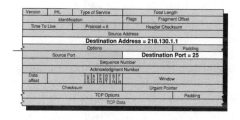

Figure 2.31 Router filter specification.

By combining classification with these functions, various differentiated service responses can be created within the network. We will examine these router functions, and their service outcomes, in further detail in Chapter 3.

Routing and Switching

As the network continues to scale, the task of the network manager is to deploy hardware platforms that can manage traffic volume. In recent years, the debate between routing and switching has achieved some attention, and various ISP networks have used both approaches to meet the demands imposed on them.

> **Routing.** Uses a forwarding decision that is local-context free, because a circuit or other soft network state does not have to be established beforehand in order for the packet to be forwarded. The forwarding decision is based on the network-level protocol destination address of the packet, rather than on either the source of the packet or the identity of the preceding router in the packet's path through the network. Routing is a Layer 3 forwarding function, using the ultimate destination address as the basis of the forwarding decision.

> **Switching.** Describes a generic technique in which the local forwarding decision is determined by attributes of the data-link layer encapsulation and in which the IP header is not used within the forwarding decision. This decision may involve some form of soft network state being set up within the switching fabric to create the appearance of end-to-end virtual circuits at the IP level. This local state is used to generate switching-level circuit descriptors that are used to make local forwarding decisions in the interior of the network. Switching is a Layer 2 forwarding function, using local information to make the next-hop forwarding decision.

Switching is intended to reduce the time and size of the forwarding function when performing routing at every forwarding point. Switching replaces the hop-by-hop IP destination address lookup with the model that uses a path through the switching fabric, which is selected upon ingress to the first switch. This technique can prove highly efficient in certain deployment environments, because the switching decision uses a smaller selection descriptor space. The size of the switch descriptor space corresponds to the size and complexity of the topology of the switched environment, whereas the IP forwarding descriptor space is the order of size of the number of distinct reachable destinations carried within the network's routing environment.

For example, in the topology in Figure 2.32, the forwarding decision is relatively simple because only two outcomes are possible. If X and Y are large-transit networks, the decision at R2 whether to forward the packet to networks X and Y involves a lookup in a table of all potential Internet destination addresses to make a choice between two outbound paths. Within a switching fabric, encompassing R1 and R2 as switches, the network feeding R1 would make the IP decision as to whether to direct the packet to X or Y at R1; then the packet would be encapsulated with a switch identification that would allow R2 to undertake a local lookup in a table of two entries, and make a switching decision based on this information.

To Switch or Not to Switch . . .

Because a path description is predictive (if the topology of the switch fabric changes, the path descriptor may be invalidated), the deployment of switching systems in a robust environment requires some containment of scale. Switching gains its ultimate strength in a high-speed local environment in which end-to-end speed is the ultimate network objective. Switching also is used within the high-speed transmission core of some Internet service provider networks in which speed and scaling are the factors that led to the decision to switch within the backbone network.

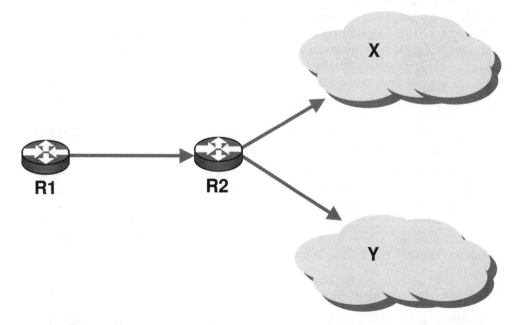

Figure 2.32 Routing and switching example.

Ultimately, the decision to switch or to route within the network is based on two very common design factors: performance and cost. Switches allow relatively simple high-capacity elements to be used within the core of the transmission network, at the cost of some increased complexity at the periphery of the switching fabric, where end-to-end switching fabric state has to be initiated and maintained. Switches also do not necessarily reduce the impact on the total routing complexity. Switching systems add an intermediate layer of switching between the IP systems and the transmission substrate that is not necessarily a cost-efficient element to add to the overall network design. Switching can allow other non-IP forms of network traffic to be injected to the network, enabling the network to be used for other purposes, as well as IP transport. Switching also can be used to support multiple IP address families on a single network, supporting various forms of private network domains simultaneously. Such additional functions can provide additional revenue streams that can offset the cost of adding this layer of functionality to the network.

Routing systems allow for higher degrees of flexibility within the choice of transmission elements and network topologies, and can allow a single packet-switching regime to operate on the underlying transmission substrate. This, in turn, may allow the network operator to realize some economies of scale of operation.

There is one other reason that attention is turning to switching systems within the network. The issue with routing systems is that the system converges to a single optimal path for each destination. Such systems are not capable of taking advantage of path diversity, and tend to concentrate traffic along a small number of heavily used paths. One of the most promising uses of switching is to create a number of diverse paths across a network, then to spread the load across these paths. This is a form of *traffic engineering*, whose intent is to make more effective use of the entire network's infrastructure to manage peak loads, rather than having a number of critical traffic congestion points within a common network backbone.

Some hybrid switching and forwarding mechanisms also are being devised, which attempt to use the outcome of a routing process that computes end-to-end paths, such as OSPF, to create path descriptors, which then are used within a packet encapsulation to allow the packet to be forwarded through a switching fabric. These Multi-Protocol Label Switching (MPLS) techniques are being considered by the IETF for adoption as standards [ID-mpls-arch].

Multi-Protocol Label Switching

The MPLS moniker describes a united effort in the IETF to blend the best of several similar switching concepts into a standardized framework and protocol suite. The IETF draft framework document says that MPLS as a "base technology (label swapping) is expected to improve the price/performance of network layer routing, improve the scalability of the network layer, and provide greater flexibility in the delivery of (new) routing services (by allowing new routing services to be added without a change to the forwarding paradigm)" [ID-mpls-frame]. MPLS does hold some interesting possibilities in the realm of QoS, and for this reason we introduce MPLS here.

---

*Y*ou can find the IETF MPLS Working Group home page, along with any associated documents, at www.ietf.org/html.charters/mpls-charter.html.

The terms *label switching* and *label swapping* are used as an attempt to convey an amazingly simple concept—although you do need to know a bit of the historical context and background information to understand why the basic routing and forwarding paradigm of today is somewhat less than ideal. To appreciate the label-swapping concept, you must examine the current *longest-match* routing and forwarding paradigm used today. The current method of routing and forwarding is referred to as longest match because a router references a routing table of variable-length prefixes, and installs the "longest," or most specific, prefix as the preference for subsequent forwarding mechanisms. Consider an example of a router that receives a packet destined for a 199.1.1.1 host address. Suppose that the router has routing table entries for both 199.1.0.0/24 and 199.1.0.0/16. Assuming that no administrative controls are in place that would interfere with the basic behavior of dynamic routing, the router will (based on an algorithmic lookup) choose to forward the packet on an output interface toward the next-hop router from which it received an announcement for 199.1.0.0/24. Why? Because it is a "longer" and more specific prefix than 199.1.0.0/16.

At first, the significance of this longest-match paradigm may not seem important. But given the number of prefixes in the global Internet (currently hovering in the neighborhood of about 60,000 variable-length unique prefixes), and the realization that the growth trend most likely will continue to increase, it is important to note that the amount of time and computational resources required to make path calculations is directly proportional to the number of prefixes and possible paths. One of the expected results of label swapping is to reduce the amount of time and computational resources required to make these decisions. The size of the label space is loosely bounded by the number of paths through the network. Given that at any routing point within the network the number of paths always is far fewer than the number of address prefixes, the goal here is to reduce the cumulative switching decision in an n node configuration from a calculation in a space of order (n times the number of unique prefixes) to a calculation in a space that can be bounded approximately by order ($n!$). For closed systems in which the number of unique end-to-end paths is well constrained (which implies relatively small values of n), this labeling technique can facilitate a highly efficient forwarding process within the router.

To a certain extent, the concept of label swapping involves replacing the need to do longest-match, by inserting a fixed-length label between the network-layer IP header (Layer 3) and the link-layer header (Layer 2), which can be used to make path and forwarding decisions (see Figure 2.33). Determining a best path based on a set of fixed-length values, as opposed to the same number of variable-length values, is computationally less intensive and thus takes less time.

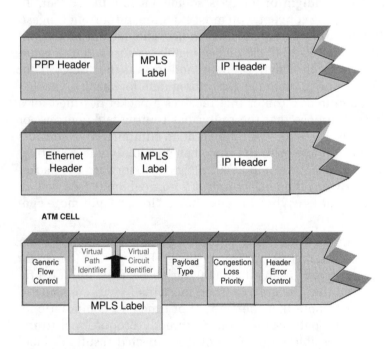

Figure 2.33 Inserting an MPLS label.

MPLS has jokingly been referred to as Layer 2.5, because it is neither Layer 3 nor Layer 2, but is inserted somewhere in the middle between the network layer and the link layer. The network layer could be virtually any of the various network protocols in use today, such as IP, IPX, AppleTalk, and so on; thus, the significance of *multiprotocol* in the MPLS technology framework. Because of the huge deployed base in the global Internet and the growing base of IP networks elsewhere, however, the first and foremost concern—and application—for this technology is to accommodate the IP.

Alternatively, MPLS can be implemented natively in ATM switching hardware, where the labels are substituted for (and situated in) the VP/VC identifiers.

Conceptually, the labels are distributed in the MPLS network by a dynamic label-distribution protocol, and they are bound to a prefix or set of prefixes. The association and prefix-to-tag binding are done by an MPLS network edge node, a router that interfaces other nodes that are not MPLS-capable. The MPLS edge node exchanges routing information with non-MPLS-capable nodes, locally associates and binds prefixes learned via Layer 3 routing to MPLS labels, and distributes the labels to MPLS peers.

The effective result is that MPLS maintains a set of label-switched paths across the MPLS network (see Figure 2.34). In the basic mode of operation, these paths are linked to the state of the IP routing system, so that a change of routing state will generate a change of the ingress prefix-to-tag binding. This allows the interior of the network to undertake forwarding decisions based only on the local label address space. The key aspect of this technology is that the ingress switching node can select which path is used to traverse the MPLS network. This allows considerable latitude for the construction of traffic engineering systems, where an ingress system shares its transit load across multiple paths, and for QoS support, where a path is selected according to the desired service characteristics.

QoS and MPLS

QoS has several possibilities with MPLS. One of the most straightforward is a direct mapping of the 3 bits carried in the IP precedence of the incoming IP packet headers to a Label CoS field, as proposed in Cisco Systems' contribution to the MPLS standardization process, *Tag Switching.* For all intents and purposes, the terms tag and label can be considered interchangeable. As IP packets enter an MPLS domain, in addition to the functions mentioned earlier, the edge MPLS router is responsible for mapping the bit settings in the IP packet header into the CoS field in the MPLS header, as shown in Figure 2.35.

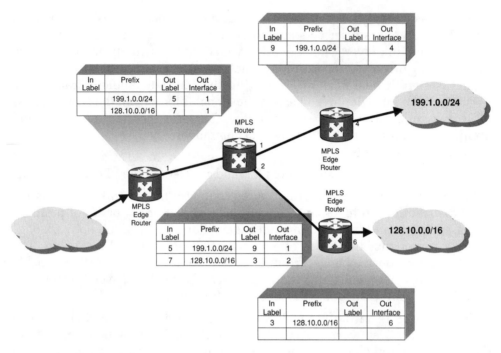

In Label	Prefix	Out Label	Out Interface
9	199.1.0.0/24		4

In Label	Prefix	Out Label	Out Interface
	199.1.0.0/24	5	1
	128.10.0.0/16	7	1

In Label	Prefix	Out Label	Out Interface
5	199.1.0.0/24	9	1
7	128.10.0.0/16	3	2

In Label	Prefix	Out Label	Out Interface
3	128.10.0.0/16		6

Figure 2.34 The label-swapping process.

Several parallel paths may exist from one end of an MPLS network domain to another, for example, each of varying bandwidth and utilizations. It certainly is possible to choose explicit ingress-to-egress paths for each specific CoS type, each path offering a distinct differentiated characteristic. Traffic labeled with higher CoS values could be forwarded along a higher-speed,

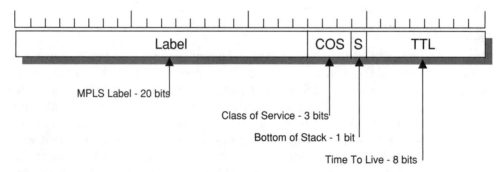

Figure 2.35 CoS bits carried in the MPLS label.

shorter-delay path, whereas traffic labeled with a lower CoS value could be forwarded on a lower-speed, longer-delay path. This example illustrates one method of providing differentiated service with MPLS; different approaches certainly exist. In fact, traffic engineering could be performed without consideration of the CoS designation altogether. It could be based on other criteria, such as source address, destination address, or per-flow characteristics; or it could be in response to RSVP messages.

One of the most compelling uses for MPLS is traffic engineering for IP networks, allowing traffic to use multiple paths, rather than the single optimal path used by conventional routing-based networks. This approach is based on the capability of MPLS to create explicit *label-switched paths* from one edge node of an MPLS network domain to another. It is envisioned that label-switched paths could be determined, and perhaps modeled, with a collection of traffic engineering tools that reside on a workstation and then downloaded to network devices. One approach is to maintain a *path state* with each path, along with a record of available capacity within the state information. Each ingress node can use this information to place traffic across multiple paths, attempting to balance the available capacity on each path [ID-mpls-te]. The intent of such measures is to avoid single-path congestion events where the underlying transmission infrastructure does support a rich mesh of connectivity. This is an important aspect for traffic-engineering purposes, giving network administrators the ability to define explicit paths through an MPLS cloud based on any arbitrary criteria. Traffic-engineering functions such as this may contribute significantly to enhanced service quality. As a result, some method of offering differentiated services may be possible using this type of explicit path switching technology.

MPLS traffic engineering in response to RSVP messages is one interesting possibility—the MPLS framework draft suggests that any MPLS-compliant implementation *must* be interoperable with RSVP [ID-mpls-frame]. Although this mandate does not mean that MPLS would provide an explicit forwarding path through the network in response to RSVP messaging, it certainly is a confident expectation for MPLS networks. The envisaged mode of operation is that RSVP would be used to establish a path across an MPLS network, using a best match between the MPLS node resource availability and the flow's resource requirements to establish the path. For this to operate correctly, each MPLS switch must maintain a record of the available RSVP capacity per egress interface, and the amount of this capacity already committed via RSVP to active QoS-mediated flows. A new QoS flow can only be installed on an interface when there is available capacity from the reserved RSVP allocation. On establishment of the path, the MPLS switches can associate the reserved resource with the explicit per-flow state, as described by the incoming label and interface. At this stage, the QoS flow is associated with a discrete *label-switched path* (LSP), and the interior switches along the LSP do not require an additional traffic classification step to determine whether a packet is part of a QoS flow. If the incoming packet or cells are part of a known QoS flow, the incoming label and interface will identify this to the switch. The label-swapping process not only can determine the next hop interface and label, but it can also trigger a desired queuing response to meet the end-to-end service conditions.

Router Requirements

The Internet document RFC1812, "Requirements for IP Version 4 Routers" (commonly referred to as the Router Requirements RFC), specifies much of the functional requirements for IP routers [Baker 1995]. The document describes the Internet architecture and the role of the router within this architecture, then discusses specific router requirements at each layer of the OSI seven-layer reference model.

> **Link layer.** The document specifies that for point-to-point lines, a conforming router must support the PPP protocol. The protocol includes the operation of the PPP Link Control Protocol (LCP), which can negotiate a number of options, including header compression, asynchronous character maps, maximum packet size negotiation, link-quality monitoring, and link-integrity monitoring via magic number exchange.

> **Internet layer.** At the Internet layer, the router may support the IP options of IP source routing, record route, and timestamp options. Support for source routing may have implications for the network's security policy, as source routing can be used in conjunction with source address spoofing to subvert simple filter sets. A router must verify the IP checksum and discard packets that fail the check. The router must implement fragmentation; and where required and permitted by the packet, the fragmentation should ensure that the least number of fragments are generated. The router also must check and decrement the TTL field on forwarding. Routers also must support the generation of ICMP messages.

The Router Requirements RFC indicates that routers should not generate ICMP Source Quench messages. In the event that it does, it should provide some mechanism for limiting the rate at which such messages are sent. This area is one of continuing research; mechanisms to inform flow-controlled sessions of the onset of router resource contention and the most effective means of host response to such messages remain an active topic of consideration. ICMP Source Quench messages can be regarded as a means of explicit congestion notification, and can be used by the sender as a signal to reduce the congestion window. A more effective approach is the use of Explicit Congestion Notification [RFC2481]. We will examine this approach in further detail in Chapter 3.

The Router Requirements RFC also mandates the support of statically entered routes, as well as support for OSPF as a routing protocol for interior routing applications. It also indicates that if support for exterior routing protocols is provided, BGP must be supported as a candidate exterior routing protocol. Finally, the router must be manageable by SNMP, and all SNMP operations must be supported, with a core set of management information bases (MIBs) included within this support.

Transmission Systems and Performance

Constructing an IP transmission infrastructure for performance is more of an art than a well-understood network engineering task. There are as many designs for IP transmission capacity as there are IP networks. Rather than selecting various transmission options from a rich selection of available alternatives, the approach to date has been characterized by adapting what is available in transmission options to meet the requirements of the IP network.

The basis for this approach is largely historical. For the last 50 years the major consumer of the world's communications facilities, both in volume and revenue terms, has been telephony. Not surprisingly, these communications systems have been constructed with a primary objective of meeting the requirements of the telephone network. Data transmission services have traditionally been a smaller-scale activity, and to date, have been provisioned on the margins of over-supply of infrastructure constructed to support voice. Data systems have been constructed as an adjunct to a well-provisioned and much larger voice transmission system. North American 56K, T1, and T3 leased circuit services, and 64K, E1, and E3 leased circuit services elsewhere, are the resale of components of the voice transmission hierarchy.

The data market is now growing very rapidly in size, and the sustained exponential growth experienced with data over the past decade is now challenging the far slower linear growth trends of telephony. As Internet systems continue to grow in size, they will start to exert significant levels of pressure on the design of the underlying transmission systems, creating a transmission infrastructure attuned to their requirements. However, a communications infrastructure designed solely to support data still remains in the future in most parts of the world, and the more common task is to adapt transmission systems that are part of the telephony infrastructure into useful data conduits.

Defining Transmission Systems

Transmission systems have three major characteristics: *bandwidth, delay,* and *cost:*

Bandwidth. Commonly interpreted to mean the capacity of a line, as measured in bits per second. But strictly speaking, bandwidth, capacity, and bit rate are different concepts. Bandwidth is the information-carrying capacity of a link; capacity is the number of analog symbols per second; and bit rate is the product of capacity and the mean number of bits per symbol.

Delay. A signal propagation metric. Electromagnetic radiation propagates at a speed of some 3.34µs/km in a vacuum. The speed of propagation in optical fiber depends on the fiber's refractive index, and is typically 5.13µs/km microseconds per kilometer. This implies that a transcontinental link of some 3900km has a one-way propagation delay of 20ms, while an undersea cable circuit may extend over 8500km of cable and have a one-way propagation delay of 44ms. Geostationary satellite systems have an altitude of 35,784km, which corresponds to a 239ms one-way (corresponding to the path up to the spacecraft and back to Earth) propagation delay for stations directly under the satellite and a propagation delay of 281ms for a more typical configuration of Earth stations at elevated latitudes.

Cost. Link cost is not a physical characteristic, but one that nevertheless is very relevant to transmission choices. In general, link cost is an expression of the original price of infrastructure installation, the level of demand for the facilities, and the anticipated rate of financial return by the original investors in the facility. Infrastructure installation may include more cost factors than the price of the conductor and any required signal amplifiers or regenerators. The cost of right of access for terrestrial systems, the cost of geostationary orbital slots, or the cost of restoration capacity and repair for undersea cable systems also may be significant factors in the total cost of the infrastructure. The rule of thumb is that the longer the link, the greater the cost, although for some satellite-based systems this is not necessarily the case. Also, within the same area of general observations, the higher the bandwidth, the greater the cost, although this is not usually in direct proportion, and the unit cost of bandwidth may decline as link bandwidth increases.

The Carrier Hierarchy

Much of the world's current communications system is based on the premise that the human spoken word uses a limited range of frequencies and has limited dynamic range. The world's telephone network is attuned to reproduce the spoken word with acceptable clarity; to do so, it uses a system that can carry analog signals of between 350Hz and 3400Hz. However, the world's telephone network is largely a digital network, where the analog signal is carried within the telephone network as a digital data stream.

The technique used to convert this analog signal to a digital signal is Pulse Code Modulation (PCM). The first step is to transform a continuous analog signal into a sequence of samples. The voice carriers standardize on a rate of 8000 samples per second to enable clear voice reproduction (sampling at more than twice the highest frequency that must be reproduced), effectively choosing a network clock base of 125μs. The number of bits used to encode the amplitude of each sample is the next conversion issue. The voice carriage industry standardized on an encoding that uses 256 levels, called *quantization levels*. Conveniently, this maps to an 8-bit encoding value. The result is that a voice call is mapped to a 64Kbps stream, and 64Kbps is the base building block of the voice carrier hierarchy.

To take individual 64Kbps streams and allow them to be carried within larger bearers across the network requires the use of a multiplexing technology. The most common multiplexing technology used in today's carrier networks is time division multiplexing (TDM).

Time Division Multiplexing

The basic problem addressed by TDM is shown in Figure 2.36. Multiplexing takes a number of discrete inputs and multiplexes them into a single higher-capacity data stream. This stream can be transmitted over a higher-capacity link and then demultiplexed back into the original discrete channels. Where the inputs are constant-rate signals, the multiplexing creates a single signal whose rate is no less than the sum of the component rates.

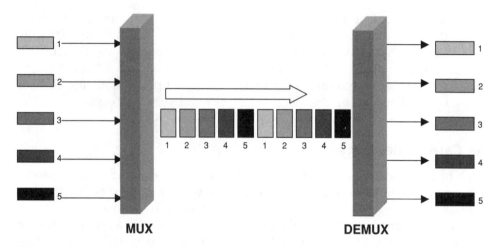

Figure 2.36 Multiplexing.

Time division multiplexing takes a frame of data from each input in turn and transmits these frames across a common link. No data is lost in this transmission model, as the speed of the common link is the sum of the component links. This is achieved within the multiplexor (MUX) by using a frame buffer for each input line. Incoming bits are loaded into the frame buffer. At every scan interval, the common line scheduler empties the frame buffer and loads it into the link driver. The common line scheduler scans each frame buffer in turn, where the complete scan of all input lines takes one scan interval. The output operation is similar. Each frame is assembled in the common input driver and then placed in the link driver buffer for the output line.

The frame buffer can be of any length, although the longer the frame buffer, the greater the latency of the multiplexing operation. The scan interval of the carrier hierarchy is typically based at the sample interval of PCM encoding of voice circuits, which is a scan time interval of 125µs, or 8000 scan intervals per second.

TDM is inherently very simple in its operation, but such simplicity is not without cost. TDM allocates a fixed amount of capacity to each input channel, whether the channel is active in transmitting a data element or not. A simple TDM MUX cannot operate on an adaptive basis where the common channel capacity is less than the sum of the input capacities, and each channel is allocated resources on the basis of data activity. Nor can TDM allow uncontrolled clock slippage of any of the input channels. The assumption in a simple TDM model is that each input line operates within a synchronized clocked mode, so that packing the common multiplexed data stream with control information is not required.

TDM systems often use a basic framing protocol, in which each scan is terminated with a single frame control bit. This bit alternates on each scan, creating a recognizable bit pattern "01010 . . . ". This pattern can be used to synchronize the demultiplexing unit to the input unit, so that the demultiplexing unit can recognize the alignment of each scan frame within the bit stream.

Perfect synchronized clocking across multiple sources is quite rare. The TDM system must allow the various clock sources to drift, within certain limits. There is no single 125µs clock pulse driving the entire voice network. The system was constructed to allow some level of longer-term clock drift between the various component PCM clocks. The TDM system compensates by providing overflow space within each scan frame, so that a clock operating at a slightly faster rate can insert additional bits into the overflow space, with an associated timeslot label to indicate which input line has generated the additional bits. This allows the TDM MUX to correct the overrunning source at periodic intervals. This is the foundation of the *Plesiochronous Digital Hierarchy* (PDH) (where plesiochronous means almost synchronous).

The Plesiochronous Digital Hierarchy

In a TDM digital switched hierarchy, the 64Kbps PCM encoded data streams are termed DS0 circuits. From this point, the carrier hierarchy is constructed (see Table 2.2).

In North America and Japan, the first level of carrier multiplexing is to take 24 of these DS0 circuits to create a DS1 circuit, operating with the inclusion of an 8Kbps framing signal stream at a data rate of 1.544Mbps (commonly termed a T1 circuit). Elsewhere, the first level of the hierarchy uses 30 DS0 circuits to create the CEPT-1 circuit, clocked at 2.048Mbps, allowing for 128Kbps framing and control signal streams, in addition to the 30 PCM streams (commonly termed an E1 circuit).

The next level of the hierarchy, DS2, is four multiplexed DS1 streams. This point in the hierarchy is not used in most carrier systems.

The next level is that of a DS3 bearer. In the North American system, this takes 28 DS1 groups and maps them into a 44.736Mbps bearer, termed a T3 circuit. In the CCITT bearer system, a CEPT-3 is a mapping of 16 E1 circuits, which is a 34.368Mbps bearer, termed an E3 circuit.

Table 2.2 Plesiochronous Digital Hierarchy

Hierarchical Level DS-x	North American CEPT-x (Kbps)	European (Kbps)	Japanese (Kbps)	International (Kbps)
0	64	64	64	64
1	1544	2048	1544	2048
2	6312	8448	6312	6312
3	44736	34368	32064	44736
4	139264	139264	97728	139264

The additional space allocated in these higher-order points of the carrier hierarchy is used to allow for slightly different clocking speeds of the individual streams, so that space is allocated for *overflow bits* to allow for clock alignment at the start of every frame. This is because the different DS1 bearers are not necessarily tightly time-synchronized with each other. If one stream is running slightly faster than the other multiplexed streams, it can overrun into the overflow bits. This technique spreads each component stream over the entire aggregate frame. This spreading of a single stream across the entire frame makes the task of extracting or inserting a single component stream into a multiplexed group without disturbing the remainder of the multiplexed streams very challenging. The more common approach is to break down the multiplexed frame into its component DS0 data frames. For this reason, every voice switch within the PDH carrier hierarchy must demultiplex the trunk signal down to the level of DS0 streams before a call can be switched.

The PDH carrier system allows for point-to-point DS0 data links operating at speeds of 56Kbps in North America, or at 64K elsewhere, to be used for private point-to-point circuits. Typically, carriers also allow for a number of DS0 circuits to be provided in a composite bundle, normally by provisioning a DS1 bearer and then marking out a number of timeslots that are available for use within the data circuit. The PDH also allows for the provisioning of DS1 point-to-point circuits: at 1.544Mbps or E1 at 2.048Mbps, depending on the locally used hierarchy, and where available, from the transmission operator. The next available service in this hierarchy is a DS3 circuit, operating at 44.736Mbps (T3), or 34.368Mbps (E3). No composite DS1 circuits are available from the transmission operators due to the operation of PDH framing, although the use of inverse multiplexors can take a number of DS1 circuits and create composite data rates between DS1 and DS3 speeds.

The Synchronous Digital Hierarchy

The issue of making the hierarchy fit a set of imprecisely synchronized clocks can be eliminated if all component data streams are synchronized to the same clock. In a precisely synchronized environment, every component stream will occupy a fixed area of the multiplexed data frame, allowing streams to be added or removed without a complete demultiplexing operation as a prerequisite. This method is used in the Synchronous Optical Network (SONET) to define a carrier hierarchy. SONET is the North American standard; the Synchronous Digital Hierarchy (SDH) standard is used in other parts of the world.

The defined speeds within the hierarchy are shown in Table 2.3. Not every speed within the hierarchy is supported by carriers, and the bolded entries in the table indicate those points in the hierarchy that are available as supported services. There is no need to add overflow bits into the multiplexed frames because the data streams are synchronously clocked. The data rates are now exact multiples of the data rates of the lower-speed trunk systems. The frame format remains locked to the constant 8000 frames per second base, or 125 µs per frame. The base frame is an OC-1 frame, which is 810 bytes. Each byte within an OC-1 frame corresponds to one 64K DS0 stream. Higher-order rates are constructed by interleaving OC-1 frames using *byte interleaving*. Higher levels in the hierarchy are provided to customers either with explicit SDH internal framing or as an unframed clear channel (*concatenated*) service. The concatenated services are designated with a trailing "c," as in OC-3c, or STM-16c.

Within the carrier industry, the OC-3c (STM-1) services are commonly available. The OC-12c services are also relatively common, whereas the STM-

Table 2.3 Synchronous Digital Hierarchy

Hierarchical Level	North American Designation	International Designation	Data Rate (Mbps)
1	OC-1		51.84
2	**OC-3**	**STM-1**	**155.52**
3	OC-9	STM-3	466.56
4	**OC-12**	**STM-4**	**622.08**
5	OC-18	STM-6	933.12
6	OC-24	STM-8	1244.16
8	OC-36	STM-12	1866.24
9	**OC-48**	**STM-16**	**2488.32**
10	**OC-192**	**STM-64**	**9953.28**

4c services are a more recent carrier access service and are not so widely available. The two higher service points, OC-48 (STM-16) and OC-192 (STM-64) correspond to the individual wavelength stream capacities within an optical fiber-based wave division multiplexing (WDM) system. These two high-speed systems are currently the highest aggregation point within the SDH-based carrier systems.

Digital Circuits

A number of point-to-point digital transmission services were introduced to the data services market following the introduction of digital infrastructure into the telephone network. The earliest of these services was the Digital Data Service (or DDS). DDSes are point-to-point 56Kbps circuits. They are provisioned as 64K digital circuits within the digital transmission network. On the copper loop, the loop is groomed to remove loading coils, taps, and other sources of unwanted noise and distortion. The encoding used on DDS systems is an alternate mark inversion (AMI), in which each 1 bit is represented alternatively by a positive and a negative voltage pulse, intended to avoid a buildup of capacitance within the loop. The properties of this encoding include a net zero DC voltage and relatively easy detection of impulse noise. However, no inherent clocking of 0 bits takes place, allowing the two end points to lose clock synchronization with a long sequence of 0-valued bits. DDS encoding contains a clock signal that uses 1 bit in every 8, encoded as a 1 value, to ensure the maintenance of clock synchronization. The DDS data rate is, therefore, 7 bits of every 8-bit frame, or 56Kbps.

A variant of the encoding system addresses the potential loss of clock synchronization on long strings of 0 bits by using bipolar eighth-zero substitution (B8ZS) line encoding. This method offers clear channel capacity at 64Kbps. The encoding system requires slightly greater complexity, in which every sequence of eight 0 bits is encoded with a pair of bipolar pulses on the eighth 0. This preserves net DC voltage and sufficient pulse density to maintain synchronous clocking at both ends.

Higher-speed digital circuits are based on a framed T1 (or E1) connection and make use of a number of 64K basic circuits in parallel. Commonly available carrier services use a number of 64K frames from a framed T1 (or E1) with a customer service unit (CSU) aggregating these framed segments and presenting them to the customer as a single-clocked service. Carrier services also typically include the provision of clear channel T1 (or E1) circuits, which allow the customer to use CSU equipment to clock at the effective data rate of the access circuit.

These circuits are all based on the PDH carrier hierarchy, as discussed in the previous section. They offer an end-to-end synchronized data clock that preserves bit-level and packet-level clocking between the sender and receiver.

In general, the lower-speed digital circuit services are used in the Internet network as *access circuits*, to provide a permanent connection from a customer's premises to the network service provider's equipment. Higher-speed digital circuits are often used within the service provider's network to construct the core backbone of the network.

SONET and SDH Circuits

The digital circuits just described create point-to-point circuits that operate at speeds between 64kbps and 1.544Mbps (or 2.048Mbps in an E1 hierarchy). There are three approaches to create higher-speed digital circuits. One is to use a technique known as *inverse multiplexing*, to bond a number of parallel T-1 (or E-1) circuits and present the bonded aggregate of these circuits as the new circuit clock. This technique can provide circuits that operate at speeds of up to 16Mbps. For higher speeds, the T-3 (45Mbps) or E-3 (34Mbps) services can be used. For still higher circuit speeds, it is necessary to use different carrier services, and it is here that the Synchronous Digital Hierarchy (SDH) is used.

The SDH uses a tightly synchronized clocking environment to create a digital transmission hierarchy. Framing techniques within SDH can compensate for clock drift within the various constituent circuits. This allows for the construction of high-speed digital transmission hierarchies that can be easily combined or split. Individual circuit groups can be peeled off the aggregate circuit or readily inserted into a vacant circuit slot without disassembling the entire timeslot. The SONET Add/Drop Multiplexor (ADM) can insert and remove a data flow into an SDH stream. This capability does not eliminate the requirement to set up a channel within an SDH bearer system manually, but it does allow the use of individual channel manipulation without having to demultiplex the entire circuit hierarchy. SDH systems are usually configured in a dual ring structure, using the architecture of a working and a protection ring. The ADM automatically switches data into the protect ring in the event of failure of the working ring.

The synchronized clocking also allows the carrier speed to be increased well beyond the 34Mbps, 45Mbps, and 140Mbps speeds, which are the typical ceiling of transmission services provided from the PDH hierarchy. The services emerging from the SDH bearer system are STM-1 at 155Mbps and STM-4c at 622Mbps and STM-16c circuits at 2.5Gbps. The comparable systems within the SONET designation are OC-3c circuits (155Mbps), OC-12c

(622Mbps), and OC-48c (2.5Gbps). The more recent digital offerings in the North American environment also include the provision of OC-192c circuits, with a clock rate of some 10Gbps.

Such high-speed circuits are commonly used to support backbone circuits within a service provider's network, where the demand for ever-increasing circuit speeds is a by-product of the escalating demand for Internet services.

ISDN

The digital circuit services discussed so far share a second common attribute, in addition to synchronized data clocking: that services are statically configured within the bearer network on an end-to-end basis. If the customer wants to change the location of either end of the circuit, the carrier must reconfigure the digital circuit service network to relocate the circuit. This administrative overhead is an additional factor to the cost of operating point-to-point digital circuits. The intent of the Integrated Services Digital Network (ISDN) is to combine the utility of digital end-to-end circuits with the switching systems used in the Public Switched Telephone Network (PSTN) environment, allowing digital calls to be dynamically created and torn down by the customer.

The ISDN system is part of the switched telephone network and uses the same switching services. Telephone and data services can be accessed via ISDN. The primary difference between the two is that in the ISDN architecture, the local loop is not an analog signal. The local loop is a collection of 64K data channels, which are effectively extensions of the internal digital channels used to carry voice circuits within the PSTN.

Two access services are defined within the ISDN architecture: a Basic Rate Interface (BRI), and a Primary Rate Interface (PRI). A BRI uses three separate channels into the network, a 16Kbps signaling channel (the *D channel*), and two independent 64Kbps data channels (*B channels*), that can be used for data or voice. The D channel is used to control the B channel connections, via a set of call control messages (the format of these messages is defined in the ITU-T standard Q.931), to initiate, and terminate, B channel calls. Each B channel call is a clear channel 64Kbps clear channel circuit. In this architecture, the customer equipment is more complex than that used to terminate digital end-to-end circuits. The customer equipment now has to manage the three circuits and use Q.931 signaling to manage calls on the two B channels. ISDN BRI interfaces can be used to initiate calls or answer incoming calls.

A PRI uses a configuration of 23 B channels in North America (30 B channels elsewhere) and a 64Kbps D channel. Again, the D channel is used to control the calls made on the B channels; the operation is similar to that of the BRI.

The use of ISDN for infrastructure transmission is dictated by the prevailing carrier tariff. It may entail a slightly greater customer investment in equipment, as the switching control is now a function of the attached equipment, but this additional cost may be offset by slightly lower tariffs for the ISDN service, as compared to the lease costs of a 64Kbps point-to-point digital circuit. ISDN is not a high-speed transmission system; the ISDN service network treats each 64K circuit as an independent call. It is the responsibility of the customer's equipment to use inverse multiplexing (or *bonding*) to group bundles of B channel calls and create the functional equivalent of a larger-capacity aggregate channel.

Currently, the major use of ISDN within the Internet environment is to terminate modem and ISDN-based access calls from service provider clients. The widespread use of V.90 high-speed analog modems by the dial-up customer base assumes that the service provider end of the call uses a digital interface to the PSTN network. This is often implemented using an ISDN PRI, where the dial-in network access server terminates the incoming access calls using onboard digital signal processors. At one stage in the evolution of ISDN, it was envisaged that the entire analog access medium would be replaced by ISDN, allowing the consumer to make both voice calls and data calls from the same network termination. This has not eventuated in most markets to date, and it is becoming evident that the digital access market is already shifting its focus to high-speed access systems using either hybrid fiber coax (HFC) common broadband systems or high speed digital subscriber loop (DSL) techniques, where the copper loop is operated with a digital encoding that allows a data transfer rate of some megabits per second. Within this environment, ISDN is having some problems in finding a natural and viable market niche.

X.25

From the synchronously clocked end-to-end digital circuits, our overview of transmission technologies and services moves into switched services. These services do not provide end-to-end synchronous clocking of data; they operate in a manner similar to IP itself; carrying discrete packets of data through a switching environment, usually mimicking a circuit-switching environment. A call creates an end-to-end sequence of local switching decisions that are set up as the call is initiated; the data packets then are carried along this virtual circuit (VC) for the duration of the call.

X.25 is a specification of an interface into a packet-switched transport technology developed by the telephony carriers in the 1970s. It was designed to address emerging corporate data communications needs through a common data-switching infrastructure that had many technical parallels with telephony architecture. The technology model used for the X.25 protocol was much like traditional telephony, using many of the well-defined constructs within telephony networks. The X.25 specification does not specify the precise internal characteristics of the packet-switched network, as it is limited to describing the external interfaces to such a packet switched network. X.25 supports switched virtual circuits, in which one connected computer can establish a point-to-point dedicated connection with another computer (equivalent to a call or a virtual circuit).

*T*he X.25 model essentially defines a reliable telephone network for computers. After setting up a virtual circuit, the interface specification assumes that packets, once accepted into the network from an interface, are reliably passed along the associated virtual circuit and are correctly delivered, in the correct order, to the output circuit interface. The only characteristic of the network that is not held constant is the interpacket timing, and the relative timing of packets accepted by the network may not be reproduced as the packets leave the network.

The X.25 protocol is not a complete end-to-end transport protocol, as it only specifies the interface between the X.25 network and the X.25 client. The network boundary point is the *data communications equipment* (DCE), or the boundary network switch, and the customer premise equipment (CPE) is the *data termination equipment* (DTE), where the appropriate equipment is located on the customer premises. X.25 specifies the DCE/DTE interaction in terms of framing and signaling. A complementary specification, X.75, specifies the switch-to-switch interaction for X.25 switches within the interior of the X.25 network.

*M*any computer and network protocol descriptions suffer from "acronym density." The excuse given is that using more generic words would be too imprecise for protocol definition. Therefore, the designers of each protocol proudly develop their own terminology and related acronym set.

An alternative view is that each protocol demonstrates its uniqueness solely through the invention of yet more acronyms. X.25 and Frame Relay are no exception to this general rule.

This book stays with the protocol-defined terminology and acronyms for reasons of precision and brevity of description, and we ask for your patience as we work through the various acronym-dense protocol descriptions.

The major control operations in X.25 are *call setup*, *data transfer*, and *call clear*.

Call setup. Refers to the creation of a virtual circuit between two DTEs. The circuit operates as a reliable flow-controlled circuit. The call setup operation consists of a handshake: One computer initiates the call, and the call is answered from the remote end by returning a signal that confirms receipt of the call. Each DCE /DTE interface then has a locally defined local channel identifier (LCI) associated with the call. All further references to the call use the LCI as the identification of the call instead of referring to the called remote DCE. The LCI is not a constant value within the network, and each data-link circuit within the end-to-end path uses a locally unique LCI. The initial DTE/DCE LCI is mapped to successive channel identifiers in each X.25 switch along the end-to-end call path, using a simple LCI swapping algorithm. Thus, when the client computer refers to an LCI as the prefix for a data transfer, it is only a locally significant reference. When a frame is transmitted out on a local LCI, the DTE simply assumes that it is correctly being transmitted to the appropriate switch, because of the configured association between the LCI and the virtual circuit. The X.25 switch receives the HDLC (high-level data link control) frame and then passes it on to the next hop in the path, using local lookup tables to decide how to similarly switch the frame out on one of various locally defined channels. Because each DCE uses only the LCI as the identification for this particular virtual circuit, there is no need for the DCE to be aware of the remote LCI.

Data transfer. Takes place across this reliable network transport protocol, so all data frames are sequenced and checked for errors.

Call clear. Removes the sequence of LCIs from the network, essentially, tearing down the connection. In contrast to TCP, a call clear is bidirectional, and the implication of a clear signal is that no further packet data can flow along the circuit in either direction, regardless of which system elected to initiate the call clear.

X.25 and QoS

The engineering implication for reliable quality in X.25 flows is that X.25 switch implementations require some form of flow control. Each interior switch within the network does not consider that a frame has been transmitted successfully to the next X.25 switch in the path until the next switch explicitly acknowledges the packet transfer. The end-to-end transfer is verified by each intermediate switch for dropped and duplicated packets by a sequence number check. Because all packets within an X.25 virtual circuit must follow the same internal path through the network, out-of-sequence packets are readily detected within the network's packet switches by this simple sequence number check. The X.25 network switches implement switch-to-switch flow control and switch-to-switch error detection and retransmission; they also preserve packet sequencing and integrity.

This level of network functionality allows relatively simple end systems to make a number of assumptions about the transfer of data. In the normal course of operation, all data passed to the network will be delivered to the call destination, in order and without error, with the original framing preserved.

In the same way that telephony uses simple peripheral devices (dumb handsets) and a complex interior switching system (smart network), X.25 attempts to place the call- and flow-management complexity into the interior of the network. X.25 creates a simple interface with minimal functional demands on the connected peripheral devices. In contrast, the Transmission Control Protocol (TCP) implements the opposite technology model, with a simple best-effort datagram delivery network (dumb network) and flow control and error detection and recovery in the end system (smart peripherals).

Such reliability within the network comes at a price. X.25 networks are generally slow-speed because the reliable, flow-controlled, hop-by-hop protocol is not conducive to high-speed switching. This functionality is also unnecessary for IP networks, as the end-to-end control algorithms in the IP protocol are more than adequate for high reliability and performance. Furthermore, X.25 networks are not time-synchronized. An X.25 network will accept a sequence of packets at ingress, and reproduce precisely the same packet sequence at egress, but the interpacket timings are not preserved by the X.25 network, and any implicit timing within the data signal is lost. Finally, end-to-end latency is not managed within an X.25 network; consequently, jitter levels can be high within such a network.

No special mechanisms exist to provide QoS within the X.25 protocol. Therefore, any differentiation of services has to be at the network layer (such as IP) or by preferential queuing (such as priority, class-based, or weighted-fair queuing). Of course, all the external prioritization one wishes to place on IP packets at the IP level will have no impact on the underlying X.25 service, and efforts to create Internet service differentiation where the IP network is layered upon an X.25 platform may be frustrated by the operational characteristics of the X.25 network. This is a compelling reason to consider, at a minimum, Frame Relay services instead of X.25, as a wide-area technology for implementing IP QoS-based services.

Frame Relay

What is Frame Relay? As Cisco describes it, Frame Relay is an "industry-standard, switched data link-layer protocol that handles multiple virtual circuits using HDLC encapsulation between connected devices. Frame Relay is more efficient than X.25, the protocol for which it is generally considered a replacement" [Cisco 1995].

So how does Frame Relay differ from X.25? Frame Relay has been described as faster, more streamlined, and more efficient as a transport protocol than X.25. The reality is that Frame Relay removes the switch-to-switch flow control, sequence checking, and error detection and correction from X.25, while preserving the connection orientation of data calls as defined in X.25. This allows for higher-speed data transfers with a lighter-weight transport protocol. In addition, the removal of some of the switch-processing functions of X.25 allows for cheaper, faster Frame Relay switches, as compared to their X.25 counterparts.

You can find approved and pending Frame Relay technical specifications at the Frame Relay Forum Web site, located at www.frforum.com.

Frame Relay's origins lie in X.25, but were influenced by the development of ISDN technology, where Frame Relay originally was seen as a packet-service technology for ISDN networks. The rationale behind Frame Relay was the perceived need for the efficient relaying of HDLC-framed data across ISDN networks. With the removal of the X.25 protocol features of data link-layer error detection, retransmission, and flow control, Frame Relay opted for a model of unreliable frame delivery, using end-to-end signaling at the transport layer of the protocol stack model to undertake error recovery and flow control functions (the similarity to Internet protocols is not entirely accidental). This allows the network switches to consider data-link frames as being forwarded without waiting for positive acknowledgment from the next switch. This in turn allows the switches to operate with less memory and to drive faster circuits with the reduced switch functionality required by Frame Relay. The outcome of this is cheaper, faster switches.

Like X.25, Frame Relay has a definition of the interface between the client and the Frame Relay network called the User-to-Network Interface (UNI). Switches within the confines of the Frame Relay network may use any form of packet-switching technologies, including ATM cell relay switches (a technology commonly used to support carrier Frame Relay services), or dedicated circuits with HDLC-framed packets passing along such circuits. While interior Frame Relay switches have no requirement to undertake error detection and frame retransmission, the Frame Relay specification does indicate that frames must be delivered in their original order, which is most commonly implemented using a connection-oriented interior switching structure. Figure 2.37 illustrates a simple Frame Relay network.

*C*urrent Frame Relay standards address only permanent virtual circuits (PVCs) that are administratively configured and managed in the Frame Relay network. However, Frame Relay Forum standards-based work currently is underway to support switched virtual circuits (SVCs). The Frame Relay Forum recently completed the definition of Frame Relay high-speed interfaces at HSSI (52Mbps), T3 (45Mbps), and E3 (34Mbps) speeds, augmenting the original T1/E1 specifications.

The original framing format of Frame Relay is defined by CCITT Recommendation Q.921/I.441 [ANSI-frame], shown in Figure 2.38. The minimum, and the default, frame address field is 16 bits. In this field, the Data Link Connection Identifier (DLCI) is addressed using 10 bits; the extended address field is 2 bits, the Forward Explicit Congestion Notification (FECN) is 1 bit, the Backward Explicit Congestion Notification (BECN) is 1 bit, and the Discard Eligible (DE) field is 1 bit. These final 3 bits within the Frame Relay header are perhaps the most significant components when examining QoS possibilities.

Committed Information Rate

The Discard Eligible (DE) bit, in tandem with the fact that Frame Relay does not provide end-to-end synchronized data clocking, is used to support the engineering notion of a *bursty connection*. Frame Relay defines this type of connection using the concepts of Committed Information Rate (CIR) and traf-

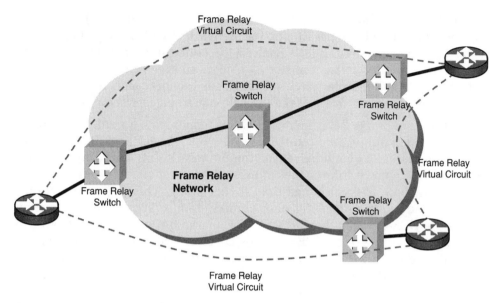

Figure 2.37 A Frame Relay network.

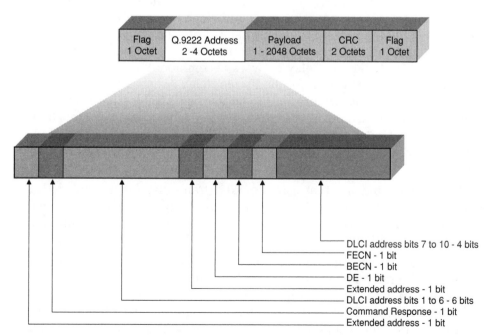

Flag 1 Octet	Q.9222 Address 2 -4 Octets	Payload 1 - 2048 Octets	CRC 2 Octets	Flag 1 Octet

DLCI address bits 7 to 10 - 4 bits
FECN - 1 bit
BECN - 1 bit
DE - 1 bit
Extended address - 1 bit
DLCI address bits 1 to 6 - 6 bits
Command Response - 1 bit
Extended address - 1 bit

Figure 2.38 The Frame Relay-defined frame format.

fic bursts, and applies these concepts to each virtual circuit (VC) at the interface between the client (DTE) and the network (DCE). (Frame Relay could never be called acronym-light!) Each VC is configured with an administratively assigned information transfer rate, or *committed rate*, which is referred to as the CIR of the virtual circuit. The CIR is an admission filter. All traffic entering the VC that is in excess of the CIR is marked as discard eligible (DE).

The first-hop Frame Relay switch (DCE) is responsible for enforcing the CIR at the ingress point of the Frame Relay network. When the CIR is exceeded, frames are marked as such by setting the frame's DE bit. This allows the network to subsequently enforce the committed rate at some point internal to the network, if there is local congestion at that point. This ingress function is implemented using a rate filter on incoming frames. When the frame arrival rate at the DCE exceeds the CIR, the DCE marks the excess frames with the Discard Eligible bit set to 1 (DE = 1). The DE bit instructs the interior switches of the Frame Relay network to select those frames with the DE bit set as discard eligible in the event of switch congestion and discard these frames in preference to frames with their DE field set to 0 (DE = 0).

As long as the overall capacity design of the Frame Relay network is sufficiently robust to enable the network to meet the sustained requirements for all VCs operating at their respective CIRs, bandwidth in the network may be consumed above the CIR rate up to the port speed of each network-attached DTE device. Using the DE bit to discard frames as a congestion management response within the Frame Relay network accommodates traffic bursts while providing capacity protection for the Frame Relay network. This method can be regarded as a *soft partitioning* of the network, in which clients' traffic may exceed the base levels allocated to them if there is no contention for the network resource. Under conditions of contention, each client is allocated its respective CIR rates.

No signaling mechanism (to speak of) is available between the network DCE and the DTE to indicate that a DE marked frame has been discarded (see Figure 2.39). This is an extremely important aspect of Frame Relay. The job of recognizing that frames somehow have been discarded in the Frame Relay network is left to higher-layer protocols, such as TCP.

The architecture of this ingress rate *tagging* is a useful mechanism. The problem with Frame Relay, however, is that the marking of the DE bit is not well integrated with the higher-level protocols. Frames normally are selected for DE tagging by the DCE switch without any signaling from the higher-level application or protocol engine that resides in the DTE device. This is particu-

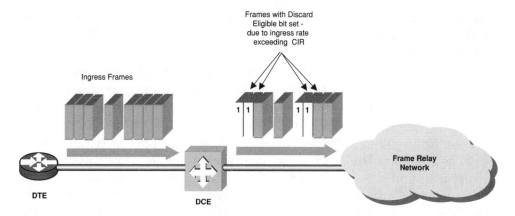

Figure 2.39 Frames with the DE bit set.

larly apparent in the interaction of TCP and Frame Relay. The worst case for TCP-over-Frame Relay occurs when CIRs are being enforced within the Frame Relay network, and the sender is in slow start mode. As the sender attempts to opens its congestion window to a rate that exceeds the CIR, the tailing sequence of packets in the transmission burst will have their Frame Relay frames marked as discard eligible. When these packets are discarded within the Frame Relay network, the sender may not receive the necessary duplicate ACKs to drop back into congestion avoidance mode, and instead may have to time-out and recommence slow start with an initial congestion window of one segment. The result is that instead of operating at a data transfer rate near to the CIR rate, the data transfer rate may be substantially lower, governed by the sender's RTT estimate and the associated retransmit timers.

Frame Relay Congestion Management

There is more to Frame Relay congestion management than the single response of CIR rate enforcement. Frame Relay congestion control is handled in two ways: *congestion avoidance* and *congestion recovery*. Congestion avoidance consists of a Backward Explicit Congestion Notification (BECN) bit and a Forward Explicit Congestion Notification (FECN) bit (yes, more acronyms!). The BECN bit provides a mechanism for any switch in the Frame Relay network to notify the originating node (sender) of potential congestion when there is a build-up of queued traffic in the switch's queues. This notice informs the sender that the transmission of additional traffic (frames) into the network should be restricted. The FECN bit notifies the receiving node of potential future delays, informing the receiver to use possible mechanisms available in a higher-layer protocol to alert the transmitting node to restrict the flow of frames.

Congestion management within a Frame Relay switch is based on the level of occupancy of the switch queues holding frames awaiting access to the transmission scheduling resource. The management response is governed by three queue occupancy threshold values.

When the frame queue length exceeds the first threshold, the switch sets the FECN and BECN bits of all frames. The bits are not set simultaneously; the precise action of whether the notification is forward or backward is admittedly somewhat arbitrary, and appears to depend on whether the notification is generated at the egress from the network (FECN) or at the ingress (BECN).

The intended result is to signal the sender or receiver on the UNI interface that there is congestion in the interior of the network. No specific action is defined for the sending or receiving node on receipt of this signal, although the objective is for the node to recognize that congestion may be introduced if the present traffic level is sustained, and that some avoidance action may be necessary to reduce the level of transmitted traffic. The avoidance mechanism does not explicitly call for the switch to discard packets, nor does it explicitly notify only those clients whose transfer rates exceed their CIR. The ECN mechanism is advisory and informs the current senders that the frame network is under stress and that CIR rates may be enforced. An appropriate response for a sender may be to back off the transmission rate to the CIR level. If the sender backs off to the CIR, the consequent packet loss rate will be minimized.

If the queue length continues to grow past the second threshold, the switch then discards all frames that have the Discard Eligible (DE) bit set. At this point, the switch is functionally enforcing the CIR levels on all VCs that pass through the switch, in an effort to reduce queue depth. The intended effect is that the sending or receiving nodes recognize that traffic has been discarded and that they subsequently throttle traffic rates to operate within the specified CIR level, at least for some period before probing for the availability of burst capacity. The higher-level protocol is responsible for detecting lost frames and retransmitting them. It is also responsible for using this discard information as a signal to reduce transmission rates to help the network back off from the congestion point.

The third threshold is the queue size itself, and when the frame switch queue length reaches this threshold, all further frames are discarded (see Figure 2.40).

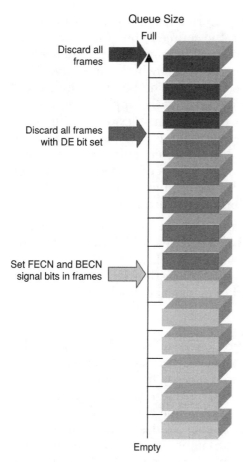

Figure 2.40 Frame Relay switch queue thresholds.

Frame Relay and QoS

Frame Relay has gained popularity in the data networking market because it allows a form of over-subscription of the underlying network through a burst rate. Therefore, subscriber traffic can exceed the sustained (and presumably tariffed) committed rate when the network has excess capacity, up to the port speed of the local circuit loop. (The concept of a free lunch always has been a powerful marketing incentive.)

Frame Relay adds a layer of switching and queuing, further increasing the delay, and possibly increasing the level of jitter imposed upon data packets in transit, which in turn may have negative performance implications for application traffic. Given the capability for a traffic source to burst into the Frame Relay network, the queuing structures within the Frame Relay switches will congest from time to time. CIR levels are enforced when the Frame Relay switch invokes packet discard on marked frames. The burst capabilities of Frame Relay may not be available when required by the application; accordingly, the opportunities presented by this form of transmission sharing using soft partitioning may not be achievable by the client. Beyond CIR, there are no guarantees within Frame Relay; even CIR is not a technologically enforced guarantee, but rather a universally enforced first level of load shedding. If the network remains over-committed, load will continue to be shed until the switches regain a state of dynamic equilibrium between load and capacity.

The important questions are: Can Frame Relay congestion management mechanisms enable the end user to set IP QoS policies, which in turn can provide some direction to the congestion management behavior of the underlying Frame Relay network? Can the Frame Relay congestion signals (FECN and BECN) be used to trigger IP layer congestion management behavior?

VC CONGESTION SIGNALS

When considering such questions, the first observation is that Frame Relay uses connection-based VCs. Frame Relay per-VC congestion signaling flows along these fixed paths, whereas IP flows do not use fixed paths through the network. Given that Frame Relay signals take some time to propagate back through the network, there is always the risk that the end-to-end IP path may be dynamically altered before the signal reaches its destination. As a result, the consequent sender's action may be completely inappropriate for handling the current situation. However, from a practical standpoint, the larger the network, the greater the pressure to dampen the dynamic nature of routing-induced, logical topology changes. The resulting probability of a topology change occurring within any single TCP end-to-end session then becomes very low indeed. It is important to determine whether any translation between IP QoS and Frame Relay QoS signaling is feasible.

FECN AND BECN SIGNALS

The FECN and BECN signaling mechanisms are intended to notify the end-to-end transport protocol mechanisms about the likely onset of congestion and potential data loss. Selectively discarding frames using a DE flag is intended to allow an additional mechanism to reduce load *fairly*, as well as to provide a secondary signaling mechanism to indicate the possibility of more serious congestion. Though it first appears that there are effective places where signals can be generated into the IP protocol stack, this is not the case. BECN and FECN signaling are not explicitly recognized by the protocols in the TCP/IP protocol suite, nor is the discarding of DE frames explicitly signaled into a TCP/IP protocol implementation as a congestion indicator.

The BECN and FECN signals are analogous to the ICMP (Internet Control Message Protocol) Source Quench signal. They are meant to inform the transmitter's protocol stack that congestion is being experienced in the network and that some reduction of the transmission rate is advisable. However, this signal is not used in the implementation of IP-over-Frame Relay for good reason. As noted in the Router Requirements document: "Research seems to suggest that Source Quench consumes network bandwidth but is an ineffective (and unfair) antidote to congestion" [RFC1812]. In the case of BECN and FECN, no additional bandwidth is being consumed (the signal is a bit set in the Frame Relay header so that there is no additional traffic overhead), but the issues of effectiveness and fairness are relevant. Although these notifications can indeed be signaled backward (or forward, as the case may be) to the CPE, where the transmission rate corresponding to the DLCI can be reduced, such action must be done at the Frame Relay layer. To translate this notification back up the protocol stack to the IP layer, a subsequent reaction would be necessary for the Frame Relay interface equipment upon receipt of a BECN or FECN signal, to set a condition that generates an ICMP Source Quench for all IP packets that correspond to such signaled frames. In this situation, the cautionary advice of RFC1812 is particularly relevant.

DISCARDING OF NONCONFORMANT TRAFFIC

The discard of DE packets as the initial step in traffic load reduction by the frame switches allows a relatively rapid feedback to the end-user TCP stack to reduce transmission rates. The discarded packet causes a TCP NAK to be delivered from the destination back to the sender upon receipt of the subsequent packet, allowing the sender to receive a congestion-experienced signal within one round trip time (RTT). Like random early deletion schemes, the behavior of such random marking of packets leads to a higher probability of having DE-marked packets, and the enforcement of the DE discarding leads to a likely trimming of the highest-rate TCP packets as the first measure in congestion management.

The most interesting observation about the interaction of Frame Relay and IP indicates what is missing rather than what is provided. The DE bit is a powerful mechanism that allows the interior of the Frame Relay network to take rapid and predictable actions to reduce traffic load when under duress. The challenge is to relate this action to the environment of the end user of the IP network.

Within the UNI specification, the Frame Relay specification allows the DTE (router) to set the DE bit in the frame header before passing it to the DCE (switch). In fact, this action is available in a number of vendors' routers. A network administrator is then able to specify a simple binary discard priority. However, this rarely is done in a heterogeneous network, simply because if no other subscriber undertakes the same action, it is somewhat self-defeating.

EFFECTIVENESS OF FRAME RELAY DISCARD ELIGIBILITY

Arguably, in a heterogeneous network that uses a number of data-link technologies to support end-to-end data paths, the Frame Relay DE bit is not a panacea. It does not provide for end-to-end signaling, and the router is not necessarily the system that provides the end-to-end protocol stack. The router is more commonly performing IP packet into frame-relay encapsulation. With this in mind, a more functional approach to user selection of discard eligible traffic is possible, one that uses a field in the IP header to indicate a defined quality level, and allows this designation to be carried end-to-end across the entire network path. With this facility, it then is logical to allow the DTE IP router (which performs the encapsulation of an IP datagram into a Frame Relay frame) to set the DE bit according to the bit setting indicated in the IP header field, then pass the frame to the DCE, which can confirm or clear the DE bit using steps outlined earlier.

Without coherence between the data-link transport-signaling structures and the higher-level protocol stack, the result is somewhat more complex. Currently, the Frame Relay network works within a locally defined context, using selective frame discard as a means of enforcing rate limits on traffic as it enters the network. This is the primary response to congestion, without respect to any hints provided by the higher-layer protocols. The end-to-end TCP uses packet loss as the primary signaling mechanism to indicate network congestion, but it is recognized only by the TCP session originator. When the network starts to reach a congestion state, the result is that the method in which end-system applications are degraded matches no particular imposed policy; and in this current environment, Frame Relay offers no great advantage over any other data-link transport technology.

However, if the TOS field in the IP header were used to allow a change to the DE bit semantics in the UNI interface, it is apparent that Frame Relay would allow a more graceful response to network congestion by attempting to reduce load in accordance with upper-layer protocol policy directives. The advantage of pushing this setting back to the end stack is the capability of the stack to select discard eligible IP packets based on the current state of the IP stack, setting the Discard Eligible bit when the transmit window is fully open. In other words, you can construct an IP-over-Frame Relay network that adheres to QoS policies if you can modify the standard Frame Relay mode of operation.

ATM

Asynchronous Transfer Mode (ATM) has been described, possibly unfairly, as the chainsaw massacre of the data transmission industry. This depiction refers to the method of data transmission within ATM in which, on ingress to the ATM network, each data packet is segmented into a series of small cells, each of which contain a 5-byte header and a 48-byte payload, and on egress from the ATM network, the cell sequence is reassembled into the original packet.

> *This section is not intended to be a detailed description of the inner workings of ATM networking. There is a wealth of published material already available on this subject. Our purpose is to describe the underlying mechanics that link ATM and certain types of Internet QoS. The majority of information in this section is condensed from the ATM Forum Traffic Management Specification Version 4.0 [AF-tm] and the ATM Forum Private Network-Network Interface Specification Version 1.0 [AF-pnni].*

One of the major attributes of ATM as a transmission solution for Internet networks is that it provides access speeds in excess of OC-3 (155Mbps) today. This is perhaps the most compelling reason that ATM has enjoyed some degree of success in supporting the Internet backbone environment. ATM also provides a complex subset of traffic-management mechanisms, VC establishment controls, and QoS parameters. It is important to understand why these underlying mechanisms are not being exploited by a the majority of organizations that are using ATM as a data-transport mechanism for Internet networks in the wide area. The major use of ATM in today's Internet networks is simply due to its availability, the high data-clocking rate, and multiplexing flexibility available with ATM implementations, rather than to a particular ATM service function. However, there is some evidence of change in this area. There is a second factor behind ATM's continuing interest to the Internet service provider industry: the use of ATM virtual circuits as an enabling technology for virtual private networks (VPNs). The use of Multi-Protocol Label Switching (MPLS) on an ATM network is thought to hold much promise in the efficient construction of VPNs. In addition, MPLS technology is seen by many as an enabler for IP to make effective use of some of the ATM service quality mechanisms.

ATM Background

Historically, organizations have used time division multiplexing (TDM) equipment to combine, or *mux*, different data and voice streams into a single physical circuit (see Figure 2.41), and subsequently *de-mux* the streams on the receiving end, effectively breaking them into their respective connections on the remote customer premise. Placing a mux on both ends of the physical circuit in this manner was a means to an end: It was considered economically more attractive to mux multiple data streams into a single physical circuit than it was to purchase different individual circuits for each application. This economic principle still holds true today.

There are, of course, some drawbacks to using the TDM approach, but the major liability is that once multiplexed, it is impossible to manage each individual data stream. This is sometimes an unacceptable paradigm in a service-provider environment, especially when management of data services is paramount to the service provider's economic livelihood. By the same token,

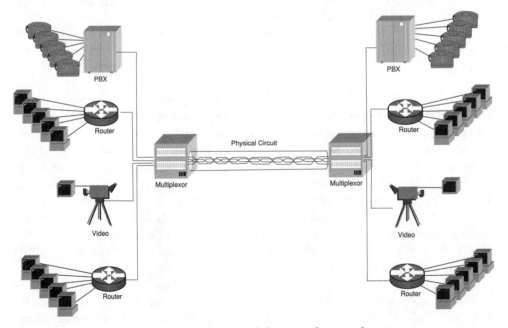

Figure 2.41 Traditional multiplexing of diverse data and voice streams.

it may not be possible to place a mux on both ends of a circuit because of the path a circuit may traverse. In the United States, for example, it is common that one end of a circuit may terminate in one Regional Bell Operating Company's (RBOC) network, while the other end may terminate in another RBOC's network on the other side of the country. Since the breakup of AT&T in the mid-1980s, the resulting RBOCs (Pacific Bell, Bell Atlantic, U. S. West, et al.) commonly have completely different policies, espouse different network philosophies and architectures, provide various services, and often deploy noninteroperable and diverse hardware platforms. Other countries have differing regulatory restrictions on different classes of traffic, particularly where voice is concerned, and the multiplexing of certain classes of traffic may conflict with such regulations. Another drawback to this approach is that the multiplexors represent a single point of failure.

Like X.25 and Frame Relay, ATM provides an alternative multiplexing technology to TDM. The origins of ATM lie in the TDM technology. The first refinement to TDM was to expand the timeslot period and then tag each TDM timeslot component with a header fragment, which enabled asynchronous time division multiplexing (ADTM). ADTM came to refer to the technology base, and ATM was adopted as the name for the set of international standards based on this technology.

This approach was further refined to allow the use of switches within the common TDM bearer, which swapped the timeslot tags at the same time the timeslot frame was swapped from one bearer to another. A set of such timeslot tag swaps can form a virtual circuit through a timeslot-switching fabric. This virtual circuit is similar in function to the X.25 and Frame Relay virtual circuit models, providing an end-to-end path across a switching fabric. In this scenario, an input circuit can be mapped to an output circuit in a fashion very similar to TDM, but with the additional flexibility of being switched within the bearer framework as a timeslot frame. As a result, the switching of circuits does not require the complete regression of the aggregated high-speed bearers back into their individual circuits at every switch. The economic attractions of this feature to the carrier industry underpin much of the industry's enthusiasm for ATM-based digital service systems.

These switched timeslots form a soft partitioning of the network, as distinct from the fixed partitioning undertaken by TDM. Each timeslot stream is not synchronously clocked, nor is it necessarily allocated a fixed amount of resources from the underlying common platform. The asynchronous, switched timeslot is mapped to a circuit through the set of timeslot switches, termed a virtual circuit. These concepts were first proposed in 1974 by Fraser, at Bell Labs [Fraser 1974]. When this ATM technology was presented to the telephone operating companies as a scalable switching architecture that had far greater flexibility and efficiency than TDM-based architectures in the mid-1980s, the response was very positive. In fact, a migration to ATM within the carriers' networks was anticipated to be completed by 2020!

The computing industry had a different response. Having just completed a transition from Ethernet to FDDI, the prospect of requiring a new generation of LAN technology for every new generation of processor was not an inviting prospect. The promise offered by ATM was a scalable architecture that was not constrained to any particular bandwidth or distance limitation. With this impetus, ATM was rapidly embraced by the computing industry; from there, ATM was reintroduced into the carrier world with a new sense of urgency.

The Virtual Circuit

The basic concept within ATM is that of the virtual circuit. In the datagram model of switching, the sender places the destination address in the datagram and passes the datagram into the network. Each switch must match the datagram destination within a table of all possible destinations, creating a switching complexity that scales as the size of the network grows, regardless of the complexity of the topology. A virtual circuit is usually implemented by an encapsulation layer, where the encapsulation contains a virtual circuit identification field. Each switch looks up the virtual circuit identifier within the local virtual circuit table, and based on the table entry, the switch selects a forwarding interface, and potentially rewrites the virtual circuit identifier with a new value for the next hop. Here the switch does not have to compute a forwarding decision based on all potential protocol-based destination addresses, but instead can perform a forwarding decision based on the number of concurrent virtual circuits.

Virtual circuits can be *static* or *dynamic*. Static virtual circuits are manually configured by the network manager, using a provisioning process that attempts to ensure that no points of the network are over-subscribed. Dynamic virtual circuits are created automatically, in response to a user request. A setup message is passed to the destination address. The return message accepting the call has the additional function of setting up a circuit state within the network. The path chosen by the network is the outcome of a routing protocol, where the protocol has the role of determining the most appropriate path for the circuit, given the service requirements and existing commitments. Once the path has been selected, subsequent packets can reference the circuit identifier rather than the destination address. The lookup space at each switch now scales back to the number of active circuits that traverse this switch, which is generally considerably fewer than the number of possible destinations within the network.

The task of assigning unique circuit identifiers across the network, which have switching significance for those switches along the chosen path, is one that cannot easily be undertaken in a sufficiently stable or robust manner. Instead, a local label is used, and the end-to-end path is constructed using a technique known as *label swapping*. In label swapping, each virtual circuit label has only local significance, and the same label can be used elsewhere in the network with impunity. Each switch operates a translation table in which the incoming interface and circuit identifier of a cell determine a mapping to an outgoing interface, together with the allocation of a new circuit identifier, which has significance for the next switch in sequence. The new circuit identifier is written into the cell header, and the cell is forwarded on the selected output interface.

This switching algorithm is very simple, and can be readily constructed using fast-switching hardware. Proponents of ATM argue that the datagram-switching algorithm cannot operate as efficiently due to the size of the routing table lookup undertaken on a per-datagram basis. On the other hand, proponents of a datagram-based switch argue that the VC identifiers are a slower and more cumbersome version of a datagram router's switching cache. One way to increase the speed of certain types of VC setup is to use the concepts of virtual paths and virtual circuits. A virtual path (VP) is a set of hops and associated label swaps that traverse the switching fabric from an entry point (the source), to an egress point (the destination). A VC in this structure is a logical end-to-end channel on a path, so that many VCs can be delivered on a single VP. Multiple traffic paths can be delivered across a single switching fabric. The overhead of path creation is limited to some extent through the segregation of an end-to-end state into paths and logical flow circuits. The mapping of a path through the network can either be performed statically by the network administrator or dynamically via an interswitch protocol that has a similar function in the discovery of the minimal cost path to that of the interior routing protocols in the IP domain. The switch also can generate additional state information on a per-VC basis, including various QoS parameters, which can be used to create various service characteristics that apply to the VC as a complete entity.

ATM Switching

The introduction of ATM in the 1980s provided an alternative method to traditional multiplexing, in which the basic concept (similar to its predecessor,

Frame Relay) is that multiple virtual channels (VCs) or virtual paths (VPs) now could be used for multiple data streams. Many VCs can be delivered on a single VP, and many VPs can be delivered on a single physical circuit (see Figure 2.42). This is very attractive from an economic, as well as a management, perspective. The economic allure is obvious: Multiple discrete traffic paths can be configured and directed through a wide-area ATM switched network, while avoiding the monthly costs of several individual physical circuits. Traffic is switched end-to-end across a network consisting of several ATM switches. Each VC or VP can be mapped to a specific path through the network, either statically by a network administrator or dynamically via a switch-to-switch routing protocol used to determine the best path from one end of the ATM network to the other, as simplified in Figure 2.43.

The primary purpose of ATM is to provide a high-speed, short-delay, low-jitter multiplexing and switching environment that can support virtually any type of traffic, such as voice, data, or video applications. ATM segments and multiplexes user data into 53-byte cells. Each cell is identified with a VC and

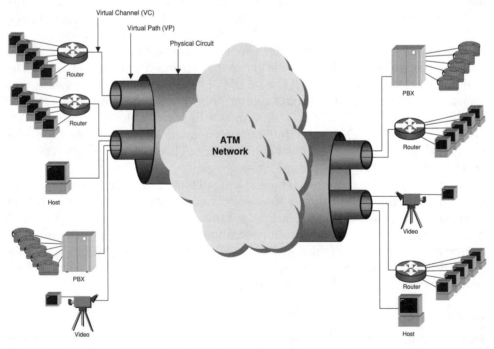

Figure 2.42 ATM assuming the role of multiplexor.

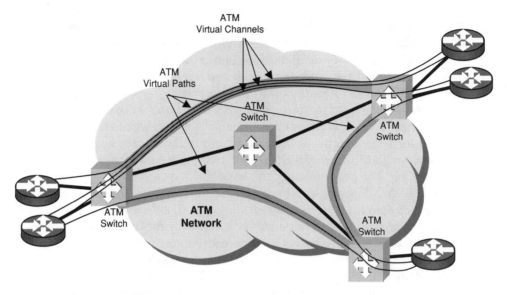

Figure 2.43 A simplified ATM network.

a VP identifier (VCI and VPI, respectively), which indicate how the cell is to be switched from its origin to its destination in the ATM switched network.

The ATM switching function is fairly straightforward. Each device in the ATM end-to-end path rewrites the VPI/VCI value, because it is only locally significant. That is, the VPI/VCI value is used only on a switch to indicate on which local interface and/or VPI/VCI a cell is to be forwarded. An ATM switch, or router, receives a cell on an incoming interface with a known VPI/VCI value, looks up the value in a local translation table to determine the outbound interface and the corresponding VPI/VCI value, rewrites the VPI/VCI value, then switches the cell onto the outbound interface for retransmission with the appropriate connection identifiers.

Frame-based traffic, such as Ethernet-framed IP datagrams, is segmented into 53-byte cells by the ingress (entrance) router and transported along the ATM network until it reaches the egress (exit) router, where the frames are reassembled and forwarded to their destination. The segmentation and reassembly (SAR) process requires substantial computational resources; and in modern router implementations, this process is done primarily in silicon on specially designed application-specific integrated circuit (ASIC) firmware.

Why 53 Bytes per ATM cell? *The 53 bytes are composed of a 5-byte cell header and a 48-byte payload. The original committee work refining the ATM model resulted in two outcomes, with one group proposing 128 bytes of payload per cell and the other 16 bytes of payload per cell. Further negotiation within the committee brought these two camps closer, until the proposals were for 64 and 32 bytes of payload per cell. The proponents of the smaller cell size argued that it reduced the level of network-induced jitter and the level of signal loss associated with the drop of a single cell, making it suitable for voice transmission. The proponents of the large cell size argued that it permitted a higher data payload in relation to the cell header, making it efficient for data transmission. Interestingly, both sides were proposing data payload sizes that were powers of 2, allowing for relatively straightforward memory mapping of data structures into cell payloads with associated efficiency of payload descriptor fields. The committee resolved this apparent impasse simply by taking the median value of 48 for the determined payload per cell. The compromise of 48 really suited neither side; it is considered too large for voice use and too small for data use. Current measurements indicate that there is roughly a 20 percent efficiency overhead in using ATM as the transport substrate for an IP network. This is a good example of why technology choices should never be made by a committee vote. All too often, technology committees can concentrate too heavily on reaching a consensus and lose sight of their major responsibility to define rational technology.*

ATM Connections

ATM networks essentially are connection-oriented; a virtual circuit must be set up and established across the ATM network before any data can be transferred across it. There are two types of ATM connections: *permanent virtual connections* (PVCs) and *switched virtual connections* (SVCs). PVCs generally are configured statically by some external mechanism, usually, a network-management platform of some sort. PVCs are configured by a network administrator. Each incoming and outgoing VPI/VCI, on a switch-by-switch basis, must be configured for each end-to-end connection.

Obviously, when a large number of VCs must be configured, PVCs require quite a bit of administrative overhead. SVCs are set up automatically by a signaling protocol—or rather, the interaction of different signaling protocols. There are also *soft PVCs,* the end points of the soft PVC; that is, the segment of the VC between the ingress or egress switch to the end system or router remains static. However, if a VC segment in the ATM network (between switches) becomes unavailable, experiences abnormal levels of cell loss, or becomes overly congested, an interswitch-routing protocol reroutes the VC within the confines of the ATM network. Thus, to the end user, there is no noticeable change in the availability of the local PVC.

ATM Traffic Management Functions

As mentioned several times earlier, certain architectural choices in any network design may impact the success of a network. The same principles ring true with regard to ATM as with other networking technologies. Being able to control traffic in the ATM network is crucial to ensuring the success of delivering differentiated QoS to the various applications that request and rely on the controls themselves. The primary responsibility of traffic management mechanisms in the ATM network is to promote network efficiency and prevent congestion situations so that the overall performance of the network does not degenerate. It is also a critical design objective of ATM that the network utilization imposed by transporting one form of application data does not adversely impact the ability to efficiently transport other traffic in the network. It may be critically important, for example, to ensure that the transport of bursty traffic does not introduce an excessive amount of jitter into the transportation of constant bit rate, real-time traffic for video or audio applications.

To deliver this stability, the ATM Forum has defined the following set of functions to be used independently or in conjunction with one another to provide for traffic management and control of network resources:

Connection admission control (CAC). Actions taken by the network during call setup to determine whether a connection request can be accepted or rejected.

Usage parameter control (UPC). Actions taken by the network to monitor and control traffic and to determine the validity of ATM connections and the associated traffic transmitted into the network. The primary purpose of UPC is to protect the network from traffic misbehavior that can adversely impact the QoS of already established connections. UPC detects violations of negotiated traffic parameters and takes appropriate actions—either tagging cells as CLP = 1 or discarding cells altogether.

Cell loss priority (CLP) control. If the network is configured to distinguish the indication of the CLP bit, the network may selectively discard cells with their CLP bit set to 1 in an effort to protect traffic with cells marked as a higher priority (CLP = 0). Different strategies for network resource allocation may be applied, depending on whether CLP = 0 or CLP = 1 for each traffic flow.

Traffic shaping. ATM devices may control traffic load by implementing leaky-bucket traffic shaping to control the rate at which traffic is transmitted into the network. A standardized algorithm called Generic Cell Rate Algorithm (GCRA) is used to provide this function.

Network-resource management. Allows the logical separation of connections by virtual path (VP), according to their service criteria.

Frame discard. A congested network may discard traffic at the ATM Adaptation Layer (AAL) frame level, rather than at the cell level, in an effort to maximize discard efficiency.

ABR flow control. In an effort to maximize the efficiency of available network, you can use the Available Bit Rate (ABR) flow-control protocol to adapt subscriber traffic rates resource utilization. (You can find ABR flow-control details in the ATM Forum Traffic Management Specification 4.0 [AF-tm].) ABR flow control also provides a crankback mechanism to reroute traffic around a particular node when loss or congestion is introduced, or when the traffic contract is in danger of being violated as a result of a local connection admission control (CAC) determination. With the crankback mechanism, an intervening node signals back to the originating node that it no longer is viable for a particular connection and no longer can deliver the committed QoS.

The major strengths of any networking technology are simplicity and consistency. Simplicity produces scalable implementations that can readily interoperate. Consistency results in a set of capabilities that are complementary. The preceding list of ATM functions may look like a grab-bag of fashionable tools for traffic management, without much regard for simplicity or consistency across the set of functions. This is no accident. Again, as an outcome of the committee process, the ATM technology model is inclusive, but without the evidence of operation of a filter of consistency. It is left to the network operator to take a subset of these capabilities and create a stable set of network services.

ATM ADMISSION CONTROL AND POLICING

Each ingress ATM switch provides the functions of admission control and traffic policing. The admission control function is called connection admission control (CAC); it is the decision process an ingress switch executes when determining whether an SVC or PVC establishment request should be honored, negotiated, or rejected. Based on this process, a connection request is entertained only when sufficient resources are available at each point within the end-to-end network path. The CAC decision is based on various parameters, including service category, traffic contract, and requested QoS parameters.

The ATM policing function is called usage parameter control (UPC); it, too, is performed at the ingress ATM switch. Although connection monitoring at the public or private UNI is referred to as UPC, and connection monitoring at a NNI (Network-to-Network Interface) can be called NPC (network parameter control), UPC is the generic reference commonly used to describe either one. UPC is the activity of monitoring and controlling traffic in the network at the point of entry.

The primary objective of UPC is to protect the network from malicious, as well as unintentional, misbehavior that can adversely affect the QoS of other, already established connections in the network. The UPC function checks the validity of the VPI and/or VCI values and monitors the traffic entering the network to ensure that it conforms to its negotiated traffic contract. The UPC actions consist of allowing the cells to pass unmolested, tagging the cell with CLP = 1 (marking the cell as discard eligible), or discarding the cells altogether. No priority scheme to speak of is associated with ATM connection services. However, an explicit bit in the cell header indicates when a cell may be dropped, usually, in the face of switch congestion. This bit is called the Cell Loss Priority (CLP) bit. Setting the CLP bit to 1 indicates that the cell may be dropped in preference to cells with the CLP bit set to 0. Although this bit may be set by end systems, it is set predominantly by the network in specific circumstances. This bit is advisory, not mandatory. Cells with the CLP set to 1 are not dropped when switch congestion is not present. Cells with CLP set to 0 may be dropped if there is switch congestion. The function of the CLP bit is a two-level prioritization of cells used to determine which cells to discard first in the event of switch congestion.

ATM SIGNALING AND ROUTING

There are two basic types of ATM signaling: the User-to-Network Interface (UNI) and the Network-to-Network Interface (NNI); the latter is sometimes referred to as the Network-to-Node Interface. UNI signaling is used between ATM-connected end systems, such as routers and ATM-attached workstations, as well as between separate, interconnected private ATM networks. A public UNI signaling is used between an end system and a public ATM network or between different private ATM networks. A private UNI signaling is used between an end system and a private ATM network. NNI signaling is used between ATM switches within the same administrative ATM switch network. A public NNI signaling protocol called B-ICI (BISDN or Broadband ISDN) Inter-Carrier Interface, which is shown in Figure 2.44, is used to communicate between public ATM networks [AF-b-ici].

The UNI signaling request is mapped by the ingress ATM switch to NNI signaling, and then is mapped from NNI signaling back to UNI signaling at the egress switch. An end system UNI request, for example, may interact with an interswitch NNI signaling protocol, such as Private Network-to-Network Interface (PNNI).

PNNI is a dynamic signaling and routing protocol that is run within the ATM network between switches; it sets up SVCs through the network. PNNI uses a complex algorithm to determine the best path through the ATM switch network and to provide rerouting services when a VC failure occurs [AF-pnni]. The original specification of PNNI, Phase 0, also is called IISP, for Interim Inter-Switch Signaling Protocol. The name change is intended to prevent confusion between PNNI Phase 0 and PNNI Phase 1.

PNNI Phase 1 introduces support for QoS-based VC establishment (routing) and crankback mechanisms. This refinement to PNNI does not imply that IISP is restrictive in nature, because the dynamics of PNNI Phase 1 QoS-based routing actually are required only to support ATM variable bit rate (VBR) services.

PNNI provides highly complex VC path-selection services that calculate paths through the network based on the cost associated with each interswitch link. The costing can be configured by the network administrator to indicate preferred links in the switch topology. PNNI is similar to Open Shortest Path First (OSPF) in many regards: Both are fast-convergence link-state protocols. However, whereas PNNI is used only to route signaling requests across the

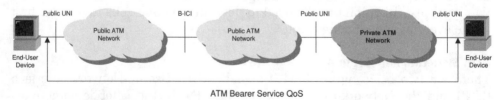

Figure 2.44 ATM signaling reference.

ATM network and, ultimately, provide for VC establishment, OSPF is used at the network layer in the Open Systems Interconnection (OSI) reference model to calculate the best path for packet forwarding. PNNI does not forward packets, nor does it forward cells; it simply provides routing and path information for VC establishment.

Importantly, PNNI does enable an aspect of QoS within ATM. Unlike other link-state routing protocols, PNNI not only advertises the link metrics of the ATM network, it also advertises information about each node in the ATM network, including the internal state of each switch and the transit behavior of traffic between switches in the network. PNNI also performs *source routing* (also known as *explicit routing*), in which the ingress switch determines the entire path to the destination, as opposed to path calculation being done on a hop-by-hop basis. This behavior is one of the most attractive features of ATM dynamic VC establishment and path calculation: the capability to determine the state of the network, to determine the end-to-end path characteristics (such as congestion, latency, and jitter), and to build connections according to this state. With the various ATM service categories (listed in the following section), as well as the requested QoS parameters (for example, cell delay, delay variance, and loss ratio), PNNI also provides an admission control function. When a connection is requested, the ingress switch determines whether it can honor the request based on the traffic parameters included in the request. If it cannot, the connection request is rejected.

---

*Y*ou can find approved and pending ATM technical specifications, including the Traffic Management 4.0 and PNNI 1.0 specifications, at the ATM Forum Web site, located at www.atmforum.com.

ATM Service Categories

One of the more unexploited features of ATM is its capability to request different service levels during SVC connection setup. Of course, a VC also can be provisioned as a PVC with the same traffic class. However, the dynamics of SVCs appear to be more appealing than the static configuration and provisioning of PVCs. Although PVCs must be provisioned manually, it is not uncommon to discover that SVCs also are manually provisioned, to a certain extent, in many instances. Although it certainly is possible to allow PNNI to determine the best end-to-end paths for VCs in the ATM network, it is common for network administrators to manually define administrative link parameters, called *administrative weights,* to enable PNNI to favor a particular link over another.

Currently, five ATM Forum-defined service categories exist (as shown in Table 2.4):

- ◆ Constant bit rate (CBR)
- ◆ Real-time variable bit rate (rt-VBR)
- ◆ Nonreal-time variable bit rate (nrt-VBR)
- ◆ Available bit rate (ABR)
- ◆ Unspecified bit rate (UBR)

The basic differences among these service categories are described in the following sections.

CONSTANT BIT RATE

The CBR service category is used for connections that transport traffic at a consistent bit rate, where there is an inherent reliance on time synchronization between the traffic source and destination. CBR is tailored for any type of data for which the end systems require predictable response time and a static amount of bandwidth continuously available for the lifetime of the connection.

Table 2.4 ATM Forum Traffic Services

ATM Forum Traffic Management 4.0 ATM Service Category	ITU-T I.371 ATM Transfer Capability	Typical Use
Constant Bit Rate (CBR)	Deterministic Bit Rate (DBR)	Real-time; QoS guarantees
Real-Time Variable Bit Rate (rt-VBR)	(For further study)	Statistical mux, real-time
Non-Real-Time Variable Bit Rate (nrt-VBR)	Statistical Bit Rate (SBR)	Statistical mux
Available Bit Rate (ABR)	Available Bit Rate (ABR)	Resource exploitation, feedback control
Unspecified Bit Rate (UBR)	(No equivalent)	Best-effort; no guarantees
(No equivalent)	ATM Block Transfer (ABT)	Burst-level feedback control

The amount of bandwidth is characterized by a peak cell rate (PCR). These applications include services such as video conferencing; telephony (voice services); or any type of on-demand service, such as interactive voice and audio. For telephony and native voice applications, the combination of ATM Adaptation Layer 1 (AAL1) and CBR service is best suited to low-latency traffic with predictable delivery characteristics. In the same vein, the CBR service category typically is used for circuit emulation. For multimedia applications, such as video, you might want to choose the CBR service category for a compressed, frame-based, streaming video format over AAL5 for the same reasons.

REAL-TIME VARIABLE BIT RATE

The rt-VBR service category is used for connections that transport traffic at variable rates, traffic that relies on accurate timing between the traffic source and destination. An example of traffic that requires this type of service category are variable rate, compressed video streams. Sources that use rt-VBR connections are expected to transmit at a rate that varies with time (traffic that can be considered bursty, for example). Real-time VBR connections can be characterized by a peak cell rate (PCR), sustained cell rate (SCR), and maximum burst size (MBS). Cells delayed beyond the value specified by the maximum cell transfer delay (CTD) are assumed to be of significantly reduced value to the application.

NONREAL-TIME VARIABLE BIT RATE

The nrt-VBR service category is used for connections that transport variable bit rate traffic for which there is no inherent reliance on time synchronization between the traffic source and destination, but there is a need for a guaranteed bandwidth or latency. An application that might require an nrt-VBR service category is Frame Relay interworking, where the Frame Relay Committed Information Rate (CIR) is mapped to a bandwidth guarantee in the ATM network. No delay bounds are associated with nrt-VBR service.

You can use the VBR service categories for any class of applications that might benefit from sending data at variable rates to most efficiently use network resources. You could use rt-VBR, for example, for multimedia applications with *lossy* properties, applications that can tolerate a small amount of cell loss without noticeably degrading the quality of the presentation. Some multimedia protocol formats may use a lossy compression scheme that provides these properties. Or you could use nrt-VBR for transaction-oriented applications, such as interactive reservation systems, where traffic is sporadic and bursty.

AVAILABLE BIT RATE

The ABR service category is similar to nrt-VBR, because it also is used for connections that transport variable bit rate traffic for which there is no reliance on time synchronization between the traffic source and destination and for which no required guarantees of bandwidth or latency exist. ABR provides a best-effort transport service in which flow-control mechanisms are used to adjust the amount of bandwidth available to the traffic originator. The ABR service category is designed primarily for any type of traffic that is not time-sensitive and expects no guarantees of service. ABR service generally is considered an option for supporting TCP/IP traffic and other LAN-based protocols that can modify their transmission behavior in response to the ABR's rate-control mechanics. However, performance issues are raised by using TCP over ABR, due to the differing control and signal systems. High-frequency oscillation of the ABR rate will cause TCP systems to take an extremely conservative view of available throughput capacity of the ABR circuit.

ABR uses resource management (RM) cells to provide feedback that controls the traffic source in response to fluctuations in available resources within the interior ATM network. The specification for ABR flow control uses these RM cells to control the flow of cell traffic on ABR connections. The ABR service expects the end system to adapt its traffic rate in accordance with the feedback, so that it may obtain its fair share of available network resources. The goal of ABR service is to provide fast access to available network resources up to the specified peak cell rate (PCR).

UNSPECIFIED BIT RATE

The UBR service category also is similar to nrt-VBR, because it is used for connections that transport variable bit rate traffic for which there is no reliance on time synchronization between the traffic source and destination. However, unlike ABR, there are no flow-control mechanisms to dynamically adjust the amount of bandwidth available to the user. UBR generally is used for applications that are very tolerant of delay and cell loss. It has enjoyed success in the Internet LAN and WAN environments for store-and-forward traffic, such as file transfers and e-mail. Similar to the way in which upper-layer protocols react to ABR's traffic-control mechanisms, TCP/IP and other LAN-based traffic protocols can modify their transmission behavior in response to latency or cell loss in the ATM network.

These service categories relate traffic characteristics and QoS requirements to network behavior. ATM network functions, such as VC/VP path establishment, CAC, and bandwidth allocation, are structured differently for each category. The service categories are characterized as real-time or nonreal-time. There are two real-time service categories: CBR and rt-VBR, both of which are distinguished by whether the traffic descriptor contains only the PCR or both the PCR and the sustained cell rate (SCR) parameters. The remaining three service categories are considered nonreal-time: nrt-VBR, UBR, and ABR. Each differs in its method of obtaining service guarantees provided by the network, and relies on different mechanisms implemented in the end systems and the higher-layer protocols to realize them. Selection of an appropriate service category is application-specific.

ATM Traffic Parameters

Each ATM connection contains a set of parameters, called *source traffic parameters,* that describe the traffic characteristics of the source. Source traffic parameters, coupled with another parameter called the cell delay variation tolerance (CDVT), and a conformance-definition parameter, characterize the traffic properties of an ATM connection. Not all these traffic parameters are valid for each service category. When an end system requests an ATM switched virtual connection (SVC) to be set up, it signals to the ingress ATM switch the type of service it requires, the traffic parameters of each data flow (in both directions), and the QoS parameters requested in each direction. These parameters form the *traffic descriptor* for the connection. The traffic parameters for the various service categories consist of the following:

Peak cell rate (PCR). The maximum allowable rate at which cells can be transported along a connection in the ATM network. In an effort to minimize jitter, the PCR is the determining factor in how often cells are sent in relation to time. PCR generally is coupled with the CDVT, which indicates how much jitter is allowable.

Sustainable cell rate (SCR). A calculation of the average allowable, long-term cell transfer rate on a specific connection.

Maximum burst size (MaxBS). The maximum allowable burst size of cells that can be transmitted contiguously on a particular connection.

Minimum cell rate (MinCR). The minimum allowable rate at which cells can be transported along an ATM connection.

QoS parameters. These parameters are discussed in the following section.

ATM Topology Information and QoS Parameters

As mentioned earlier, ATM service parameters are negotiated between an end system and the ingress ATM switch prior to connection establishment. This negotiation is accomplished through UNI signaling, and, if negotiated successfully, is called a *traffic contract*. The traffic contract contains the traffic descriptor, which also includes a set of QoS parameters for each direction of the ATM connection. This signaling is done in one of two ways. In the case of SVCs, these parameters are effected through UNI signaling, and are acted on by the ingress switch. In the case of PVCs, the parameters are configured statically through a network management system (NMS) when the connections are established. SVCs are set up dynamically, then torn down in response to signaling requests. PVCs are considered either permanent or semi-permanent, because once they are configured and set up, they are not torn down until manual intervention. (Of course, there also are soft PVCs, but for the sake of brevity of this overview, they are not discussed here.)

Topology attributes and topology metrics are two other important aspects of topology information carried around in PNNI routing updates. A *topology metric* is the cumulative information about each link in the end-to-end path of a connection. A *topology attribute* is the information about a single link. The PNNI path-selection process determines whether a link is acceptable or desirable for use in setting up a particular connection based on the topology attributes of a particular link or node. It is within these parameters that the mechanics of ATM QoS start to appear.

The topology metrics are as follows:

Cell delay variation (CDV). An algorithmic determination for the variance in the cell delay, primarily intended to determine the amount of jitter. The CDV is a required metric for CBR and rt-VBR service categories. It is not applicable to nrt-VBR, ABR, and UBR service categories.

Maximum cell transfer delay (maxCTD). A cumulative summary of the cell delay on a switch-by-switch basis along the transit path of a particular connection, measured in microseconds. The maxCTD is a required topology metric for CBR, rt-VBR, and nrt-VBR service categories. It is not applicable to UBR and ABR services.

Cell loss ratio (CLR). CLR is the ratio of the number of cells unsuccessfully transported across a link, or to a particular node, compared to the number of cells successfully transmitted. CLR is a required topology attribute for CBR, rt-VBR, and nrt-VBR service categories; it is not applicable to ABR and UBR. CLR is defined for a connection as:

CLR = Lost Cells / Total Transmitted Cells

Administrative weight (AW). The AW is a value set by the network administrator to indicate the relative preference of a link or node. The AW is a required topology metric for all service categories; when one is not specified, a default value is assumed. A higher AW value assigned to a particular link or node is less preferable than one with a lower value.

The topology attributes consist of the following:

Maximum cell rate (maxCR). The maximum capacity available to connections belonging to the specified service category. The maxCR attribute is required for ABR and UBR service categories; it is an optional attribute for CBR, rt-VBR, and nrt-VBR service categories. The maxCR attribute is measured in units of cells per second.

Available cell rate (AvCR). A measure of effective available capacity for CBR, rt-VBR, and nrt-VBR service categories. For ABR service, AvCR is a measure of capacity available for minimum cell rate (MCR) reservation.

Cell rate margin (CRM). The difference between effective bandwidth allocation and the allocation for sustained cell rate (SCR) measured in units of cells per second. CRM is an indication of the safety margin allocated above the aggregate sustained cell rate. CRM is an optional topology attribute for rt-VBR and nrt-VBR service categories; it is not applicable to CBR, ABR, and UBR.

Variance factor (VF). A variance measurement calculated by obtaining the square of the cell rate normalized by the variance of the sum of the cell rates of all existing connections. VF is an optional topology attribute for rt-VBR and nrt-VBR service categories; it is not applicable to CBR, ABR, and UBR.

Figure 2.45 illustrates the PNNI topology state parameters. Figure 2.46 shows a matrix of the various ATM service categories and how they correspond to their respective traffic and QoS parameters.

The ATM Forum's Traffic Management Specification 4.0 specifies six QoS service parameters that correspond to network-performance objectives. Three of these parameters may be negotiated between the end system and the network, and one or more of these parameters may be offered on a per-connection basis.

The following three negotiated QoS parameters were described earlier; they are repeated here because they also are topology metrics carried in the PNNI topology state packets (PTSPs). Two of these negotiated QoS parameters (CDV

PNNI Topology State Information		
Topology Metrics	**Topology Attributes**	
	Performance/Resource-Related	Policy-Related
Cell Delay Variation (CDV)	Cell Loss Ratio for CLP=0 (CLR_0)	Restricted Transit Flag
Maximum Cell Transfer Delay (maxCTD)	Cell Loss Ratio for CLP=0+1 (CLR_{0+1})	
Administrative Weight	Maximum Cell Rate (maxCR)	
	Available Cell Rate (AvCR)	
	Variance Factor (VF)	
	Restricted Branching Flag	

Figure 2.45 PNNI topology state parameters.

and maxCTD) are considered *delay parameters*, and one (CLR) is considered a *dependability parameter*:

◆ Peak-to-peak cell delay variation (peak-to-peak CDV)
◆ Maximum cell transfer delay (maxCTD)
◆ Cell Loss Ratio (CLR)

The following three QoS parameters are not negotiated:

Cell error rate (CER). Successfully transferred cells and errored cells contained in cell blocks counted as severely errored cell block rate (SECBR) cells should be excluded in this calculation. The CER is defined for a connection as:

CER = Errored Cells / (Transferred Cells + Errored Cells)

Severely errored cell block rate (SECBR). A cell block is a sequence of consecutive cells transmitted on a particular connection. A severely errored cell block determination occurs when a specific threshold of errored, lost, or misinserted cells is observed. The SECBR is defined as:

SECBR = Severely Errored Cell Blocks / Total Transmitted Cell Blocks

Cell misinsertion rate (CMR). The CMR most often is caused by an undetected error in the header of a cell being transmitted. This performance parameter is defined as a rate rather than a ratio, because the mechanism that produces misinserted cells is independent of the number of

Attribute	ATM Layer Service Categories				
	Constant Bit Rate	Real-Time Variable Bit Rate	Nonreal-Time Variable Bit Rate	Unspecified Bit Rate	Available Bit Rate
	CBR	**rt-VBR**	**nrt-VBR**	**UBR**	**ABR**
Traffic Parameters					
Peak Cell Rate (**PCR**) and Cell Delay Variation Tolerance (**CDVT**) [1,2]	specified			specified [3]	specified [4]
Sustainable Cell Rate (**SCR**), Maximum Burst Size (**MBS**), Cell Delay Variation Tolerance (**CDVT**) [1,2]	n/a	specified		n/a	
Minimum Cell Rate (**MCR**) [1]	n/a				specified
QoS Parameters					
peak-to-peak Cell Delay Variation (**ptpCDV**)	specified		unspecified		
Maximum Cell Transfer Delay (**maxCTD**)	specified		unspecified		
Cell Loss Ratio (**CLR**) [1]	specified			unspecified	Network-specific [5]
Other Attributes					
Feedback	unspecified				specified [6]

Notes:
1. These parameters are either explicitly or implicitly specified for permanent virtual circuits or switched virtual circuits.
2. Cell Delay Variation Tolerance (CDVT) is not signaled. In general, CDVT need not have a unique value for a connection. Different values may aply at each interface along the path of a connection.
3. May not be subject to Connection Admission Control (CAC) and Usage Parameter Control (UPC) procedures.
4. Represents the maximum rate at which the Available Bit Rate may send. The actual rate is subject to control information.
5. Cell Loss Ratio (CLR)is low for sources that adjust cell flow in response to control information. Whether a quantitative value for CLR is specified is network-specific.
6. See [AF 1996b]

Figure 2.46 ATM Forum service category attributes.

transmitted cells received. The SECBR should be excluded when calculating the CMR. The CMR can be defined as:

CMR = Misinserted Cells / Time Interval

Table 2.5 lists the cell-transfer performance parameters and their corresponding QoS characterizations.

ATM QoS Classes

There are two types of ATM QoS classes: one that explicitly specifies performance parameters (*specified QoS class*) and one for which no performance parameters are specified (*unspecified QoS class*). QoS classes are associated with a particular connection, and provide a set of performance parameters and objective values for each performance parameter specified. Examples of performance parameters that could be specified in a given QoS class are CTD, CDV, and CLR.

An ATM network may support several QoS classes. At most, however, only one unspecified QoS class can be supported by the network. It also stands to reason that the total performance provided by the network should meet or exceed the performance parameters requested by the ATM end system. The ATM connection indicates the requested QoS by a particular class specification. For PVCs, the Network Management System (NMS) is used to indicate the QoS class across the UNI signaling. For SVCs, a signaling protocol's information elements are used to communicate the QoS class across the UNI to the network.

A correlation for QoS classes to ATM service categories results in a general set of service classes:

Service class A. Circuit emulation, constant bit rate video.

Service class B. Variable bit rate audio and video.

Table 2.5 **Performance Parameter QoS Characterizations**

Cell-Transfer Performance Parameter	QoS Characterization
Cell Error Ratio (CER)	Accuracy
Severely Errored Cell Block Rate (SECBR)	Accuracy
Cell Loss Ratio (CLR)	Dependability
Cell Misinsertion Rate (CMR)	Accuracy
Cell Transfer Delay (CTD)	Speed
Cell Delay Variation (CDV)	Speed

Service class C. Connection-oriented data transfer.

Service class D. Connectionless data transfer.

Currently, the following QoS classes are defined:

QoS class 1. Supports a QoS that meets service class A performance requirements. This should provide performance comparable to digital private lines.

QoS class 2. Supports a QoS that meets service class B performance requirements. Should provide performance acceptable for packetized video and audio in teleconferencing and multimedia applications.

QoS class 3. Supports a QoS that meets service class C performance requirements. Should provide acceptable performance for interoperability connection-oriented protocols, such as Frame Relay.

QoS class 4. Supports a QoS that meets service class D performance requirements. Should provide for interoperability of connectionless protocols, such as IP.

The primary difference between specified and unspecified QoS classes is this: In an unspecified QoS class, no objective is specified for the performance parameters. However, the network may determine a set of internal QoS objectives for the performance parameters, resulting in an implicit QoS class being introduced. For example, a UBR connection may select best-effort capability, an unspecified QoS class, and only a traffic parameter for the PCR with a CLP = 1. This criteria then can be used to support data capable of adapting the traffic flow into the network based on time-variable resource fluctuation.

ATM and IP Multicast

Although it does not have a direct bearing on IP unicast QoS issues, it is nonetheless important to touch on the basic elements that provide for the interaction of IP multicast and ATM. These concepts will surface again later and be more relevant when we discuss the IETF Integrated Services architecture, the Resource Reservation Protocol (RSVP), and ATM.

Of course, there are no issues outstanding when IP multicast is run with ATM PVCs, because all ATM end systems are static and generally available at all times. Multicast receivers are added to a particular multicast group, as they normally would be in any point-to-point or shared media environment.

The case of ATM SVCs is a bit more complex. There are two methods for using ATM SVCs for IP multicast traffic. The first is the establishment of an end-to-end VC for each sender-receiver pair in a multicast group. This is fairly straightforward; however, depending on the number of nodes participating in a particular multicast group, this approach has obvious scaling issues associated with it. The second method uses ATM SVCs to provide an ingenious mechanism to handle IP multicast traffic by *point-to-multipoint* VCs. As multicast receivers are added to the multicast tree, new branches are added to the point-to-multipoint VC tree (see Figure 2.47).

When Protocol Independent Multicast (PIM) is used, certain vendor-specific implementations may provide for dynamic signaling between multicast end systems and ATM UNI signaling to build point-to-multipoint VCs [ID-pim-arch, RFC2363].

ATM-based IP hosts and routers may alternatively use a Multicast Address Resolution Server (MARS) to support RFC1112-style Level 2 IP multicast over the ATM Forum's UNI 3.0/3.1 point-to-multipoint connection service [RFC2022]. The MARS server is an extension of the ATM Address Resolution Protocol (ARP) server described in RFC2225. For practical reasons, the MARS functionality can be incorporated to the router to facilitate multicast-to-ATM host address resolution services. MARS messages support the distribution of multicast group membership information between the MARS server and multicast end systems. End systems query the MARS server when an IP address needs to be resolved to a set of ATM end points that make up the multicast group. End systems inform MARS when they need to join or leave a multicast group.

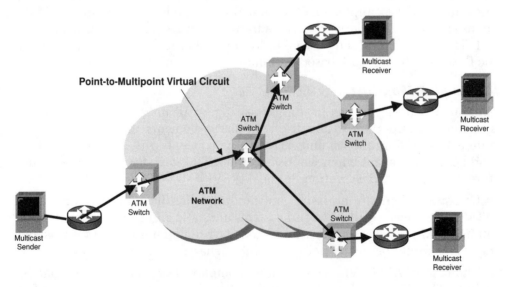

Figure 2.47 Point-to-multipoint VCs.

Factors That May Affect ATM QoS Parameters

It is important to consider factors that may have an impact on QoS parameters, factors that may be the result of undesirable characteristics of a public or private ATM network. There are several reasons that QoS might become degraded [AF-sc], and certain network events may adversely impact the network's capability to provide qualitative QoS. One of the principal reasons that QoS might become degraded is because of the ATM switch architecture itself. The ATM switching matrix design may be suboptimal, or the buffering strategy may be shared across multiple ports, as opposed to providing per-port or per-VC buffering. Buffering capacity may be less than satisfactory; as a result, congestion situations may be introduced into the network. Other sources of QoS degradation include media errors; excessive traffic load; excessive capacity reserved for a particular set of connections; and failures introduced by port, link, or switch loss. Table 2.6 lists the QoS parameters associated with particular degradation scenarios.

General ATM Observations

Several noteworthy issues make ATM somewhat controversial as a method of networking. Among them are excessive signaling overhead, encapsulation inefficiencies, and inordinate complexity.

THE CELL TAX

Quite a bit of needless controversy has arisen over the overhead imposed on frame-based traffic by using ATM. In some cases, the overhead can consume more than 20 percent of the available bandwidth, depending on the encapsulation method and the size of the packets. The controversy has centered

Table 2.6 QoS Degradation

Attribute	CER	SECBR	CLR	CMR	CTD	CDV
Progagation delay					X	
Media error statistics	X	X	X	X		
Switch architecture			X		X	X
Buffer capacity		X	X		X	X
Number of tandem nodes	X	X	X	X	X	X
Traffic load			X	X	X	X
Failures	X	X	X			
Resource allocation			X		X	X

around the overhead imposed because of segmenting and reassembling variable-length IP packets in fixed-length ATM cells for IP traffic in the Internet.

The last cell of an AAL5 frame, for example, will contain anywhere between 0 to 39 bytes of padding, which can be considered wasted bandwidth. Assuming that a broad range of packet sizes exists in the Internet, you could conclude that the average waste is about 20 bytes per packet. Based on an average packet size of 200 bytes, for example, the waste caused by cell padding is about 10 percent. However, because of the broad distribution of packet sizes, the actual overhead may vary substantially. Note that this 10 percent is in addition to the 10 percent overhead imposed by the 5-byte ATM cell headers (5 bytes subtracted from the cell size of 53 bytes is approximately a 10 percent overhead) and various other overhead (some of which also is present in frame-over-SONET, Synchronous Optical Network, schemes).

Suppose that you want to estimate the ATM bandwidth available on an OC3 circuit. With OC-3 SONET, 155.520Mbps is reduced to 149.760Mbps due to section, line, and SONET path overhead. Next, you reduce this figure by 10 percent, because an average of 20 bytes per 200-byte packet is lost (due to ATM cell padding), which results in 134.784Mbps. Next, you can subtract 9.43 percent due to ATM headers of 5 bytes in 53-byte packets. Thus, you end up with a 122.069Mbps available bandwidth figure, which is about 78.5 percent of the nominal OC-3 capacity. Of course, this figure may vary depending on the size of the packet data and the amount of padding that must be done to segment the packet into a 48-byte cell payload and fully populate the last cell with padding. Additional overhead is calculated for AAL5 (the most common ATM adaptation layer used to transmit data across ATM networks), framing (4 bytes length and 4 bytes CRC), and Link Layer Control (LLC) SubNetwork Access Protocol (SNAP) (8 bytes) encapsulation of frame-based traffic.

When you compare this scenario to the 7 bytes of overhead for Point-to-Point Protocol (PPP) encapsulation, which is run on point-to-point circuits, you can see how this produces a philosophical schism between IP engineering purists and ATM proponents. Although conflicts in philosophies regarding engineering efficiency clearly exist, once you get beyond the *cell tax*, as ATM cell overhead is called, ATM does provide interesting traffic-management capabilities. After accepting the fact that ATM does indeed consume a significant amount of overhead, ATM does provide service benefits for certain application classes.

MORE ROPE, PLEASE

Developing network technologies sometimes is jokingly referred to as being in the business of selling rope: The more complex the technology, the more rope with which to hang oneself.

In this vein, you can see that virtual multiplexing technologies such as ATM and Frame Relay also provide the necessary tools that enable people to build sloppy networks. These poor designs attempt to create a *flat* network in which all end points are virtually one hop away from one another, regardless of how many physical devices are in the transit path. This design approach is not a reason for concern in ATM networks, in which an insignificantly small number of possible ATM end points exists. However, in networks with a large number of end points, this design approach presents a reason for serious concern over scaling the Layer 3 routing system. Many Layer 3 routing protocols require that routers maintain adjacencies or peering relationships with other routers to exchange routing and topology information. The more peers or adjacencies, the greater the computational resources consumed by each device. Therefore, in a flat network topology, which has no hierarchy, a much larger number of peers or adjacencies exists. Failure to introduce a hierarchy into a large network, in an effort to promote scaling, can be suicidal.

ATM QoS Observations

This section attempts to present an objective view of relying solely on ATM to provide quality of service on a network. It is difficult to quantify the significance of some issues because of the complex nature of the ATM QoS delivery mechanisms and their interactions with higher-layer protocols and applications. The inherent complexity of ATM and its associated QoS mechanisms may be a big reason that many network designers are reluctant to implement QoS mechanisms based on ATM mechanisms.

Many people feel that when ATM is tested against the principle of Occam's Razor, ATM by itself would not be the choice for QoS services. The complexity involved, compared to other technologies that provide similar service outcomes, remains a significant factor.

---

*W*illiam of Occam (or Ockham) was a philosopher (presumed dates 1285-1349) who coined Occam's Razor, which states that "Entities are not to be multiplied beyond necessity." The familiar modern English version is "Make things as simple as possible, but no simpler." A translation frequently used in the engineering community is "All things being equal, choose the solution that is simpler." The popular version of this principle is yet another acronym: KISS, for keep it simple, stupid!

ATM enthusiasts correctly point out that ATM is complex for good reason. In order to provide predictive, proactive, and real-time services, such as dynamic network resource allocation, resource guarantees, virtual circuit rerouting, and virtual circuit path establishment to accommodate subscriber QoS requests, ATM's complexity is unavoidable.

Higher-layer protocols, such as TCP/IP, provide the end-to-end transportation service in most cases. Although it is possible to create QoS services in a lower layer of the protocol stack, namely ATM in this case, such services may cover only part of the end-to-end data path. This gets to the heart of the problem in delivering QoS with ATM: The true end-to-end bearer service is not pervasive ATM. Such partial QoS measures often have their outcomes masked by the effects of the traffic distortion created from the remainder of the end-to-end path in which they do not reside. The overall outcome of a partial QoS structure often is ineffectual.

In other words, if ATM is not pervasively deployed end-to-end in the data path, efforts to deliver QoS using ATM can be ineffectual. Traffic distortion is introduced to the ATM landscape by traffic-forwarding devices that service the ATM network and upper-layer protocols such as IP, TCP, and UDP, as well as other upper-layer network protocols. Queuing and buffering introduced to the network by routers and non-ATM-attached hosts skew the accuracy with which the lower-layer ATM services calculate delay and delay variation. Routers also may introduce needless congestion states, depending on the quality of the hardware platform and the network design.

LINES OF REASONING

An opposing line of reasoning suggests that end stations simply could be ATM-attached. However, this suggestion introduces several new issues, such as the inability to aggregate downstream traffic flows and provide adequate bandwidth capacity in the ATM network. Efficient utilization of bandwidth resources in the ATM network continues to be a primary concern for network administrators.

Yet another line of reasoning suggests that upper-layer protocols are unnecessary, because they tend to render ATM QoS mechanisms ineffectual by introducing congestion bottlenecks and unwanted latency into the equation. The flaw in this line of thinking is that native ATM applications do not exist for the majority of popular, commodity, off-the-shelf software applications, and even if they did, the ability to build a hierarchy of separate administrative domains into the network system is diminished severely. Scalability, such as exists in the global Internet, is impossible with native ATM. Even in the local network environment, ATM to the desktop has fallen victim to the far more pervasive switched Ethernet office environment.

On a related note, some have suggested that most traffic on ATM networks would be primarily UBR connections, because higher-layer protocols and applications cannot request specific ATM QoS service classes, and therefore cannot fully exploit the QoS capabilities of the VBR service categories. A quick take of deployed ATM networks and their associated traffic profiles reveals that this is indeed the case, except in the rare instance when an academic or research organization has developed its own native ATM-aware applications that can fully exploit the QoS parameters available to the rt-VBR and nrt-VBR service categories. Real-world experience reveals that this scenario is the proverbial exception and not the rule.

It is interesting to note that it is not sufficient to have a lossless ATM subnetwork from the end-to-end performance point of view. This observation is due to the fact that two distinct control loops exist: ABR and TCP (see Figure 2.48). Although it generally is agreed that ABR can effectively control the congestion in the ATM network, ABR flow control simply pushes the congestion to the edges of the network (in other words, the router), where performance degradation or packet loss may occur as a result. Arguably, the reduction in buffer requirements in the ATM switch caused by using ABR flow control may be at the expense of an increase in buffer requirements at

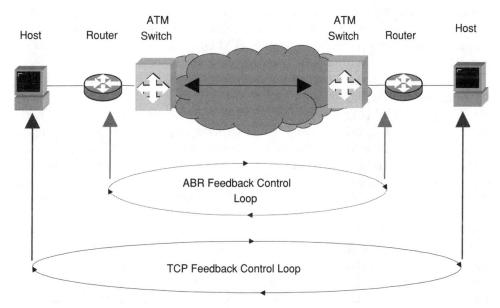

Figure 2.48 ABR and TCP control loops.

the edge device (for example, ATM router interface, legacy LAN-to-ATM switches). Because most applications use the flow control provided by TCP, you might question the benefit of using ABR flow control at the subnetwork layer, as UBR (albeit with early packet discard) is equally effective and much less complex. ABR flow control also may result in longer feedback delay for TCP control mechanisms, and as a result may exacerbate the overall congestion problem in the network.

Aside from traditional data services that may use UBR, ABR, or VBR, circuit-emulation services that may be provisioned using the CBR service category clearly can provide the QoS necessary for telephony communications. But this becomes an exercise in comparing apples and oranges. Delivering voice services on virtual digital circuits using circuit emulation is quite different from delivering packet-based data found in local area and wide area networks.

By the same token, providing QoS in these two environments is substantially different. It is substantially more difficult to deliver QoS for data, because the higher-layer applications and protocols do not provide the necessary hooks to exploit the QoS mechanisms in the ATM network. As a result, an intervening router must make the QoS request on behalf of the application, and the ATM network really has no way of discerning what type of QoS the application truly requires. Recent research and development efforts have attempted to address this shortcoming and to investigate methods of allowing the end systems to request network resources using RSVP, and then map these requests to native ATM QoS service classes as appropriate [RFC2205].

CLARIFICATION AND UNDERSTANDING

Finally, you should understand that ATM QoS commitments are probable estimates, intended only to provide a first-order approximation of the performance the network expects to offer over the duration of the ATM connection. Because there is no limit to the duration of connections, and the ATM network can make decisions based only on the information available to it at the time the connection is established, the actual QoS may vary over the duration of the connection's lifetime. In particular, transient events (including uncontrollable failures in the transmission systems) can cause short-term performance to be worse than the negotiated QoS commitment. Therefore, the QoS commitments can be evaluated only over the long term and with other connections that have similar QoS commitments. The precision with which the various QoS values can be specified may be significantly greater than the accuracy with which the network can predict, measure, or deliver a given performance level.

This leads to the conclusion that the use of the term guarantee is misleading, and should not be taken literally. Although ATM certainly is capable of delivering QoS when dealing with native cell-based traffic, the introduction of packet-based traffic (that is, IP) and Layer 3 forwarding devices (routers) to this environment may have an adverse impact on the ATM network's capability to properly deliver QoS, and certainly may produce unpredictable results. With the upper-layer protocols, there is no equivalent of a guarantee. In fact, packet loss is expected to occur to implicitly signal the traffic source that errors are present or that the network or the specified destination is not capable of accepting traffic at the rate at which it is being transmitted. When this happens, the discrete mechanisms that operate at various substrates of the protocol's stack (for example, ATM traffic parameter monitoring, TCP congestion avoidance, random early detection, ABR flow control) may well demonstrate self-defeating behavior because of these internetworking discrepancies and the inability of these different mechanisms to explicitly communicate with one another. ATM provides desirable properties with regard to increased speed of data-transfer rates, but in most cases, the underlying signaling and QoS mechanisms are viewed as excess baggage when the end-to-end bearer service is not ATM.

ATM AND IP DESIGN

Given these observations about QoS and ATM, the next question is: What sort of QoS can be provided by ATM in a heterogeneous network such as the Internet, which uses (in part) an ATM transport level?

When considering this question, the basic differences in the design of ATM and IP become apparent. The design philosophy for the Internet is based on coherent end-to-end data delivery services that can function across a path that uses a diverse collection of transport technologies, rather than a single technology. To achieve this functionality, the TCP/IP signaling mechanism uses two very basic parameters for end-to-end characterization: a dynamic estimate of end-to-end round trip time (RTT) and packet loss. If the network exhibits a behavior in which congestion occurs within a window of the RTT, the end-to-end signaling can accurately detect and adjust to the dynamic behavior of the network.

ATM, like many other data-link layer transport technologies, uses a far richer set of signaling mechanisms. The objective is to support a wider set of data-transport applications including a wide variety of real-time applications and traditional nonreal-time applications. This enhanced signaling capability is available to ATM because of its homogenous nature, and it can be used to support a wide variety of traffic-shaping profiles that are available in ATM switches. However, this richer signaling environment, together with the use of a profile adapted toward real-time traffic with very low jitter tolerance, can create a somewhat different congestion paradigm. For real-time traffic, the response to congestion is immediate load reduction, on the basis that queuing data can dramatically increase jitter and lengthen the congestion event duration. The design objective in a real-time environment is the immediate and rapid discard of cells to clear the congestion event. Assuming that integrity of real-time traffic is of critical economic value, data that requires integrity will use end-to-end signaling to detect and retransmit the lost data. Thus, the longer recovery time for data transfer is not a significant economic factor to the service provider.

The result of this design objective is that congestion events in an ATM environment occur and are cleared (or at the very least, are attempted to be cleared) within time intervals that generally are well within a single end-to-end IP round-trip time. Therefore, when the ATM switch discards cells to clear local queue overflow, the resultant signaling of IP packet loss to the destination system (and the return signal of a NAK for missing a packet) takes a time interval of up to one RTT. By the time the TCP session reduces the transmit window in response to this signaling, the ATM congestion event is cleared. It is a design challenge to define the ATM traffic-shaping characteristics for IP-over-ATM traffic paths for end-to-end TCP sessions to sustain maximal data-transfer rates. This, in turn, impacts the overall expectation that ATM provides increased cost efficiency through multiplexing different traffic streams over a single switching environment, when it is countered by the risks of poor payload delivery efficiency.

For networks like the Internet, the QoS objective is to direct the network to alter the switching behavior at the IP layer so that certain IP packets are delayed or discarded at the onset of congestion. The goal is to reduce the impact of congestion on other classes of IP traffic. When looking at IP-over-ATM, the issue (as with IP-over-Frame Relay) is that there is no mechanism for mapping such IP service directives to the ATM level, nor is it desirable, given the small size of ATM cells and the requirement for rapid processing or discard. Attempting to increase the complexity of the ATM cell discard mechanics to the extent necessary to preserve the original IP QoS directives by mapping them to the ATM cell is counterproductive.

It appears that the default IP QoS approach is best suited to IP-over-ATM. It also stands to reason that if the ATM network is adequately dimensioned to handle burst loads without the requirement of large-scale congestion avoidance at the ATM layer, there is no need for the IP layer to invoke congestion-management mechanisms. It appears that the discussion comes full circle to an issue of capacity engineering, and not necessarily one of QoS within ATM.

SMDS

Switched Multi-Megabit Data Service (SMDS) is a carrier offering that has been the predecessor to ATM services in many markets. SMDS was designed to provide a switched high-speed carriage system with the economies of a large public data service offering and transmission properties that resembled a high-speed LAN.

It is important to note that X.25, Frame Relay, and ATM are packet-switched virtual circuit environments. In these environments, packet flows are supported by the addition of virtual circuit states in which the packet headers reference the circuit, rather than the identity of the ultimate destination point. SMDS retains the packet-switched structure, but eliminates the imposition of a circuit state from the carriage environment.

The core service is a central hub intended to serve a metropolitan area. The switching hub is a connectionless switching system, rather than a virtual circuit-switching environment. Packets, up to 9188 bytes in length, can be addressed to any other connected interface; or, using a multicast group address, they can be addressed to a group of interfaces. Closed user groups are implemented through packet filters, which define the set of interface locations that can address packets to the customer egress port.

Connection speeds to the central hub can vary from 64Kbps DS0 circuits to 45Mbps DS3 access circuits, and the media can vary from twisted pair copper loops to fiber access tails. The central hub uses an internal cellular switch, which switches 53 octet cells. The internal formatting of these cells is different from ATM; the cellular format is a 7-byte header, a 44-byte payload, and a 2-byte trailer.

The major application of SMDS appears to be in servicing the market for LAN-level interconnection of metropolitan offices. It is unclear whether

SMDS will enjoy significant investment by carriage providers given that other high-speed carriage services, based on ATM and SONET technologies, provide a broader solution space.

Wavelength Division Multiplexing

In contrast to switching systems that attempt to impose end-to-end virtual circuits that span multiple media segments and share the medium between multiple uses via some form of time division multiplexing, wavelength division multiplexing (WDM) appears to be a return to frequency division multiplexing. Current modulation of signals into fiber optic cable is undertaken at 622Mbps, 2.5Gbps, and 10Gpbs. WDM can be used to increase the capacity of the fiber by multiplexing a number of data streams.

In WDM, each signal is encoded into a different wavelength (or *lambda*), and different, highly stable, lasers inject each signal into a common fiber. These signals are passed through the fiber together. Signal regeneration is undertaken using optical amplification with Erbium Doped Fiber Amplification (EDFA), which provides equal gain to all optical wavelengths. At the remote end, the light is passed through a grating to separate each wavelength.

Recently, there has been a spate of WDM capacity announcements, extending to a capability to support more than 150 wavelengths, each of which can carry an OC-192c (10Gbps) data payload. In this fashion, a single strand of fiber can be configured to carry more than a terabit per second of data. Although it is unlikely that the data service carriage market will see the retail offering of a complete WDM channel to each customer, it does offer the interesting opportunity to shift into supporting high-capacity trunk IP systems using optical multiplexing. Add optical switching to this scenario, and there is the potential to construct an all-optical trunk carriage switching system.

This leads to the potential of a network of abundant capacity, whose interior is capable of being dimensioned to a level well in excess of the imposed load. If this potential is realized, there will be some surprising consequences. The first is that the price of transmission capacity will become vanishingly small. More relevant to Internet performance, in such an environment of abundant capacity, it is questionable whether any performance controls will be necessary at all. If each application can make use of as much network capacity as it requires, there is no demonstrable need for the contention control that is at the heart of network performance management. Of course, underlying this is the proposition of a network of abundant capacity. Given the increasing speed, processing capability, and number of computing devices connected to public IP service networks, it remains to be seen whether the environment of abundant networking will become a reality.

Observations

In this chapter we have examined the basic building blocks of networks: the Internet Protocol, routers, and transmission elements. The interaction between an Internet application and these three elements results in differentiated performance levels within a network.

The UDP protocol will not impose any additional constraints on the performance of the application. The TCP protocol is used to create reliable data transfer; and the overhead of reliability imposes considerable performance constraints on the application. TCP uses a positive feedback control loop to implement an adaptive performance framework; and the protocol is intended to dynamically oscillate its rate of transfer around the point of maximal network efficiency.

Routers can distort the actions of the protocol in four basic ways: discard, delay, fragment, and forwarding. In addition, the router can operate in a selective fashion by applying a filter to the traffic, so that some level of differentiated response of these four fundamental actions can be applied.

Transmission systems can be synchronous or asynchronous. Synchronous systems use a synchronized data clock, so that each packet will sustain a constant propagation delay factor. The fixed clocking rate of the circuit also imposes a bandwidth delay to each packet, equal to the packet size divided by the data clock rate. Asynchronous systems, such as Frame Relay or ATM, use some form of Layer 2 switching within the network. These switches operate like IP routers. To some extent, all such transmission systems can add both variable delay and an incremental discard probability to the IP packet. Asynchronous systems can also support filtering functions, allowing some level of differentiated performance response to transmitted packets.

Currently, the debate within the carrier industry is over how many levels of switching and protection are necessary for a competitively priced and sufficiently reliable QoS IP platform. In some carrier systems, the IP network may be constructed using Frame Relay virtual circuits, which in turn may be constructed from ATM virtual circuits. The ATM system may be constructed using SDH rings, constructed on top of a WDM fiber system. Other carrier systems are looking to place the IP datagram directly into a point-to-point WDM channel, using a form of PPP framing of the IP packet. Another IP-over-fiber framing approach being offered is the Gigabit Ethernet protocol, used in a point-to-point configuration. Part of the substance of the debate over the number of switching levels is the intended architecture of the multiservice network.

If IP is seen as offering sufficiently rich functionality to support all forms of multiservice applications, then some rationalization of the number of levels of switching within the carrier system will improve the cost efficiency of the network. Each level of switching introduces capital cost for the switching equipment, operational cost in configuring and managing the network environment at that switching level, and further overhead in terms of additional framing and addressing bits surrounding the data payload. Eliminating a switching layer from the carriage infrastructure does admit the potential for lowering the carriage cost base. The ultimate end point of this approach of reduction of switching layers is that of IP-over-WDM. The differing service qualities required by the various applications can be provided through management of differential responses from the packet switches, potentially coupled with the use of different wavelengths, where there is a need for strong isolation of traffic systems.

On the other hand, if a multiservice network is seen as requiring a collection of distinct switching and transport systems, then the approach of layering multiple switching systems above the basic WDM optical transport does give the carrier greater flexibility to offer different services to different markets; it also enables the placement of some level of resiliency into the lower levels of switching. A typical architecture using this approach is a layering of IP-over-ATM, which in turn is layered on SDH rings. SDH is the framing protocol used to transmit across a wavelength of a WDM system.

Both approaches are being implemented in the carrier industry at present. Recent entrants are evidently concentrating on the cost efficiencies available through a minimal layering of IP over WDM, while the more established carriers are looking to a richer layering to support a broader set of carriage service requirements, allowing the existing carriage service portfolio to coexist with an IP platform on a common layered switching fabric.

References

[AF-b-ici] *BISDN Inter-Carrier Interface (B-ICI) Specification, Version 2.0 (Integrated)*, af-bici-0013.003. The ATM Forum Technical Committee, December 1995.
The standard specification of the interfaces used within the ATM architecture.

[AF-lane] *LAN Emulation over ATM Version 1.0 Specification*, af-lane-0021.000. The ATM Forum Technical Committee, January 1995.

The standard specification of LAN emulation layered above ATM networks. The most common application of this approach is to logically segment a LAN into a number of distinct security domains, although it also has potential application in the area of virtual private networks.

[AF-tm] *Traffic Management Specification, Version 4.0,* af-tm-0056.000. The ATM Forum Technical Committee, April 1996.

ATM has a rich set of traffic management control tools. This specification details the functionality of these tools.

[AF-pnni] *Private Network-Network Interface Specification Version 1.0*, af-pnni-0055.000. The ATM Forum Technical Committee, March 1996.

PNNI is a dynamic signaling and routing protocol that is run within the ATM network between switches, and sets up SVCs through the ATM network. This is the reference specification for version 1.0 of this protocol.

[AF-sc] *ATM Service Categories: The Benefit to the User*, Lambarelli, L., ed., CSELT, for the ATM Forum. www.atmforum.com/atmforum/library/service_categories.html

A useful description of the ATM service categories and their intended applicability.

[ANSI-frame] *Core Aspects of Frame Relay*, ANSI 1S1 DSSI , March 1990.

The standard specification of the framing format of Frame Relay protocol data units.

[Cisco 1995] *Internetworking Terms and Acronyms*, Text Part Number 78-1419-02. Cisco Systems, Inc., September 1995.

A useful and complete communications acronym decoder.

[Clark 1988] *The Design Philosophy of the DARPA Internet Protocols*, D.D. Clark. Proc SIGCOMM 88, ACM CCR, vol 18, no. 4, August 1988. (Reprinted in *ACM CCR*, vol 25, no. 1, January 1995).

The original paper describing the end-to-end design philosophy used within the Internet protocols.

[Jacobson 1988] *Congestion Avoidance and Control*, V. Jacobson. *Computer Communication Review*, vol. 18, no. 4, August 1988.

A landmark paper on end-system flow control mechanisms for TCP, to avoid network congestion collapses.

[Jacobson 1990] *Berkeley TCP Evolution from 4.3 Tahoe to 4.3 Reno*, V. Jacobson. Proc. 18th Internet Engineering Task Force, University of British Colombia, Vancouver, BC, September 1990.

A description of a refinement in the retransmission timers from the original congestion paper.

[Saltzer 1984] *End-to-End Arguments in System Design*, J.H. Saltzer, D.P. Reed, and D.D. Clark. *ACM TOCS*, vol. 2, no. 4, November 1984.

The original paper describing how network performance can be managed through the interaction between the end host systems and the network.

[Stevens 1994] *TCP/IP Illustrated, Volume 1*, W.R. Stevens, Addison-Wesley, 1994.

An excellent reference volume on the dynamic behavior of TCP.

RFCs

Request for Comments documents (RFCs) are documents published by the RFC editor. They are available online at www.rfc.editor.org.

[RFC791] *Internet Protocol, DARPA Internet Program Protocol Specification*, J. Postel, ed. RFC791, Standard RFC, September 1981.

The specification of version 4 of the Internet Protocol.

[RFC793] *Transmission Control Protocol, DARPA Internet Program Protocol Specification,* J. Postel, ed. RFC793, Standard RFC, September 1981.

The specification of the Transmission Control Protocol component of version 4 of the Internet Protocol.

[RFC896] *Congestion Control in IP/TCP Internetworks*, J. Nagel. RFC896, January 1984.

The original description of what is now know as the Nagle Algorithm, where congestion load on the network is relieved by delayed packet transmission.

[RFC1112] *Host Extensions for IP Multicast*, S. Deering. RFC1112, Standard RFC, August 1989.

A description of the host functionality required to support IP multicasting.

[RFC1122] *Requirements for Internet Hosts: Communications Layers*, R. Braden. RFC1122, Standard RFC, October 1989.

A compendium of requirements gathered from Internet standards relating to host functionality on the Internet. This RFC exmaines the functionality at the communications level.

[RFC1123] *Requirements for Internet Hosts: Application and Support*, R. Braden. RFC1123, Standard RFC, October 1989.

A compendium of requirements gathered from Internet standards relating to host functionality on the Internet. This RFC exmaines the functionality at the applications level.

[RFC1191] *Path MTU discovery*, J. Mogul and S. Deering. RFC1191, Draft Standard RFC, November 1990.

This document describes a technique for dynamically discovering the maximum transmission unit (MTU) of an arbitrary internet path. The intent of this technique is to avoid network fragmentation of IP packets.

[RFC1323] *TCP Extensions for High Performance*, V. Jacobson, R. Braden, and D. Borman. RFC1323, Proposed Standard RFC, May 1992.

This RFC presents a set of TCP extensions to improve performance over large-bandwidth delay product paths and to provide reliable operation over very high-speed paths. It defines TCP options for scaled windows and timestamps.

[RFC1349] *Type of Service in the Internet Protocol Suite*, P. Almquist. RFC1349, Proposed Standard RFC, July 1992.

This RFC defines some aspects of the semantics of the Type of Service octet in the Internet Protocol header. The handling of IP Type of Service by both hosts and routers is specified in some detail. This document is superseded by the IETF Differentiated Services (DS) field definition.

[RFC1644] *T/TCP—TCP Extensions for Transactions Functional Specification*, R. Braden. RFC1644, Experimental RFC, July 1994.

This RFC specifies T/TCP, an experimental TCP extension for efficient transaction-oriented (request/response) service. The document notes that this backward-compatible extension could fill the gap between the current connection-oriented TCP and the datagram-based UDP.

[RFC1700] *Assigned Numbers*, J. Reynolds and J. Postel. RFC1700, Standard RFC, October 1994.

Many protocols have defined fields that have associated values and semantic associations. The listing of assigned protocol values is contained in the online updates of this document, held at ftp://ftp.isi.edu/in-notes/iana/ assignments.

[RFC1715] *The H Ratio for Address Assignment Efficiency*, C. Huitema. RFC1715, Informational RFC, November 1994.

A recurring concept in protocol design and deployment is that of "assignment efficiency," a term that is commonly expressed as the ratio of the effective number of systems in the network over the theoretical maximum. The document examines this ratio for certain applications and protocols.

[RFC1809] *Using the Flow Label in IPv6*, C. Partridge. RFC1809, Informational RFC, June 1995.

The purpose of this document is to distill various opinions and suggestions of the End-to-End Research Group regarding the handling of flow labels into a set of suggestions for IPv6.

[RFC1812] *Requirements for IP Version 4 Routers*, F. Baker. RFC1812, Proposed Standard RFC, June 1995.

As a companion to the host requirements documents, this is a compendium of requirements gathered from Internet standards relating to router functionality on the Internet.

[RFC1883] *Internet Protocol, Version 6 (IPv6) Specification*, S. Deering and R. Hinden. RFC1883, Obsoleted by RFC2460, December 1995.

The original IPv6 specification document, superseded by RFC2460.

[RFC1889] *RTP: A Transport Protocol for Real-Time Applications*, H. Schulzrinne, S. Casner, R. Frederick, and V. Jacobson. IETF Audio-Video Transport Working Group, RFC1889, Proposed Standard, January 1996.

This RFC describes RTP, the Real-Time Transport Protocol. RTP provides end-to-end network transport functions suitable for applications transmitting real-time data, such as audio, video, or simulation data, over multicast or unicast network services. RTP does not address resource reservation and does not guarantee quality-of-service for real-time services. The data transport is augmented by a control protocol (RTCP) to allow monitoring of the data delivery in a manner scalable to large multicast networks, and to provide minimal control and identification functionality. RTP and RTCP are designed to be independent of the underlying transport and network layers. The protocol supports the use of RTP-level translators and mixers.

[RFC1958] *Architectural Principles of the Internet*, B. Carpenter, ed. IAB, RFC1958, Informational RFC, June 1996.

This RFC describes the principles of the Internet architecture. This is a general document rather than a description of a formal reference model.

[RFC2022] *Support for Multicast over UNI3.0/3.1-based ATM Networks*, G. Armitage. RFC2022, Proposed Standard RFC, November 1996.

Mapping the connectionless IP multicast service over the connection-oriented ATM services provided by UNI 3.0/3.1 is a nontrivial task. This RFC describes a mechanism for ATM networks to support the needs of IP multicasting.

[RFC2131] *Dynamic Host Configuration Protocol*, R. Droms. RFC2131, Draft Standard, March 1997.

The Dynamic Host Configuration Protocol provides a mechanism for the automatic allocation of network addresses and additional configuration options for IP hosts.

[RFC2205] *Resource ReSerVation Protocol (RSVP), Version 1 Functional Specification*, R. Braden, ed., L. Zhang, S. Berson, S. Herzog, and S. Jamin. RFC2205, Proposed Standard RFC, September 1997.

This RFC describes a resource reservation setup protocol designed for support of the Integrated Services Internet architecture. RSVP provides setup of resource reservations for unicast or multicast data flows.

[RFC2225] *Classical IP and ARP over ATM*, M. Laubach and J. Halpern. RFC2225, Proposed Standard, April 1998.

This RFC defines an initial application of classical IP and ARP in an Asynchronous Transfer Mode (ATM) network environment. This memo considers only the application of ATM as a direct replacement for transmission circuits, operating in the "classical" IP paradigm.

[RFC2328] *OSPF Version 2*, J. Moy. RFC2328, Standard RFC, April 1998.

Version 2 of the OSPF protocol. OSPF is a link-state routing protocol, designed to operate internal to a single autonomous system. Each OSPF router maintains an identical database describing the autonomous system's topology. From this database, a routing table is calculated by constructing a shortest-path tree.

[RFC2330] *Framework for IP Performance Metrics*, V. Paxson, G. Almes, J. Mahdavi, and M. Mathis. RFC2330, Informational RFC, May 1998.

The purpose of this RFC is to define a general framework for particular metrics to be developed by the IETF's IP Performance Metrics effort. The study has been characterized by a very disciplined approach to metrics, and the methodologies described in this document are well thought out.

[RFC2362] *Protocol-Independent Multicast-Sparse Mode (PIM-SM): Protocol Specification*, D. Estrin, D. Farinacci, A. Helmy, D. Thaler, S. Deering, M. Handley, V. Jacobson, C. Liu, P. Sharma, and L. Wei. RFC2362, Experimental RFC, June 1998.

This RFC describes a protocol for efficiently routing to multicast groups that may span wide-area internets, Protocol Independent Multicast-Sparse Mode (PIM-SM).

[RFC2414] *Increasing TCP's Initial Window*, M. Allman, S. Floyd, and C. Partridge. RFC2414, Experimental RFC, September 1998.

A description of an experimental proposal to increase from one segment to roughly 4K bytes in the permitted initial window for TCP. The document discusses the advantages and disadvantages of such a change, outlining experimental results that indicate the costs and benefits of such a change to the TCP specification.

[RFC2460] *Internet Protocol, Version 6 (IPv6) Specification*, S. Deering and R. Hinden. RFC2460, Draft Standard, December 1998.

The draft standard specification of Version 6 of the IP, designed as the successor protocol to IP version 4.

[RFC2474] *Definition of the Differentiated Services Field (DS Field) in the IPv4 and IPv6 Headers*, K. Nichols, S. Blake, F. Baker, and D. Black. RFC2474, Proposed Standard, December 1998.

A description of the differentiated services changes to the IP, specifying a change in interpretation of the Type of Service field of the IPv4 header and the Traffic Class field of the IPv6 header to allow the specification of a per-hop behavior in the field.

[RFC2475] *An Architecture for Differentiated Service*, S. Blake, D. Black, M. Carlson, E. Davies, Z. Wang, and W. Weiss. RFC2475, Proposed Standard, December 1998.

The description of the Differentiated Services architecture, allowing service responses to be generated from the network in a stateless fashion using packet marking and per-hop behaviors in the network's routers. The architecture is intended to offer a broad range of service responses with a very low protocol overhead.

[RFC2481] *A Proposal to Add Explicit Congestion Notification (ECN) to IP*, K. Ramakrishnan and S. Floyd. RFC2481, Experimental RFC, January 1999.

This RFC document describes a proposed addition of Explicit Congestion Notification (ECN) to IP. The document describes TCP's use of packet drops as an indication of congestion. The document argues that, with the addition of active queue management (e.g., RED) to the Internet infrastructure, where routers detect congestion before the queue overflows, routers are no longer limited to packet drops as an indication of congestion. Routers can instead set a Congestion Experienced (CE) bit in the packet header of packets from ECN-capable transport protocols. It describes when the CE bit would be set in the routers and what modifications would be needed to TCP to make it ECN-capable.

[RFC2581] *TCP Congestion Control*, M. Allman, V. Paxson, and W. Stevens. RFC2581, Proposed Standard RFC, April 1999.

This RFC documents TCP's four intertwined congestion control algorithms: slow start, congestion avoidance, fast retransmit, and fast recovery. In addition, the document specifies how TCP should begin transmission after a relatively long idle period, and discusses various acknowledgment-generation methods.

IETF Internet Drafts

IETF Internet drafts are work-in-progress documents. They documents are valid within the IETF process for a maximum period of six months from the date of submission. Internet drafts are not normally referenced, but those cited here are the best pointers to current research and developmental efforts, and so have relevance to this topic of quality of service and network performance. Typically, these documents follow the Internet Standards process and are published as RFCs sometime in the future. The current collection of Internet drafts, along with pointers to the RFC documents, can be found at www.ietf.org references.

[ID-ippm-btc] *Empirical Bulk Transfer Capacity*, M. Mathis and M. Allman. IETF Internet Draft, draft-ietf-ippm-btc-framework-00.txt, November 1998.

Bulk transport capacity is meant to define a measure of a network's capability to transfer data using an elastic transport protocol. The intention of the metric is to provide the expected long-term average data rate of a single ideal TCP implementation over the path being examined.

[ID-ippm-delay] *A One-Way Delay Metric for IPPM*, G. Almes, S. Kalidindi and M. Zekauskas, IETF Internet Draft, draft-ietf-ippm-delay-07.txt, May 1999.

This document defines a metric for one-way delay of packets across Internet paths. It builds on concepts introduced and discussed in the IPPM Framework document, RFC2330.

[ID-ippm-ipdv] *Instantaneous Packet Delay Variation Metric for IPPM*, C. Demichelis, IETF Internet Draft, draft-ietf-ippm-ipdv-02.txt, November 1988.

This document describes a metric for variation in delay of packets across Internet paths. The metric is based on statistics of the difference in one-way-delay of consecutive packets. This particular definition of variation is called *instantaneous packet delay variation* (IPDV).

[ID-ippm-loss] *A One-Way Packet Loss Metric for IPPM*, G. Almes, S. Kalidindi, and M. Zekauskas, IETF Internet Draft, draft-ietf-ippm-loss-07.txt May 1999.

This document defines a metric for one-way loss of packets across Internet paths. It builds on concepts introduced and discussed in the IPPM Framework document, RFC2330.

[ID-mpls-arch] *Multiprotocol Label Switching Architecture*, E. Rosen, A. Viswanathan, and R. Callon, IETF Internet Draft, draft-ietf-mpls-arch-05.txt, April 1999.

The formal description of the Multiprotocol Label Switching architecture, a technique of path computation within a switching domain that is primarily intended to allow efficiencies of operation through constraint of the size of the per-packet per-hop forwarding lookup function.

[ID-mpls-frame] *A Framework for Multiprotocol Label Switching*, R. Callon, P. Doolan, N. Feldman, A. Fredette, G. Swallow, and A. Viswanathan, IETF Internet Draft, draft-ietf-mpls-framework-05.txt, September 1999.

This document discusses technical issues and requirements for the IETF Multiprotocol Label Switching working group. The purpose of this document is to produce a coherent description of all significant approaches under consideration by the working group.

[ID-mpls-te] *Requirements for Traffic Engineering over MPLS*, D. Awduche, J. Malcolm, J. Agogbua, M. O'Dell, and J. McManus, IETF Internet Draft, draft-mpls-traffic-eng-01.txt, June 1999.

Multiprotocol Label Switching (MPLS) allows a greater degree of control over traffic flows within an IP network than is normally provided by destination address routing. This document presents a set of requirements for traffic engineering over Multiprotocol Label Switching. It identifies the functional capabilities required to implement policies that facilitate efficient and reliable network operations in an MPLS domain. These capabilities can be used to optimize the utilization of network resources and to enhance traffic-oriented performance characteristics.

[ID-pim-arch] *Protocol-Independent Multicast-Sparse Mode (PIM-SM): Motivation and Architecture*, S. Deering, D. Estrin, V. Jaconson, D. Farinacci, L. Wei, M.

Handley, D. Thaler, C. Liu, P. Sharma, and A. Helmy, IETF Internet Draft, draft-ietf-idmr-pim-arch-05.txt, August, 1998.

The formal description of the Protocol-Independent Multicast-Sparse Mode (PIM-SM) protocol architecture.

chapter three

Performance Tuning Techniques

Brute strength and ignorance is one alternative.
Of course, sometimes it's better to use precision and subtlety.

In Chapter 2, we examined the basic actions of network components, focusing on the characteristics of the IP protocol, routers, and transmission elements. In this chapter, we will take a closer look at the router, paying particular attention to router behaviors that can be adjusted to provide different service outcomes.

The service performance-related functions that can be undertaken within the router when handling packets include:

- Classification of packets
- Admission control, using traffic profiling and shaping
- Queuing and scheduling systems
- Congestion management systems
- Forwarding actions (QoS routing)
- Packet fragmentation

We will address each of these functions in turn in this chapter, to discover how they can affect the service delivery profile of the network.

Packet Classification

The basic action of an IP router is to treat all IP packets in precisely the same fashion. Each packet is processed in the order of arrival; the same service response is applied to each packet. In the general Internet model, this function is implemented without recourse to any local state memory, so that each packet is treated as a new event in terms of the router's response to the packet.

One of the basic building blocks for differentiated services within a network is the ability to partition network traffic into different service classes. To accomplish this, the router must be equipped with some form of classification function. Classification involves the use of some form of a packet filter to assign each packet to a service class, subsequently enabling differentiated handling of the packet. The classification function can also be used to police the traffic entering the network, to ensure that the traffic pattern conforms to an agreed profile of average and burst load. The classification can also allow a class of traffic to be passed through a traffic marking function, identifying traffic as *in* or *out* of an agreed service profile.

Packet Service Classification

Packet service classification means that the packet itself carries an explicit service classification field, thus creating a self-classification function. The precedence subfield of the Type of Service IP header field is an example of self-classification. In this case, the packet is classified in up to seven differentiated service classes (allowing for a distinct base service class, using a 0 precedence value). The 4-bit service selector subfield is also usable as a service classification field. Of the 16 possible service selector values, five are defined by RFC1349; the remaining 11 values can be defined by the network operator as a service classification field (see Figure 3.1).

Other mechanisms configured within the network can then use the value in the header field to determine how to respond to this particular packet. This can include congestion management, bandwidth allocation and queuing priority, or service class weighting. The advantage of this approach is that the packet does not have to be reclassified at every router along the end-to-end network path. The packet's service classification in the Type of Service header drives the routers' service responses along the packet's path.

This approach of per-packet service classification is being adopted within the IETF's efforts in standardizing a Differentiated Services architecture. The Type of Service header is being redefined as a Differentiated Service header, and the meaning of the individual values within this field include a broader range of service responses that can be triggered by the packet. We will examine the Differentiated Service architecture in greater detail in Chapter 4, "Quality of Service Architectures."

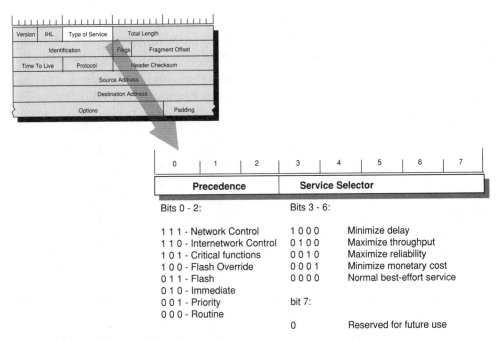

Figure 3.1 IP Type of Service field.

Admission Classification Controls

It is possible for the network to accept an application's packet service marking or a client network's packet service marking without any further admission controls. Such an unconditional acceptance of externally defined premium service packets does allow the uncontrolled use of any associated premium network services, but with the consequent risk of service overload. For any network service response to be effective, the amount of traffic admitted to the service class must be within the bounds of the amount of network resource devoted to providing the service. An overloaded premium service will deliver a worse response than a lightly loaded best-effort service. For a differentiated service network to deliver predictable service responses, the amount of traffic admitted into each service class within the network must be controlled. This is accomplished using an admission control function. Normally, such marked packets are passed though an admission filter, where an incoming classification can be checked against the network's admission policies.

The admission function may include service code translation, in which case the encoding for a particular service class may use a different code point within the network from that used in the client network. The admission filter must rewrite the packet Type of Service field (or Differentiated Service field in a Differentiated Services network) to match the network's service policies.

The admission classification policy may use some level of feedback relating to the current load conditions within the network, where the amount of traffic admitted within a premium service class may rely on the current load conditions within the network.

Admission packet classification schemes may use other packet attributes. The packet's IP source and destination addresses may form part of the classification policy; or possibly the packet's UDP or TCP source and destination port addresses, because the port addresses can correspond to defined applications. The physical interface that received the packet may be a classification condition or the packet's MAC-level address. Alternatively, the classification can correspond to the Internet Protocol field, so that, for example, IMCP packets can be identified. Note that, at times, the number of fields available to the classification function are deliberately restricted. In the case of the IP Security Architecture (IPSEC) it is possible for the external host (or firewall) to deliberately mask the packet's actual IP header and any associated TCP or UDP header information [RFC2401]. Even in the more general case of IP tunneling, some potential IP packet classification information is embedded deeper within the encapsulating IP packet's payload. The most generic packet classification mechanisms should only assume that the basic IP headers are visible, not that UDP and TCP header information is visible to the classification operation.

A common form of network abuse is the so-called ping flood denial-of-service (DoS) attack, where a target system is attacked by a flood of IMCP ping packets. A QoS form of response to such attacks is to configure a classification rule for IMCP ping packets, together with a corresponding service class that uses rate limitation. Conventional ping applications are not affected, but the flood attacks are extensively limited by such a technique.

All these systems share a common classification method: A potentially large set of classification conditions are mapped to a smaller set of service classes at the edge of the network, while the interior of the network sees only a per-packet mark that corresponds to a selection of a single configured service class. Though this may be very efficient structure for some QoS architectures, particularly those that use aggregated service responses such as the Differentiated Services architecture, it is less appropriate for other service architectures that use a finer level of granularity. Where the service architecture is intended to offer a service response to an individual data flow or an individual host or network, the classification rules and the associated service response must be distributed across the network. Additionally, each router must classify the packet into a particular context, then apply the service profile associated with that context.

Other Forms of Classification

Many routers use an *access list*, a configuration construct that supports packet classification. An access list allows the specification of a packet's source and destination address, protocol, and, in the case of UDP or TCP, the protocol port address. Some systems include a more generic configuration mechanism that allows any field of the packet header to be identified using a specification of an offset and a field length. Access lists offer a broad spectrum of granularity, ranging from identification of traffic flows associated with a particular application on a single host through to all traffic associated with a network client, including all of the client's network applications.

Such packet classification systems are *static*—the network administrator must enter fixed classification rules into the network equipment. A change in QoS policy, the addition of a new client, or a change in the QoS profile of a client may entail a configuration change in the access lists of a number of routers. In a large network, this task may present significant operational complexity, and may be further complicated by a QoS environment that extends across a number of cooperating networks, where a change in the QoS policy in one network may involve configuration changes in all other cooperating networks.

The alternative is a classification system that uses the routing system to carry QoS classification attributes within the routing advertisements, allowing a classification condition to be dynamically associated with a network address prefix. The Border Gateway Protocol (BGP), a routing protocol, allows various attribute values to be attached to a routing advertisement. These attribute values can be used to trigger a QoS response, so that when a packet is forwarded in response to a learned BGP route, the associated BGP attribute value can be used as a classification mechanism. The advantage of this approach is that a classification condition associated with a group of network prefixes within one network can be dynamically communicated to all cooperating neighboring networks without any additional configuration activity (see Figure 3.2).

Packet classification can include path and flow identification. For example, in an MPLS network, an individual flow can be identified using a combination of the incoming interface and the incoming label. A per-flow service response can be associated with this, using a classification of interface and incoming label to determine not only the outgoing interface and label, but also the queuing precedence and drop preference.

Classification Actions

Once a packet is classified, other local QoS responses can be used on the packet, including congestion management, queuing priority, bandwidth allocation, supporting delay bounds, and supporting admission traffic profiles. End-to-end service quality demands a consistent service response from each router along the path. While it is possible to repeat the classification and response on each router in the path, this is a considerable processing over-

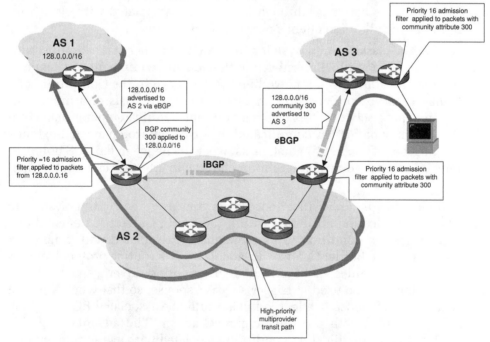

Figure 3.2 Dynamic classification with BGP.

head. The alternative approach is to classify the packet, then write a service response mark into the packet header, using the encoding scheme of RFC1349, or the encoding scheme of the Differentiated Services architecture [RFC2475], or possibly some local encoding scheme. In this packet classification scheme, the interior of the network is configured to respond to a small set of packet markings, so that the packet itself calls up the service response corresponding to the selected service class.

A common marking scheme has considerable breadth of application, spanning multiple networks. A client, or the end application, can generate the packet marking to correspond to the desired service level. The network admission function can either confirm this service request or override it with a service level based on the service contract or the prevailing level of load within the network.

Queuing Disciplines

A router is a *shared resource* within a network, where each application generates requests for a share of this resource to pass packets through the network. Each data packet may be considered a request for service from the router. At times, these requests may arrive in a well-ordered sequential fashion, allowing the router to service the requests immediately. At other times there will be contention, as multiple requests may arrive at the same time; or the time taken to service requests may overlap with the arrival of other requests. A critical aspect of performance management is how the router resolves such issues of contention for resources [RFC2309].

The idealized model used to examine contention management is that of a *server* and a sequence of service *requests*. The server has no control over the request arrival rate, and the time it takes to service each request may vary. While the server is processing one request, additional requests may arrive, so the server will need to queue these requests until the processor is available. This local queuing of requests will consume some of the server's resources. As long as the request arrival rate continues to exceed the request processing rate, the service queue will continue to expand. Given that the server has a finite resource available to hold the queued requests, if the arrival rate continues to exceed the processing rate for an extended period of time, the server's queue will ultimately fill. At that point, the server will have to discard requests.

Resource Contention and Performance Why is this aspect of contention management so central to the entire issue of performance management? One line of argument is that if the network has abundant capacity, to the extent that no two packets ever contend for network resources, then any form of performance management is irrelevant. Certainly this conclusion is valid, but the underlying requirement is that additional network transmission and switching capacity can be provided without marginal cost. In other words, engineering for abundance is only possible within a free network.

Free networks do not exist. All networks, regardless of their technology base, have some form of marginal cost of use. In the desire to achieve ever greater cost efficiency of operation, the objective is to defray this cost across the greatest possible amount of traffic. In other words, the basic engineering intent is to accept the greatest possible amount of network traffic while maintaining some basic level of service integrity. Within this engineering scheme there will be some level of contention of access to the network's transmission and switching systems at some point in time. It is possible to manipulate this contention by creating a set of different service responses from a common network, so that the resolution of contention favors one class of traffic over another. This is the very essence of the entire performance management rationale.

The server must use a *queuing discipline* to manage the request queue. The function of this discipline is to select which request to service when the server finishes processing the previous request. It may be that the chosen queuing discipline is one of preservation of the request arrival order, so that requests are processed in strict order of their arrival. Other disciplines are also possible, where some form of preemptive priority can be assigned to certain classes of request by configuring the queue manager to select such requests for service in preference to other queued requests. Accordingly, the queuing discipline that is chosen has direct implications for the level of service obtained by each client. As an example, a queuing discipline may select requests from client A in preference to client B, meaning that requests from client B will be selected only when there are no outstanding requests from client A. In this case, client A would experience a level of service that will certainly be no worse than that experienced by client B; and, when there is contention between A and B for the server, A will experience a superior level of service (see Figure 3.3).

A queuing discipline also includes management of the request discard function. When the local request queue is full, and another request arrives at the server, the queuing discipline may elect to simply discard the incoming request. Alternatively, the queuing discipline may select a previously queued request to discard, in order to make room for the new request. As an example, if the queuing discipline is configured to offer client A preemptive service priority over client B, when a request arrives from A and the queue is full, the queue manager may scan the queue for requests from B, and if successful, discard B's request and insert the new request from A into the service queue.

This model of requests and servers forms the basis for many studies of queuing theory that examine the interaction between the request arrival rates, the queuing discipline, and the server's time to service each request. It is beyond the scope of this book to study queuing theory in detail, however, so we will put aside any rigorous mathematical analysis of various queuing models, in favor of a more practical view of queuing disciplines that includes their behaviors and various outcomes.

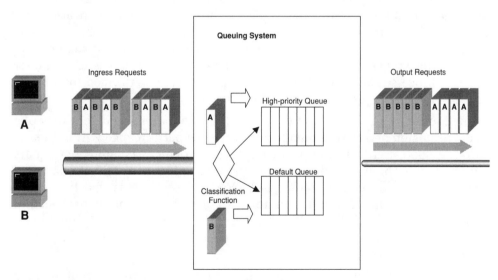

Figure 3.3 Queuing and preferential servicing.

L en Kleinrock is arguably one of the most prolific theorists on traffic queuing and buffering in networks. His books on queuing theory comprise a comprehensive study of this field [Kleinrock 1975, 1976]. You can find a bibliography of his papers at millennium.cs.ucla.edu/LK/Bib/.

Queuing disciplines can be characterized as either *work-conserving* or *non-work-conserving*. A work-conserving discipline ensure that the server is busy whenever the service queue contains queued requests, whereas a nonwork-conserving discipline may allow the server to be idle in such a situation.

In the context of network engineering, queuing is the act of storing (holding) packets or cells for subsequent processing. Typically, queuing occurs within a router when packets are received by a device's interface subsystem and are awaiting processing by the switch engine in order to make a forwarding decision. Queuing also occurs prior to transmitting the packets to the output interface on the same device. Queuing is the central component of the architecture of the router, whereby a number of asynchronous processes are linked to switch packets. A basic router is a collection of processes: There are input processes that assemble packets as they are received, checking the integrity of the basic packet framing. There are one or more forwarding processes that determine the output interface to which a packet should be passed. There are output processes that frame and transmit packets on to their next hop. Conceptually, each instance of a process operates on one packet at a time and works in parallel with all other process instances. The binding of the input processes to the forwarding processes and then to the output processes is performed by a packet queue-management process, shown in Figure 3.4.

It is important to understand the central role that queuing disciplines play in the provision of differentiated levels of service performance within the network. All QoS architectures assume deployment of an associated queuing discipline that implements the architecture's functionality. The fundamental assumption made by all differentiated QoS architectures is that the routers can exercise a classification function to correctly categorize packets into defined service groups. This allows the associated queuing discipline to allocate router resources to each service group in the relative order and priority defined by the QoS architecture.

All work-conserving queuing disciplines that implement differentiated service responses obey a law of conservation of resources. The law of conservation states that any improvement in the average delay characteristics provided

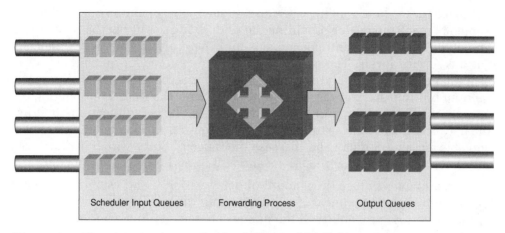

Scheduler Input Queues   Forwarding Process   Output Queues

Figure 3.4 Router queuing and scheduling: a block diagram.

to one class of traffic will be matched by a corresponding decline in the delay characteristics of other classes of traffic. In other words, the sum of the mean queuing delay for each class of traffic, multiplied by the relative share of the resource allocated to that class by the queuing discipline, will be constant, regardless of the queuing discipline in use. For nonwork-conserving queuing disciplines, this sum will be no greater than the work-conserving outcome [Kleinrock 1975].

All work-conserving queuing disciplines have the same total outcome with respect to mean queuing delay, but this should not be interpreted to mean that all such disciplines are functionally equivalent. Different queuing disciplines do have different outcomes, as you'll see when we will examine a number of specific instances of queuing disciplines and their service outcomes later in this chapter. The law of conservation simply states that any improvement in network service performance offered to one class of client will result in some form of reduced network service performance offered to other clients. Put more colloquially, there is no such thing as a free lunch.

Managing Queues

Choosing an algorithm for placing packets into a queue, and choosing the maximal length of the queue itself, may, at first blush, appear to involve relatively simple and indeed trivial configuration decisions. However, you should not make these choices lightly, because queue management is one of the fundamental mechanisms for providing the underlying quality of the service and one of the most important mechanisms for differentiating the performance of various service levels. The correct configuration choice for queue length can be extraordinarily difficult to accurately determine, because of the apparently random traffic patterns present on a network, and the interaction among the consequent delay, jitter, and loss characteristics on the applications that use the network. If you impose too long a queue, you introduce a potentially unacceptable amount of delay and increase the range of delay variance (increased jitter), which can break applications and end-to-end transport protocols.

Shorter queues reduce the amount of delay variance and establish a bound on the cumulative jitter measure. But these improvements will occur at a cost of increased loss probability. If the queues are too shallow, you may run into the problem of trying to dump data into the network faster than the network can accept it, resulting in a significant amount of discarded packets or cells. In a reliable traffic flow, such discarded packets have to be identified and retransmitted in a sequence of end-to-end protocol exchanges. The retransmission process causes the client's actual data throughput to plummet, as we described for TCP flow control behavior. For efficient operation of TCP flows, the router's queues should be of a comparable size to the delay bandwidth product of the output link to which the queue is associated. For UDP, the design choice is not so simple. While shorter queues reduce the overall jitter bound and reduce the maximum delay, they also increase the packet loss probability. Both behaviors are problematic for UDP-based real-time flows, such as audio or video applications. Packet loss is manifested as signal degradation, where parts of the signal are simply missing in the playout; jitter is manifested as a discrete break in the continuity of the playout.

Queuing within networks occurs as an outcome of a number of conditions, including the aspects of protocol behavior and the more generic issue of resource sharing. Queuing is a natural artifact of the reliable data transfer protocol, TCP. Within the TCP, the initial slow-start rate control places data packets onto the wire in a burst configuration, whereby two packets are transmitted within the ACK timing of a single packet's reception (see Figure 3.5). During this phase of flow rate control, the sender is effectively transmitting into the network at twice the data rate of packets leaving the network. The rate adaptation that the network must perform between the sender's output data rate and the maximum input data rate, as seen by the receiver, is done within the network's queues. The congestion onset rate limit that TCP is searching for is signaled by packet loss, caused by queue saturation. Within this process, the effective congestion onset point discovered by TCP is not a

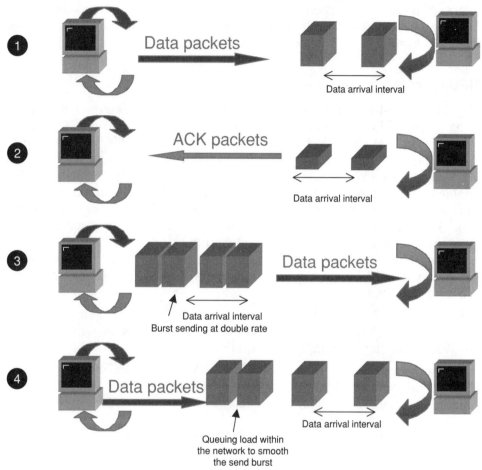

Figure 3.5 TCP slow start.

fair share of the actual available transmission capacity, but is, to be more precise, a fair share of the available queue resource.

Queuing is also a natural artifact of sharing a unique resource among multiple clients. Moreover, it is a natural result of a dynamic network of multiple flows, where the imposition of a new flow on a fully occupied link will cause queuing while the flow rates all adapt to the increased load. Queuing also can happen when data rate adaptation is required, such as when a Layer 3 device (a router) queues and switches traffic toward the next-hop Layer 3 device faster than the underlying Layer 2 network devices (such as Frame Relay or ATM switches) in the transit path can accommodate. Some intricacies in the realm of queue management, as well as their effect on network traffic, are challenging to understand. It is also challenging to deploy network architectures that minimize the related performance impacts.

FIFO Queuing

First-in, first-out (FIFO) queuing is a simple work-conserving queuing discipline in which the arrival order is the same as the servicing order. This is also referred to as first-come-first served (FCFS) queuing. FIFO is considered to be the standard method for store-and-forward handling of traffic from an incoming interface to an outgoing interface, and it is certainly the most widely deployed across Internet networks. So widely used is FIFO queuing that you can consider anything more elaborate to be somewhat exotic or even "abnormal." This is not to say that non-FIFO queuing mechanisms are inappropriate; quite the contrary. Non-FIFO queuing techniques certainly have their merits and usefulness. It is to say that you must know what their limitations are, when they should be considered, and, perhaps most important, when they should be avoided.

As Figure 3.6 shows, within a FIFO queuing discipline, packets are serviced into the appropriate output interface queue in the order in which they are received—thus the designation first-in, first-out.

FIFO queuing usually is considered standard base queue behavior. Many router vendors have highly optimized the forwarding performance of their products by making this standard behavior as fast as possible. In fact, when

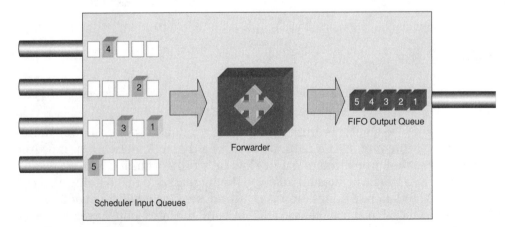

Figure 3.6 FIFO queuing.

coupled with topology-driven forwarding cache population, this particular combination of technologies could be considered the potentially fastest of the technology implementations available today as far as packets-per-second forwarding performance is concerned. This is due to the simplicity of the simple queuing algorithm and the ease at which simple tail insertion operations can be optimized into high-speed firmware. When more elaborate queuing strategies are implemented instead of FIFO, there may very well be some negative impact on forwarding performance, and an increase (sometimes dramatically) on the computational overhead of the system. This depends, of course, on the queuing discipline and the quality of the vendor implementation.

Advantages and Disadvantages of FIFO Queuing

When a network, or part of a network, operates in a mode with abundant levels of transmission capacity, so that the internal capacity of the network exceeds the network's admission capacity, and the network also possesses adequate levels of switching capacity, then queuing is necessary only to ensure that short-term, highly transient traffic bursts do not cause packet discard. If every admission system presented a packet to the switch at the same point in time, then the switch would need to operate some local queue to avoid unnecessary packet discard (see Figure 3.7). In this case, queuing undertakes the simple role of contention resolution, without performing any form of differential resource allocation. In such an over-provisioned environment, FIFO queuing is highly efficient because, as long as the queue depth

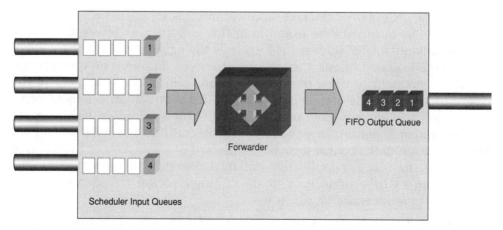

Figure 3.7 Queuing as contention resolution.

remains sufficiently short, the average packet-queuing delay is an insignificant fraction of the end-to-end packet transmission time.

Such over-provisioned environments are not common in deployed internet service networks. The more common operational environment is one where the internal capacity of the network is somewhat lower than the aggregate admission capacity. In such cases, the network relies on statistical multiplexing gain and queuing to deliver its service profile. To compound this, the nature of Internet traffic is *self-similar* in nature, where across any chosen time base, the traffic load will exhibit strong burst characteristics [Leland 1993]. Some level of contention is a constant factor in internet networks, and queuing mechanisms are necessary to resolve such local contention. As the admitted load on network increases, the load peaks, or bursts, cause significant queuing delay (significant in terms of the fraction of the total transmission time); and when the queue is fully populated, all subsequent arriving packets must be discarded. When the network operates in this mode for extended periods, the offered service level inevitably degenerates. FIFO queuing operates uniformly across all IP packet services, increasing the mean queue delay for all traffic as the load increases. Once the queue is full, all further traffic is discarded.

Under simple FIFO queuing, all traffic is offered the same network service profile but the interaction between the protocol and this FIFO-induced packet delay and discard behavior causes an outcome that generally favors nonrate-adaptive applications over those that adapt their rate to the prevailing network load. In many circumstances, rate-adaptive applications that use TCP will reduce their sending rate during periods of congestion-triggered packet loss, while externally clocked applications using UDP will continue unabated. The most extreme example of this condition is where the steady state nonadaptive UDP ingress load exceeds the egress capacity of a switch. Under such load conditions, the switch behaves as if there is no queue at all, for at any moment in time, there is either one empty slot at the tail of the queue or the queue is completely full. In such a situation, a rate-adaptive application will be unable to open its transmission window, as the attempt to send a number of packets in a packet train will result in the discard of all but the first packet of the packet train or the discard of all packets in the train. In either case, the adaptive application will never see sufficient space within the network, and will operate in a mode of single-packet transactions, interspersed with retransmission time-outs.

An alternative outcome is when a very large adaptive TCP traffic source bursts into a FIFO queue at periodic intervals. At the time of the burst, the

FIFO queue is filled with packets from the bursting application and all other traffic flows are denied service until the burst is finished. Thereafter, other flows resume their use of the queue until the next high-volume burst. In this case, the local condition is that all flows passing through the congestion point are synchronized against the burst interval of the dominant flow. Neither of these outcomes is desirable from a network performance perspective.

Different queuing strategies are intended to alter the characteristics of service-level degradation, allowing some services to continue to operate without perceptible degradation, while imposing more severe degradation on other services. Queuing strategies cannot create additional network resources; they can simply redistribute the available resources according to an adopted QoS policy. This behavior drives the use of queue management as the mechanism to provide QoS-arbitrated differentiated services.

Priority Queuing

One of the first queuing variations to be widely implemented was *priority queuing*. This is based on the concept that certain types of traffic can be identified and conceptually shuffled to the front of the output queue, so that some classes of traffic is always transmitted ahead of other types of traffic.

Priority queuing relies on the processor identifying the service class in which each packet is classified and placing the packet in the associated service queue. The scheduling algorithm is relatively straightforward: A packet is scheduled from the head of service queue q as long as all queues of higher priority are empty. The associated discard algorithm can be *preemptive* or *nonpreemptive*. Using nonpreemptive discard, a packet is placed into the relevant service queue whenever there are available queue resources; otherwise, the packet is discarded. A preemptive discard algorithm will attempt to discard a packet of a lower priority to make space to queue a higher-priority packet.

As shown in Figure 3.8, priority queuing can be visualized as a collection of FIFO output queues. As packets are passed to the output system, the particular FIFO queue to be used is based on a packet classification function. In this example, three queues are used: high-priority packets are placed in a high-priority queue and scheduled before any queued medium-priority packets; low-priority packets are held in the queue until no further high- or medium-priority packets are awaiting transmission. Typical implementations of priority queuing also include the ability to define queue lengths for each priority level. Within each priority, packets are processed in a FIFO fashion, implementing a strict arrival processing regime within the priority.

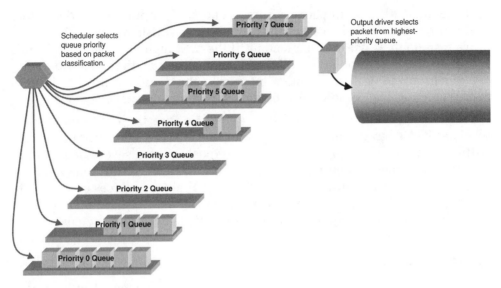

Figure 3.8 Priority queuing.

Advantages and Disadvantages of Priority Queuing

Within a priority queuing discipline, the granularity in identifying traffic to be classified into each queue is quite flexible. In a multiprotocol network, various priorities can be assigned to particular protocols to create a set of relative service priorities. For example, IPX could be queued before IP, IP before SNA, and SNA before AppleTalk. Also, specific services within a protocol family can be classified in this manner. When we discussed FIFO queuing, we noted that the common outcome of congestion in networks is that the TCP sessions back off their sending rates, while the UDP sending rate continues unaltered. In a priority queuing discipline, TCP traffic can be prioritized ahead of UDP traffic.

> *U*sing priority queuing to unconditionally elevate the relative priority of TCP over that of UDP is not generally recommended. The flow control algorithms of TCP will operate so that the TCP sessions will attempt to equilibriate, and in so doing will consume all available bandwidth, effectively starving the UDP sessions of any resources.

Priority queuing can also be used to bias application performance levels for different applications. For example, telnet (TCP port 23) can be prioritized ahead of FTP (TCP ports 20 and 21). Similarly, by assigning the routing traffic a higher priority than user traffic, priority queuing can be used to improve the stability of various routing protocols under heavy load conditions. In general, the greater the amount of differentiation attempted, the greater the impact on computational overhead.

In theory, many levels of priority are possible, while in practice a more restricted set of priorities is advisable, such as the three-level approach of high, medium, and low. Increasing the number of priority levels does not necessarily create an increased number of predictable service outcomes. As long as the higher-priority queues are serviced to exhaustion before lower-priority queues are serviced, any traffic at a certain level of priority will cause all traffic at lower priorities to experience delay. For this reason, it is common to mitigate this condition by using a relatively short high-priority queue, while using a longer low-priority queue.

The condition of service delay to low-priority traffic can be exacerbated to the extent that it poses a vulnerability in this queuing approach. If the volume of high-priority traffic is unusually high, normal traffic waiting to be queued may be dropped because of *buffer starvation*. Buffer starvation occurs because the rate of arrival of lower-priority packets exceeds the service rate available for this priority level. In such a situation, the queue associated with these lower priorities will inevitably increase to the point at which queue overflow occurs, as there are too many packets waiting to be queued and not enough room in the queue to accommodate them. Another consideration is the adverse impact that induced latency may have on applications when traffic sits in a low-priority queue for an extended period, awaiting servicing. It is challenging for end system protocol stacks to estimate how much additional latency to factor into the end-to-end round-trip time (RTT) due to the effects of such additional latency. The typical response to such latency increase is for the retransmit timers to trigger, which further exacerbates the overload condition by placing additional low-priority load into the network. In a worst-case scenario, this combination of high latency and increasing packet drop probability leads to complete congestion collapse; and throughput efficiency of the low priority traffic plummets. Even when the high-priority task clears, it takes a number of RTT intervals for the lower-priority reliable traffic to recommence flow control and resume normal operation. The point is, the effects of a high-priority burst condition may last for significantly longer intervals than that of the burst itself.

The problem is well known in operating systems design: Absolute priority scheduling systems can cause complete resource starvation to all but the highest-priority tasks. Thus, the use of priority queues creates an environment where the degradation of the highest-priority service class is delayed until the entire network is devoted to processing only the highest-priority service. The side effect of this resource preemption is that the lower levels of service are starved of system resources very quickly. Ultimately, due to the reallocation of critical resources, the total effective throughput of the network degenerates dramatically.

Nevertheless, priority queuing has been used for a number of years as a simple method of differentiating traffic into various classes of service. Over the course of time, however, it has been discovered that this mechanism simply does not scale to provide the desired performance at higher speeds. A basic step to protecting the integrity of the network when using priority queuing is to associate admission controls for high-priority traffic, ensuring that the amount of premium traffic admitted into the network will not consume all available network resources.

Class-Based Queuing

Another queuing mechanism is called *class-based queuing* (CBQ) or *custom queuing* (also referred to as *weighted round-robin*, or WRR). It is widely used within operating system design to prevent complete resource denial to a particular class of service, thereby addressing the major weakness of strict priority queuing. CBQ is a variation of priority queuing, and like priority queuing, several output FIFO queues can be defined. With CBQ, you can define the preference for each of the queues, the amount of queued traffic (typically measured in bytes) that should be drained from each queue on each pass in the servicing rotation, and the length of each queue.

CBQ is an attempt to provide some semblance of fairness by prioritizing queuing services for certain types of traffic, while not allowing any one class of traffic to monopolize system resources and bandwidth. With CBQ, at least one packet will be removed from each of the component queues within one processing round, so that all queues will receive one quantum of service. Additional packets will be dequeued from each queue until the sum of the packet sizes exceeds the queue's service amount.

The configuration in Figure 3.9, for example, has three buffers: high, medium, and low. The router could be configured to service 200 bytes from the high-priority queue, 150 bytes from the medium-priority queue, and 100 bytes from the low-priority queue within each service cycle (implementing a relative service weighting of 4, 3, and 2, respectively). As the packets in each queue are processed, packets continue to be removed from the queue until the

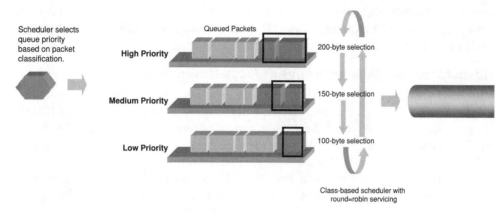

Figure 3.9 Class-based queuing (CBQ).

byte count of the dequeued packets exceeds the configured threshold for this queue, or until the queue is drained. In this fashion, traffic that has been classified into the various queues and then entered into an available slot within the queue will not be denied service. Each queue will receive a quantum of service within each service cycle, so all queued packets will, eventually, be serviced. Queue stagnation, where a low-priority queue is starved of any service whatsoever, is prevented within this queuing discipline, as is the associated condition of buffer resource shrinkage (where packets resident in stagnant queues continue to consume buffer space within the router).

CBQ's design also includes the concept that certain classes of traffic, or applications, may need minimal queuing latency to function properly. CBQ implementations typically provide mechanisms to configure the size of each service queue, as well as how much traffic can be drained off each queue in a servicing rotation.

CBQ provides a basic mechanism that ensures that a packet within a specific class does not sit in the outbound queue indefinitely. In addition, within bounds of variance determined by maximum packet sizes, the latency experienced by any queued packet in any service class is bounded to a predictable maximum delay within CBQ queuing. Of course, packet loss probability is not similarly bounded, because packet loss is a function of the ingress rate as well as the queue length and the service weighting provided to the queue. An administrator may have to tune the various queue parameters to gauge whether the desired behavior is achieved, varying the filters that assign packets to a particular service class, the length of each service class queue, and the relative service weighting assigned to each service class. The resulting configuration may benefit from observing the router and application throughput, as well as more conventional performance engineering.

Advantages and Disadvantages of Class-Based Queuing

The CBQ approach generally is perceived as a method of allocating dedicated portions of bandwidth to specific types of traffic. The major strength of CBQ is its provision of more graceful mechanisms of preemption, in which the model of dedicated service to the highest-priority queue and resource starvation to other queues in the priority-queuing model is replaced by a more equitable model of an increased level of resource allocation to the higher-precedence queues and a relative decrease to the lower-precedence queues. The fundamental assumption made by CBQ is that *resource denial* is far worse than *resource reduction*. Resource denial not only denies data but also denies any form of signaling regarding the denial state. TCP applications faced with resource denial lose the feedback control signals coming from the receiver, then shift to a time-out retransmission state, continuing to send probe packets into the congestion point. UDP flow applications typically use an external clock to determine the sending rate, and will continue to send traffic into the congestion point at the same rate. Such behaviors reduce overall network efficiency. Resource reduction is perceived as a more effective form of resource allocation, because there is still some level of low-precedence traffic throughput, allowing TCP end systems to continue to receive feedback signals. Such signals will reflect the changing state of the network, allowing the TCP sender to adapt its transmission rates accordingly.

CBQ also can be considered a basic method of differentiating traffic into various classes of service. For several years, CBQ has been considered a reasonable method of implementing a technology that provides link sharing for *classes of service* (CoS) and an efficient method for queue-resource management. That said, CBQ is not a perfect resource-sharing algorithm, due to the variable packet sizes that exist in an IP network. Figure 3.10 shows a three-queue CBQ, with relative service weightings of 3:2:1. The byte counts chosen for each queue are 600, 400, and 200. If the high-weighted queue is used for interactive applications, the average packet size will be some 60 bytes per packet, so that 10 packets are drained from the queue in each service round. If the median-weighted queue is used for streaming audio and video applications, the packet size may be larger, averaging some 150 bytes, for example. In this case, three packets will serviced per round, with a total of 450 bytes allocated to this service class. The low-weight queue is used for bulk data transfer. Our example assumes an average packet size of 1500 bytes per packet; in

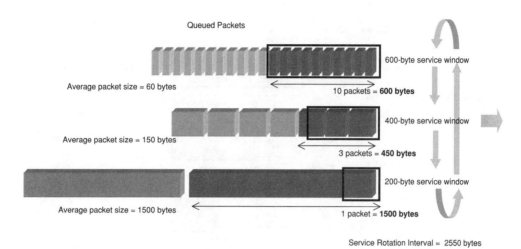

Figure 3.10 CBQ example.

each service round, one packet is processed from this queue. The resulting actual service allocations are, in bytes, 600, 450, 1500, or an actual relative service weighting of 5:2:12.

As shown by this example, CBQ can be unfair in its allocation of bandwidth; if one service class contains larger packets on average when compared to another, the larger packet class may obtain a larger service allocation than has been configured for it. Within a variable packet size regime, the CBQ service routine will continue to dequeue packets from a service class while there is still a "credit" of bytes to transmit from that class. The final packet chosen will exceed the available credit, and the service interval will therefore consume a greater proportion of bandwidth than the configured allocation. The larger the average packet size in a queue, relative to other service queues, and the smaller the service quantum (in bytes), the greater the discrepancy. A fairer output can be achieved by increasing the total number of bytes dequeued in each service rotation, or, in effect, slowing down the service rotation time. Unfortunately, while such a measure increases the accuracy of the method, so that the packet-based actual service levels correspond more accurately to the configured service weightings, the increased rotation latency adds a large jitter factor to all traffic. Figure 3.11 shows the outcome of increasing the service rotation interval by a factor of 100. In this example, the relative service outcomes are improved to within 5 percent of the desired outcome, but the jitter of the short service queue has increased dramatically.

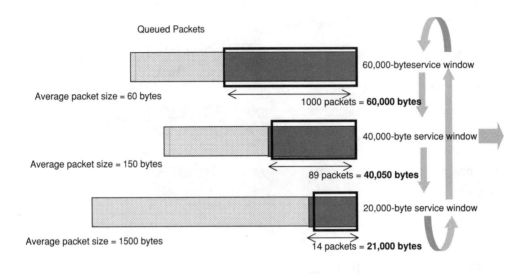

Figure 3.11 CBQ example with longer service rotation interval.

CBQ is only fair within an interval of time greater than the service rotation interval. Each queue will only receive one service quantum within each service interval. If fairness is measured at shorter intervals than this rotation time, then CBQ will show considerable unfairness. The service rotation interval is the sum of the service quantum intervals for each class, divided by the egress bandwidth. Because of the packet quantization, choosing a short service rotation interval produces service outcomes that can differ greatly from the desired policy-based outcome. Increasing the service rotation interval increases the accuracy of the method, but also the amount of jitter imposed on all traffic classes. For low-speed lines, or where a large number of service classes are used, or where large service quantum values are used, the service rotation interval increases, and the minimum time interval where fairness can be observed will extend. For this reason, when using relatively low-speed circuits, CBQ is not the optimal solution to provide network services that include factors of bounded latency or bounded jitter.

Where the speed of the circuit is considerably greater than the maximum packet size transmission time, and the number of class queues is kept reasonably small, this is less of a consideration. For example, a four-queue CBQ structure feeding a 155Mbps circuit would exhibit very low levels of jitter even with large per-queue service allocation quantities. Therefore, CBQ is a reasonable choice as a bandwidth allocation queuing mechanism when allocating absolute levels of bandwidth to aggregated service flows within the core of a high-speed backbone network. This has particular relevance when looking at support structures for Differentiated Services behaviors that assume the existence of some form of reasonably robust bandwidth allocation mechanisms. We will examine this further in Chapter 4, "Quality of Service Architectures."

A great deal of informative research on CBQ was completed by Sally Floyd and Van Jacobson while they worked at the Lawrence Berkeley National Laboratory for Network Research. For more detailed information on CBQ, see ftp.ee.lbl.gov/floyd/cbq.html.

Weighted Fair Queuing (WFQ)

In order to implement a resource allocation mechanism that avoids resource denial but implements fairness between various service classes down to a time interval of the same order as the transmission time of the maximal size packet within the network, the scheduler must use a different service algorithm to that of strict round-robin. The scheduler has to process an elevated service class queue more frequently than lower-priority service class queues. In other words, to achieve both service accuracy and acceptable performance, heavily weighted traffic should be serviced more frequently than lighter weighted traffic, allowing the higher-precedence traffic to be interleaved within the background traffic.

Fair Sharing

Before discussing weighted fair queuing in further detail it is necessary to digress slightly to *fair sharing*, as it relates to queuing disciplines. The problem is how to evenly divide a resource among competing unequal requests. When the sum of the requests is less than the available resource, all requests can be fully met. When the sum of the requests is greater than the available resource, how should the resource be divided to achieve *fairness*?

One approach is to discount all requests by the proportion of over-demand, ensuring that all competing requests are under-serviced by the same proportion. For example, if the sum of the requests is 25 percent greater than the available resource, all requests could be serviced by allocating each request 80 percent of its original request. Of course, this penalizes small requests to the same level as large requests, providing no incentive for conservatively stated requests. Alternatively, all requests could be allocated the same proportion of the resource, penalizing the large requests in direct proportion to the original amount requested. If there are five unequal requests, where the sum of these requests is greater than the available resource, then, regardless of the amount of overload of request, each request will be allocated one-fifth of the available resource. Such a mechanism is wasteful where the smaller request amounts are less than the allocated quantity, as the resource will be idle for the balance of the allocation quantity.

A commonly used fair-share algorithm is *min-max fair sharing*, where the intent is to meet the requirements of small requests and fairly distribute the remaining resource among larger requests. Min-max fair sharing is best explained by the following algorithm:

- Requests are sorted in order of increasing resource require- ments.
- Each request is serviced in order with an exact proportion of the remaining resource; if this amount is greater than the amount requested, the unused balance is placed back into the resource pool.

It is useful to define this in slightly more precise terms. Consider a resource of size R, with n requests of size s_1, s_2, ..., s_n. The requests are ordered in size to form the set of requests r_1, r_2, ..., r_n. Initially, the server will attempt to allo- cate R/n amount of resource to the first (and smallest) request (that is, grant an equal fair share of the resource). If R/n is greater than s_1 then the demand will be fully satisfied and the balance of resource remaining is $(R - s_1)$. If R/n is less than s_1 then the demand will only be partially met, and the balance of the resource remaining is $(R - R/n)$. In either case, once this request has been met, there is now a new available resource amount to be shared over the remaining $(n - 1)$ requests, and the same process is applied again. At the end of the ser- vice cycle, each request has either been granted the amount of resource requested or, if its request was not satisfied, received no less of the resource than was granted to larger requests.

This algorithm does not differentiate between requests; all are treated on an equal basis. The logical refinement to this algorithm is to accommodate *weighted* requests, where the weight reflects a relative loading to place on the request. For example, where there are two equal requests, weighted 1 and 2, respectively, weighted fair sharing will grant two units of resource to the second request for every single unit of resource granted to the first request. A *min-max weighted fair share* algorithm is as follows:

- Requests are sorted in order of increasing weighted resource requirements.
- Each request is serviced in order with a weighted proportion of the remaining resource; and if this amount is greater than the amount requested, the unused balance is placed back into the resource pool.

Again we can take this informal description and use an example to add some precision. Consider the general example of n requests, now with relative weights associated with each request. Consider a resource of size R, with n requests of size s_1, s_2, ..., s_n, and normalized weights w_1, w_2, ..., w_n. (A set of weights are normalized by multiplying all weights by a constant factor such that the smallest normalized weight is set to 1.) A set of *weighted sizes* of these requests can be generated, ws_1, ws_2, ..., ws_n, where $ws_i = (s_i/w_i)$. The requests are ordered in weighted size to form the set of requests r_1, r_2, ..., r_n. The sum of the normalized weights is W. Initially, the server will attempt to allocate ($w_1 \times R/W$) amount of resource to the first request (an equal weighted fair share of the resource). If this amount is greater than s, then the demand will be fully satisfied and the balance of resource remaining is ($R - s_1$). If this amount is less than s, then the demand will only be partially met, and the balance of the resource remaining is ($R - (w_1 \times R/W)$). In either case, there is now a new available resource amount to be shared over the remaining ($n - 1$) requests, and the same process is applied again.

Generalized Processor Sharing

We can now use this definition of a weighted fair-sharing algorithm to define queuing disciplines that exhibit this same fair-sharing property. The ideal work-conserving weighted fair sharing model is that of *generalized processor sharing* (GPS) [Keshav 1997]. Informally stated, GPS implements a min-max weighted fair-sharing resource allocation, using an infinitely small service quota.

Over any chosen time interval, this ideal model exhibits the desired property of weighted fair-share resource allocation, as the service quota is arbitrarily small. Of course, this is an ideal, and unimplementable, algorithm, in that it ignores distortion of the resource-sharing algorithm caused by data being quantized into packets. Its utility is that it defines a metric of effectiveness to examine implemented queuing disciplines, by measuring how close the queuing discipline comes to the ideal GPS behavior for any particular set of conditions.

Weighted Fair Queuing

According to Partridge [Partridge 1994], the concept of fair queuing was developed by John Nagle [RFC970], and later refined in a paper by Demers, Keshav, and Shenker [Demers 1990]. In essence, the practical problem addressed in both of these papers is a single misbehaving TCP session consuming a large percentage of available bandwidth, which causes other flows to unfairly experience elevated levels of packet loss, delay, and jitter. Fair queuing presents an alternative management discipline that relegates competitive traffic flows to their own bounded queues so they cannot unfairly consume network resources at the expense of other traffic flows.

Weighted fair queuing (WFQ) is an approximation of GPS behavior at the packet level. WFQ attempts to service a collection of queues in a min-max fair-shared manner. The relative share applied to each queue is in accordance with the weight applied to each queue. The behavioral objectives of WFQ are identical to weighted min-max fair sharing: under maximal load, each service class receives a measure of service in direct proportion to its relative weight. Where a service class requires less than its allocation, the excess is shared among all other service classes in proportion to their relative service weights. However, WFQ is a packet management process; the WFQ algorithm cannot manipulate bits or bytes, but must manage variable-sized packets.

The informal description of WFQ is that it attempts to emulate the behavior of a GPS discipline by attempting to order the queuing of packets to achieve the same order that a GPS system would use to transmit the trailing bit of each packet (see Figure 3.12). The logic of the trailing bit order can be demonstrated by the example of a GPS system transmitting arbitrarily small fragments of data to a packet reassembler at the other end of a link. In this scenario, the order in which packets are fully assembled will be determined by the order in which the trailing bit of each packet is transmitted. WFQ attempts to generate a service ordering of packets identical to this theoretical outcome of trailing bit order. WFQ also defines a packet drop policy in addition to packet order. If a packet is to be scheduled into a queue that is already full, the queue manager computes the finish order of the new packet in relation to queued packets. If the result of this computation is queued packets that would be serviced after the new packet, then enough of these packets are discarded to make space for the new packet. If this is not the result, the new packet will be discarded.

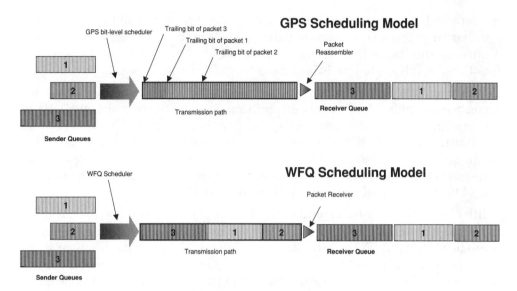

Figure 3.12 Comparing WFQ and GPS algorithms.

WFQ is min-max fair to a time level equal to the mean packet size. WFQ can be a considerable improvement in fairness accuracy and performance guarantees over CBQ. The outcome is that each service class in a WFQ environment is allocated no less than its weighted fair-share proportion of the resource. If the number of discrete service classes is bounded, then the minimum amount of resource allocated to each service class is predictable.

Predictability of service response is a critical outcome in queuing. WFQ offers a way for a network operator to guarantee the minimum level of resource that will be allocated to each defined service class. This guarantee holds regardless of the amount of activity that may occur in other service classes. It also ensures that the minimum performance level offered to one service class can be committed without reference to the cross-impacts of activity levels of other service classes.

In terms of offering performance guarantees, this guaranteed minimum resource behavior is a crucial step. To create guaranteed delay bound from this resource commitment requires the addition of traffic admission filters. If traffic admission filters are used to bound the amount of traffic entering the network in any particular service class, then the delay experienced by every packet in this service class in their transit across the network will also be bounded by a maximum delay value. This is perhaps an intuitive result—if a bounded load is processed by a committed resource, then the throughput of the system will also have a guaranteed minimum value. At this point, it appears that we have a queuing discipline that, when coupled with admission controls, provides functional solutions to providing guarantees of network service.

However, this picture is not yet complete, and as we noted in Chapter 2 when we examined the aspects of performance, the other major attribute of performance is jitter. Unfortunately, even with admission filters, WFQ will not intrinsically bound the jitter level to any useful value (the jitter bound is at the same level as the delay bound). If smaller jitter bounds are required for some service classes, then some element of priority queuing must be reintroduced back into the queuing discipline, as must admission control.

Jitter is not the only issue with WFQ. The implementation itself is not without some complexity, which will have performance implications for router implementations. WFQ is implemented using a GPS simulator, where a bitwise weighted round-robin scheduler is simulated by means of a GPS service time counter. When a packet arrives at a WFQ manager, the packet is assigned to a service class, and a GPS service time interval must be updated. Similarly, when a packet is drained from the service queue, the GPS service time interval must be updated. This update of the GPS simulator will involve an iterative scan of all the service queues to recompute the simulated GPS behavior in terms of relative packet ordering. Such complex computation may occur every few microseconds on a high-speed router. When there are a large number of active service classes and a large packet forwarding rate, the computational load of WFQ is a significant performance engineering factor. Variations of WFQ have been proposed, including *self-clocked fair queuing*, and *start-time fair queuing*, where a trade-off is made between the complexity and time it takes to perform iterative queue scans for the GPS simulator, and the resultant delay bounds. These alternate forms of WFQ can achieve considerable algorithm improvements, but such improvements are at the expense of the accuracy of the scheduling system in its adherence to the ideal GPS behavior [Keshav 1997].

WFQ uses a servicing algorithm that attempts to provide predictable response times and to negate inconsistent packet-transmission timing. WFQ does this by sorting and interleaving individual packets by flow, and by queuing each flow based on the volume of traffic in each flow (see Figure 3.13). Using this approach, the WFQ algorithm prevents larger flows (those with greater byte quantity) from consuming network resources (bandwidth), which could subsequently starve smaller flows. This is the fairness aspect of WFQ— ensuring that larger traffic flows do not arbitrarily starve smaller flows.

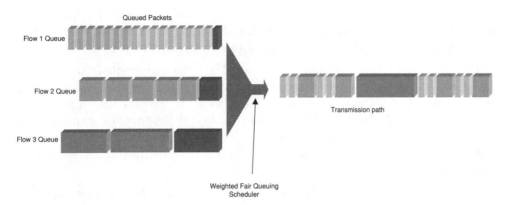

Figure 3.13 Weighted fair queuing.

The weighted aspect of WFQ is usually vendor-specific, and at least one implementation uses the IP precedence bits in the TOS (Type of Service) field in the IP packet header to weight the method of handling individual traffic flows. In this implementation, the higher the precedence value, the greater access to queue resources a flow is given, because the relative weight used by WFQ is the packet's precedence value.

Advantages and Disadvantages of Weighted Fair Queuing

WFQ does provide significant outcomes for supporting performance guarantees within a network. WFQ will ensure a minimum level of resource allocation to a service class, independent of the activity levels in other classes. When used in conjunction with some form of admission control, the result will be bounded delay as well as guaranteed throughput.

All this comes at a price. WFQ is a complex algorithm that requires per-flow (or flow-aggregate) state, and iterative scans of all per-flow states on packet arrival and packet departure. This presents considerable scaling challenges. The practical conclusion is that WFQ, when operating in a per-flow manner, simply does not scale to provide the desired performance in concentrated high-volume, high-speed routing environments. The primary reason is the computational overhead and forwarding impact that packet reordering and queue management impose on networks with significantly large volumes of data and very high-speed links. WFQ has some potential application when configured to operate within a bounded domain of a fixed number of aggregate service classes. In this case, the fixed number of service classes bounds the computational load, and WFQ will undertake fair bandwidth allocation between the service classes. In such a configuration, WFQ reorders packets to simulate per-service class fair-bandwidth allocation, and it does so in a fashion that minimizes induced jitter and rationalizes packet discard.

The associated issue is that of service class management. Bounding the number of service classes and aggregating individual flows into a single compound service class will offer some relief to scaling WFQ into high-speed, high-volume environments. The cost in this case is the level of granularity of the offered performance, because the performance profile will apply only to the aggregate service class, while individual flows within the aggregate may receive a far less determinate level of service.

The conclusion is that WFQ is not a universally applicable queuing discipline. WFQ, when configured to operate on a per-flow service classification, is most effective in low-speed access circuits, where a low-jitter form of mixing multiservice applications is critical. For example, WFQ bounds each TCP session, ensuring that any individual session does not attempt to burst across the entire queue space. WFQ can interleave voice and video streams, ensuring that each stream receives a fair share of the available bandwidth, without experiencing high jitter or unnecessary packet drops. In larger-scale environments, where the number of active flows and the available bandwidth rise dramatically, per-flow WFQ can add little value. For such reasons, per-flow WFQ is best regarded as an edge control mechanism for low-speed circuits, where bandwidth management and jitter are of paramount importance to effective multiservice delivery (see Figure 3.14).

When operating in a mode of a bounded number of aggregate service classes, WFQ offers applicability to determine per-aggregate resource allocation in larger-scale environments. The bounded service class environment allows WFQ to offer each service class a guaranteed minimum bandwidth allocation, and in a manner that minimizes per-service class jitter. The bounded number of service classes, when kept reasonably small, allows the algorithm to operate across very high-speed circuits. However, there is some level of trade-off in such high-speed environments. Within a very high-speed environment, the gains in jitter management through optimizing fairness down to a time scale of one packet transmission interval are marginal when compared to the slightly coarser multiple packet transmission interval fairness outcomes of CBQ in a similar environment. The reduced computational overhead of CBQ may be a determining factor in the decision as to which approach to adopt in such circumstances. Of course, neither approach provides management of per-flow relative fairness within each aggregate service class, and the issue of effectively managing resources at both an aggregate and

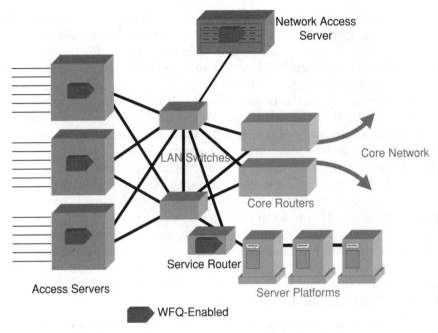

Figure 3.14 WFQ deployment in a service network POP.

a per-flow level simultaneously remains the greatest challenge in designing multiservice IP networks.

A Comparison of Queuing Disciplines

Each of these queuing disciplines has its relative advantages and areas of applicability within a multiservice network, which we review in this section.

FIFO queuing has two major attributes: First, FIFO implementations are very efficient, and scale into large queues and fast-switching environments; and second, FIFO has very predictable outcomes. FIFO does not reorder packets, and the maximum jitter added by a FIFO queue is proportionate to the configured queue size. Of course FIFO does not implement any form of service policy, so that an uncontrolled or bursting source may consume the entire queue, causing other active flows to experience varying levels of jitter and loss. And because FIFO has no capability to react to traffic priorities, a FIFO queuing system may delay or discard real-time voice or video streaming data and allow the queue to fill with nontime-critical TCP background bulk data transfers.

So where does FIFO queuing make sense? The obvious first answer is that if the network has no QoS policy, then FIFO is a very efficient queuing mechanism to use, because it also has no QoS policy. However, FIFO queuing has a role in QoS-enabled networks. If the network has an admission policy applied at the edge of the network, and the interior of the network is engineered so that the available capacity is commensurate with the amount of traffic admitted into the network, there is no need to run a policy-based queuing system throughout the interior of the network. This deployment assumes that the edge admission control functions have successfully limited the amount of traffic admitted to the network to a level commensurate with the capacity of the network, and in so doing have honored the QoS policies of the network. For very large networks, such a division of function is highly practical. The more computationally intensive service monitoring is performed at the edge of the network, where the traffic density is lower and the packet transit rates are lower, permitting a higher processing overhead per packet to be added without impairing the total capacity of the network. The interior of the network uses FIFO queuing, because it is computationally efficient within the high-speed, high-volume network core.

Priority queuing is a queuing mechanism with very coarse outcomes. Traffic within the highest-priority classification receives a low-jitter, low-loss service, as long as the amount of high-priority traffic is less than the available network capacity. Priority queuing is strictly preemptive in its service profile, so all other traffic is stalled while traffic of a higher priority is being serviced. Given that the primary objective of performance tuning is to stop uncontrolled applications from damaging the performance of other applications, priority queuing itself does little to alleviate this situation. The applicability of priority queuing is in those environments where there is a single time-critical application that has significantly higher value to the network operator than all other traffic. This may be voice or video traffic or customer-defined premium traffic passed into the network under the terms of a service level agreement (SLA). Priority queuing will allow the operator to maximize the network's response to this high-value traffic, at the potential expense of all other traffic types.

In addition, priority queuing is applicable in those environments where some form of traffic conditioning is undertaken as an admission function. If the admission control function can limit the amount of elevated priority traffic entering the network, then the subsequent high-priority service is a sustainable high-performance, low-jitter service platform. The approach to priority queuing of combining admission control to the interior priority queuing service response can form the basis of virtual circuit-based services. We will examine this in further detail in Chapter 4, "Quality of Service Architectures," when we explore the Differentiated Services Expedited Forwarding service.

Class-based queuing has similar computational overheads to priority queuing; packets are placed into a limited number of queues, based on a classification function. The basic difference between the two approaches is that CBQ offers every service class some service level (it will dequeue a minimum of one packet from each service queue within each service cycle). CBQ offers an approximation of relative bandwidth allocation to each service class, in that a minimum of a certain number of bytes will be services from each class. Given the variability of the service cycle timer, this does not correspond exactly to a minimum bandwidth allocation, but it may present an acceptable compromise. The variability of the service cycle timer also introduces an element of jitter into all traffic classes. The advantages of CBQ are, one, that is does allow some level of bandwidth allocation to each service class, and, two, with a careful selection of per-queue service amounts and service queue lengths, quite acceptable differential service outcomes can be achieved with very modest computational overhead. This is particularly the case in very high-speed environments, where CBQ's packet quantization jitter can be quite low in absolute terms. Of course, with CBQ, there is a statically configured number of service queues, and within each queue the service order remains strictly FIFO. This implies that CBQ cannot adapt to an arbitrarily fine granularity of service, and it cannot react to the service needs of individual applications. CBQ does provide the essential elements of resource allocation to a small number of aggregated service classes. Within large-scale networks, CBQ's small set of service classes is of limited utility when looking at service management of individual flows; and even with an associated collection of admission control functions at each network ingress, the potential of queue overflow within each service class still remains. CBQ is a useful low-overhead tool in corporate service networks and in the high-bandwidth core of ISP service networks. Its limited service capability restricts its adoption within the conditioning environment at the edge of the network, where a finer level of granularity of control is typically called for.

Weighted fair queuing offers a reasonable simulation of min-max fair sharing across all active flows. There is no requirement for a small set of preconfigured service classes, as the algorithm will adapt to a dynamically changing set, so that it is possible to undertake WFQ on a per-flow basis. In this mode of operation, WFQ does not intend to emulate a bandwidth allocation function, and it does not attempt to undertake allocation of fixed bandwidth to a fixed set of recognized traffic flows. WFQ attempts to fairly share the available resource across all active traffic flows, obeying any relative weighting that may be locally applied to any individual flow. This prevents any single uncontrolled application—or any collection of applications for that matter—from bursting traffic into the network, to the detriment of all other traffic. To the extent that WFQ provides a fair outcome across all active flows, WFQ does provide effective performance control. This functionality comes at a price, however. WFQ has a computational overhead that grows in direct proportion to the number of active flows. In the core of a high-speed, high-volume service network, per-flow WFQ is not an attractive option. Per-flow WFQ is a very useful performance management tool for medium- to low-speed access networks. Per-flow WFQ can also be used in conjunction with admission control functions and path reservation protocols, such as RSVP, to allow the creation of a service class that matches the RSVP service profile. Of course, such per-flow computations will not scale within the core of a large high-speed service network. WFQ is also applicable to operate across a fixed set of aggregate service classes, and in this mode, per-aggregate service class WFQ is of some potential interest. As long as the admission control function maintains accurate accounting of available network resources for premium service traffic, and does not over-commit resources to premium service requests, then careful use of service classes and relative service weightings allow per-aggregate service class WFQ within the interior of the network to ensure that the uncontrolled best-effort service class can be supported alongside rate-controlled premium services.

The conclusion from this comparison is that though queuing mechanisms are essential performance tools, queuing alone is an inadequate response to performance engineering for multiservice networks. Queuing mechanisms must be put into a structured performance architecture that includes classification functions, admission control, packet discard control, forwarding systems, and the associated path maintenance function. And it is necessary to examine these additional performance control components before we look at comprehensive performance architectures.

Packet Discard

A scheduling discipline of a router has two components: a queuing component, controlling the order in which packets are serviced, and a discard component, controlling the timing and selection of packet discard. Congestion leading to packet discard happens when there is a mismatch between input rate and output capacity. Where the input rate exceeds the output capacity for some period of time, internal queues will attempt to absorb the transient mismatch condition. That said, queues are finite structures, and, for a sufficiently large mismatch, the queue will saturate and packet discard will occur. A queue will absorb a small mismatch between input and output capacities over a long period before the queue saturates, or it will absorb a large mismatch over a very short period. In either case, once the queue is saturated, packet drops must occur. Dealing with such congestion-induced discard is a control theory problem. The desired behavior is when the discard of a packet will generate some control-level signal to the packet's sender that the path is traversing a congestion point, causing the sender to moderate its sending behavior. This section details the packet discard function.

Packet discard should be regarded as a last-resort response to congestion situations; any packet that is discarded will have already consumed some level of resources in its transit through the network, and the discard is a write-off of such resource consumption. However, discard cannot be entirely eliminated, since any finite queuing structure that supports uncontrolled variable load will inevitably experience periods of queue exhaustion and consequent packet discard.

---

*P*acket dropping can be compared to the random early detection (RED) mechanism for TCP. A common philosophy is that it is better to reduce congestion in a controlled fashion before the onset of resource saturation until the congestion event is cleared, rather than wait until all resources are fully consumed, resulting in complete discard. Controlling the quality of a service often is the task of monitoring how a service degrades in the face of congestion. It is more stable to degrade incrementally than to wait until the buffer resources degrade to complete exhaustion in a single catastrophic collapse.

It is fruitless to attempt to eliminate all possibility of packet discard from the network. The intent, similar to that of queuing, is to introduce the concept of fairness to the discard selection. An ideal discard environment will protect sources that are well behaved (in that they are consuming no more than their fair share of the resource) from those that are badly behaved (in that they are attempting to consume more than their fair share of the resource).

Tail Drop

The default packet discard algorithm, normally used in conjunction with FIFO queuing, is used to discard packets when there is no further queue space left. This is termed *tail drop*, as packets are discarded from the logical tail of the queue. Packets continue to be discarded until space is available on the queue. This discard algorithm is a simple load-shedding algorithm, but it does not exhibit fair discard behavior. A bursting source may saturate the available queues, forcing the scheduler to discard packets from well-behaved sources. Indeed, such a system exhibits a worst-case behavior where a high-volume, long-delay TCP flow can force all other simultaneous traffic to synchronize with the round-trip time of the TCP flow.

One way to protect one service class from another is to extend the service class structure used within the queuing strategy. This extension places a limit on the size of the service queue. Though placing a limit on the size of each service queue may cause packet discard even when there is available packet storage in the router, this approach will assist in ensuring fair allocation among service classes—a large burst load of traffic in one service class will not cause discard within another service class.

A refinement of this per-queue size limit approach uses a common memory pool. When a packet cannot be queued, packets are discarded from the tail of the longest service queue. In this fashion, backlogged service classes that are exceeding their service allocation will have a high discard probability, while service classes operating within their service allocation will maintain short queues, and therefore will experience a low discard probability.

A further variation of this approach can be used in the WFQ queuing discipline. In this discipline, the packets with the GPS simulator's greatest finishing order are discarded to make room for the new packet. Depending on the service weightings, this may not necessarily be the longest queue, but it will correspond to the last packet to be served, the one that was going to receive the lengthiest delay in awaiting service.

Early Drop

It was noted that if per-service class queue lengths are enforced, a packet may be dropped even while there is pool space available. This is one instance of *early drop*, where the discard is triggered prior to complete resource exhaustion.

The intent of an early drop discard strategy is to enforce a form of fair sharing. If packets from the heaviest bursting flows are discarded prior to complete queue exhaustion, then space will still remain for other, presumably smaller, flows to burst into the remaining queue space (see Figure 3.15). Early drop can be applied with varying amounts of precision. The queue manager can monitor each individual flow and its dynamic characteristics. When the available queue space falls below a certain threshold, those flows with a high, constant flow rate, and those flows with a rapidly rising flow rate are likely candidates for packet discard, as it is likely that these flows are the major contributors to continued growth in the queue size. This precise response does cause considerable overhead in terms of both classifying traffic into individual flows and in terms of monitoring the packet arrival rate within each flow in order to calculate the current flow rate and its rate of change. Is this additional processing overhead strictly necessary? If a random selection process was used to select packets for discard, statistically it is likely that the packet so chosen will be a member of one of the flows contributing to the excessive queue load. Given the scaling issues with per-flow monitoring in large networks, it is not surprising that a random early drop is the chosen approach in most environments.

Early packet drop applied to the heaviest bursting flows is also useful to signal to these heaviest flow applications that the available resource is now critically low, and that though further use can be supported, any expansion of use will cause complete exhaustion of the resource. TCP responds to this form of signaling, and we will examine early drop in further detail when we look at TCP congestion control later in this chapter, as part of our examination of random early detection.

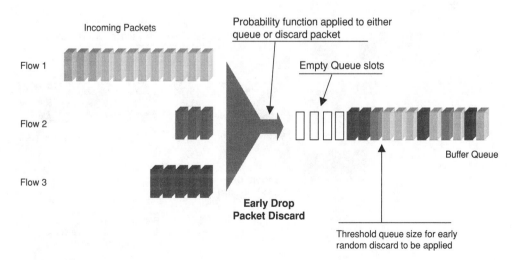

Figure 3.15 Early drop.

Head of Queue and Random Drop

The discard approaches described so far assume that the discard occurs at the tail of the service queue, either by discarding the most recently arrived packet or by discarding the packet that is last in line to receive service. The discard can occur on queue saturation, or, as noted above, the discard can occur randomly once the queue size has reached some certain threshold. Discard of the most recently arrived packet is not the only approach. Alternate approaches can be used to discard at the head of the queue or, potentially, from a random position within the queue, thereby making room for the most recently arrived packet. Of course, discarding the packet prior to placing it in the queue can be implemented very efficiently in router system designs, while head of queue drop, or random position drop, may have some router performance implications in some architectures.

This discard approach attempts to implement a fair discard function through probability. In performing a random selection from the queue for discard, a heavily bursting connection that is filling the queue will have a higher probability of having one of its packets randomly selected than a packet from a well-behaved source (see Figure 3.16). When dealing with TCP, this approach of discarding closer to the head of the queue can dramatically

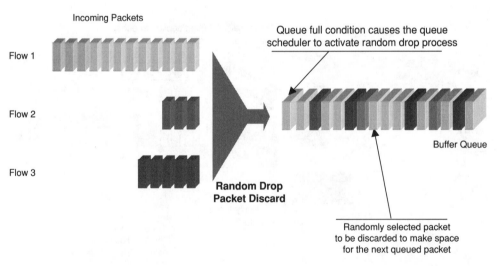

Figure 3.16 Random drop.

improve overall TCP throughput efficiency. We will look at this in greater detail as well in the discussion on TCP congestion control.

Drop Preference

The preceding approaches work on the principle that "the network knows best," and that the network selects which packet to discard based on local conditions at the point of congestion. An alternative approach, used in Frame Relay and ATM, is to use externally defined drop preference. When the local average queue load exceeds some defined proportion of available resources, all other packets with a certain drop preference (or greater) will be discarded. In this way, the congestion effects—in terms of loss—will be staged in, first affecting what could be termed "nonessential" traffic, and then more critical traffic only in the face of sustained critical traffic load. The advantage of this approach is that the definition of what is essential traffic can be left to the admission control function, to the end-host protocol driver, or, in the case of UDP-based applications, to the application itself. This is a good example of splitting the classification function from the local event. The classification of which packets are good candidates for in-transit discard is left to the admission system at the ingress to the network or even passed out to the end client. When a local congestion condition occurs, the first response can be to exercise the available option to discard all such marked packets.

The Frame Relay Discard Eligible bit and the ATM Cell Loss Priority bit are examples of a two-level drop preference. In these cases, there is a single drop preference, allowing a single internal discard threshold within the network. There is no defined field or bit within the Internet Protocol header. Consideration is being given to the use of a discard preference indicator as a part of the Differentiated Services architecture, and the option remains to use either a single discard threshold or a larger header field and allow a range of discard preferences.

In IP, the typical application of a drop preference mechanism is in conjunction with marking admission controllers. In this scenario, all traffic that is outside of the configured profile is marked with an elevated drop preference by the admission filter. If the overall network load is low, the excess traffic may be carried without any performance penalty. As the load from other traffic classes increases, the out-of-profile traffic can be discarded as an initial load-shedding measure. In essence, this defers the actual profile enforcement of the admission filter, deferring the discard decision to the interior of the network on the basis of actual current load.

Traffic Shaping and Admission Control

With the progressive refinement of queuing disciplines, the ability to define the manner in which a shared resource is distributed among competing demands is increasing. The objective of such efforts is to define the contention-resolution policies of the network's interior routers, to determine the relative priority of queued packets awaiting service, and to determine the relative priority of packets to be discarded once the queue becomes saturated. By using WFQ, the queuing discipline can define base levels of network performance for each service class by eliminating unfair resource allocation across service classes. Still, however, within each service class, there is a missing measure of necessary control. The base service level experienced by any service class is equal to the maximum load level divided by the base performance. If the load is uncontrolled, the service level is uncontrolled. Traffic shaping and admission control mechanisms are intended to complete the basic performance capability of the network by imposing a control on the load admitted into the network.

The function of such admission functions is to control the traffic actually admitted to the network and, potentially, the rate at which it is admitted. Such functions can also include service classification functions, controlling the service class or priority assigned to various types of traffic as they enter the network. Traffic shaping and admission control have very distinct differences, which we will examine in more detail. There are several schemes for both admission control and traffic shaping, some of which are used as standalone technologies and others that are used integrally with other technologies, such as the IETF Integrated Services architecture. How each of these schemes is used and the approach each attempts in conjunction with other specific technologies defines the purpose each is attempting to serve, as well as the method and mechanics by which each is being used.

It also is important to understand the basic concepts and differences between admission control, traffic shaping, and policing.

Admission control. In the most primitive sense, the practice of discriminating which traffic is admitted to the network in the first place and, by implication, which traffic is discarded.

Traffic shaping. The practice of controlling the volume of traffic entering the network, along with the rate at which it is transmitted.

Policing. The practice of determining on a hop-by-hop basis within the network beyond the ingress point whether the traffic being presented is compliant with prenegotiated traffic-shaping policies or other distinguishing mechanisms.

When determining which of these mechanisms to implement, it is necessary to consider the resource constraints placed on each device in the network, as well as the performance impact on the overall network system. For this reason, traffic-shaping and admission-control schemes need to be implemented at the network edges to control the traffic entering the network. Traffic policing obviously needs one of two factors to function properly: Each device in the end-to-end path must implement an adaptive shaping mechanism similar to what is implemented at the network edges; or a dynamic signaling protocol must exist that collects path and resource information, maintains state within each transit device concerning the resource status of the network, and dynamically adapts mechanisms within each device to police traffic to conform to shaping parameters.

Traffic shaping and admission control manage the profile of load placed on the network. It is essential to describe a load profile in terms that are simple to translate into a data model and to define profile manipulation methods in ways that correspond to manipulations of this data model. The commonly used data model is that of the *token bucket*.

The Token Bucket

Traffic shaping and admission control policing can both use a similar conceptual model, which invokes a simulation of an ideal traffic profile. In the case of shaping, packet delay is used to retime the transmission packets to approximately match the profile of the model; in policing, those packets not within the profile are discarded or marked.

The basic mechanism required to perform this simulation to support both traffic shaping and admission control policing is the *token*. In this context, a token allows the admission of a defined quantity of data into the network. The rate at which tokens are generated determines the long-term average data rate permitted by the token system. The token model in this format is a strict matching model, in that packets are passed through the filter only at the time at which a token is available for use. If no token is available, then the packet is discarded. This model is a nonbursting bounded rate admission filter, where the admitted traffic corresponds to maximum flow rate as determined by the corresponding token rate.

A model that allows a more approximate match to the profile model uses a token accumulator, or token bucket. Tokens may be assembled in a bucket and retained for later use if there is no immediate demand for packet admission. Each token admits a certain number of bytes into the network. A number of tokens may be drawn from the bucket in response to admission of a large packet or in response to a burst of packets.

In a traffic-shaping application, a packet may be delayed at the token gate until an adequate number of tokens are available to permit the packet through. In a traffic-policing application, a packet will be discarded or its priority will be reduced if there are not sufficient tokens available to permit the packet to proceed through the admission gate. Intuitively, the configured depth of the token bucket describes the permitted variance from the ideal constant rate traffic profile.

Let's look at this structure in further detail, using Figure 3.17 as a reference for the token bucket operation. Tokens are placed into the bucket at a regular rate (or *drip rate*, to use a water analogy). Each token permits the admission of a fixed amount of data. When a packet is presented for admission, the packet is permitted past the regulation point only if there are sufficient tokens already in the bucket. If there are not, the packet is discarded or queued,

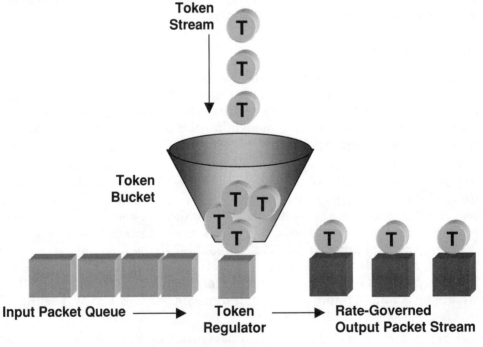

Figure 3.17 A token bucket traffic regulator.

depending on the regulation function being performed. If the packet is permitted, tokens are removed from the bucket corresponding to the size of the packet. The token bucket has a finite depth, and if tokens accumulate to the limit of the bucket, further tokens are discarded. Intuitively, the token generation rate defines the long-term average traffic rate that is enforced by the token bucket. The token bucket depth matches the largest permitted burst size. The associated packet queue defines the shaping capability of the regulator, where input traffic bursts at a rate greater than the permitted burst rate are smoothed to a permitted burst followed by an average rate stream.

If the tokens and the token bucket size are set to the equivalent of a single packet, then the regulation function performed by the token is *rate enforcement,* whereby the output rate will be no greater than the defined token generation rate. In this case, the depth of the associated packet queue defines the smoothing capability, defining the maximum size of a burst that will be smoothed by the token bucket (see Figure 3.18).

If the token bucket can hold token for more than one packet, the bucket defines the largest above-average burst that will be permitted through the rate enforcer. Don't forget that, in this case, the token bucket will still strictly enforce the long-term average data rate passing through the token bucket, so

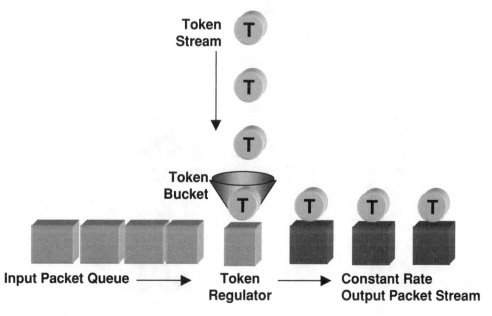

Figure 3.18 A rate-enforcing token bucket.

that a burst must be matched by a preceding period of below-average rate traffic, allowing the token bucket to accumulate sufficient credit to permit the burst. This is shown in Figure 3.19.

Token bucket regulator functions can be strung in series to perform more complex traffic-shaping functions. A buffered burst-capable token bucket operating at a token rate of ρ_1 can feed its output to a buffered rate-enforcing token bucket operating at rate ρ_2. This results in a *bounded burst* traffic-shaping function, where the long-term average traffic rate is no greater than ρ_1, while the burst rate is limited to no greater than ρ_2.

Traffic Shaping

Queuing and packet discard are defensive mechanisms used by routers when congestion is experienced. When a packet arrives at the point at which the router resource is already busy, the packet is held in a queue to await service. If the queue resource saturates, the router will need to discard packets, as it has

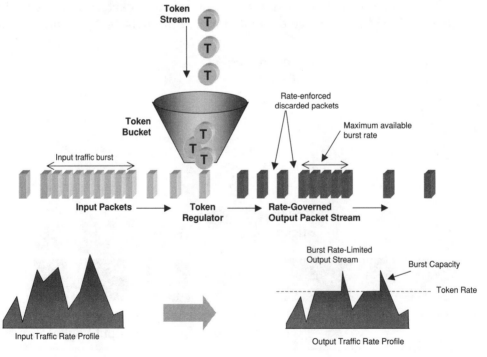

Figure 3.19 Token bucket burst rates.

no further local storage. We have examined queuing disciplines that impose some limits on the minimum amount of resource allocated to each service class, but in the face of a highly varying traffic source, we also need to control the traffic admitted from the source to create consistent performance levels.

Traffic shaping provides a mechanism to control the amount and volume of traffic being sent into the network and the rate at which the traffic is being sent. It also may be necessary to identify traffic flows at the ingress point (the point at which traffic enters the network) with a granularity that allows the traffic-shaping control mechanism to separate traffic into individual flows and shape them differently.

Two predominant traffic-shaping functions exist: one to smooth traffic bursts, so that all the traffic is bounded by a configured flow rate, the other to enforce a long-term average traffic rate and to permit bursts of a higher rate, provided that the burst is of a defined maximum size. Both these schemes have distinct properties and are used for different purposes. Discussions of scheme combinations follow. These schemes expand the capabilities of the simple traffic-shaping paradigm and, when used in tandem, provide a finer level of granularity than each method alone provides.

Traffic Smoothing

A *traffic-smoothing* function is used as a mechanism by which bursty traffic can be shaped to present a steady stream of traffic to the network, as opposed to traffic with erratic bursts of low- and high-volume flows. This is implemented using a token bucket with a depth of a single packet, with a token generation rate equal to the configured smoothed rate (see Figure 3.20). An appropriate analogy for the leaky bucket is an automobile traffic scenario in which four lanes of cars converge into a single lane. A regulated admission interval into the single lane of traffic flow helps the cars move. The benefit of this approach is that traffic flow into the major traffic circuits of the network is predictable and controlled. The major liability is that when the long-term volume of traffic is greater than the smoothing output rate, the packets will back up beyond the capacity of the regulator queue and be discarded. This smoothing function will also retime a bursty ingress data flow, an action that may have severe impacts on some forms of real-time traffic flows.

It also can be argued that a traffic-smoothing admission function does not efficiently use available network resources. Because the regulator's maximum output rate is a fixed parameter, there will be instances when the ingress traffic volume is high and large portions of network resources are not being used, but because the smoothed traffic is rate-limited it cannot take advantage of the

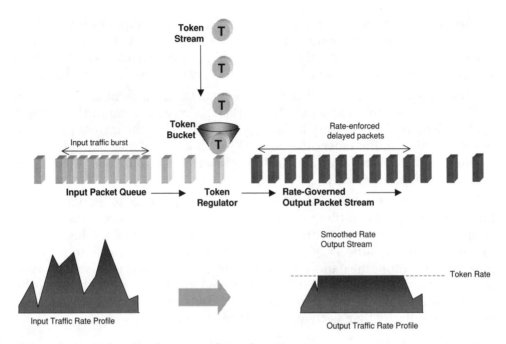

Figure 3.20 Token bucket smoothing function.

available network capacity. This illustrates a more generic problem with all traffic-shaping functions. Because such functions do not incorporate any form of control loop that has feedback of the current network conditions into the shaping function, the shaping function is incapable of taking advantage of temporarily available network capacity.

Traffic Burst Shaping

Another method of providing traffic shaping and ingress rate control is *burst shaping* (refer back to Figure 3.19). This differs from the smoothing function by increasing the depth of the token bucket traffic regulator. The resulting traffic-shaping function differs substantially from traffic smoothing. Where traffic smoothing uses a buffer to capture bursts, and uses the token mechanism to steadily transmit the traffic at a continuous fixed rate when traffic is present, the burst-shaping regulator uses a larger token bucket that can accumulate *credits* when the input is idle. This accumulation of credit is bounded, and once the token bucket depth has been reached, further tokens are simply discarded.

When a burst is presented to the regulator, the burst can be passed through without change, to the limit of the available tokens. Further packets that may form part of the burst are placed into the local packet buffer and are fed into the network at the configured average rate. In this case, the long-term average traffic rate is capped to the configured token generation rate; but bursts of a size up to the token bucket size may be passed through the regulator without imposed smoothing. The trailing edge of larger bursts is smoothed to the average traffic rate.

Combining Traffic-Shaping Mechanisms

A variation of the single traffic-shaping function is to use a number of parallel traffic shapers with different shaping parameters, in conjunction with a traffic classifier. In this case, the traffic classifier could interact with separate traffic shapers, each with a different peak-rate threshold, thus permitting different classes of traffic to be shaped independently.

The other common configuration of traffic shapers is to use two or more in sequence. The first shaping function could be configured to burst shape traffic to a long-term average load, while allowing bursts above this rate to a defined maximum size. The second shaping function could be configured as a smoothing shaper, using a smoothed rate equal to a desired peak rate. The resulting traffic profile would be one with a maximum long-term rate, a maximum peak data rate, and a maximum burst volume above the long-term rate. This function is illustrated in Figure 3.21.

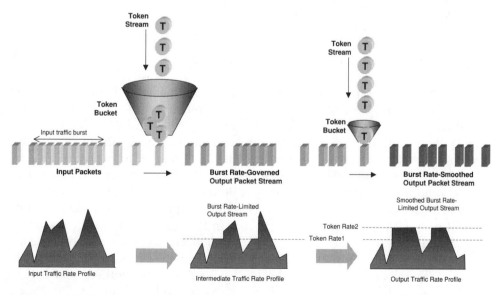

Figure 3.21 Multiple traffic shapers.

Such compound models can be further refined, using a classifier and multiple dual traffic shapers, allowing a finer granularity for controlling the types of traffic entering the network. One peak burst admission rate could be set arbitrarily low for lower-priority traffic, for example, while another could be set arbitrarily high for traffic that an administrator knows has bursty characteristics, allowing for greater consumption of available network resources.

Admission Control

Admission control is a policing function rather than a shaping function, whereby traffic that does not conform to a configured profile is not adjusted through the use of a local packet buffer, but is either discarded or marked. The traffic regulator mechanism is the same token bucket structure; but in this case, the packet buffer is eliminated and a packet "marker" or a discard function is used in its place.

The smoothing traffic shaper uses delay to smooth bursts into a rate-controlled traffic flow. The admission control variant on the smoothing function is that instead of delaying burst packets, the packets are simply discarded. Here, the peak rate of the smoothing function is strictly enforced, and any burst sequence is trimmed so that the rate does not exceed the configured peak rate. A similar variant of the burst-shaping function can be constructed by replacing the local buffer with a discard function. Once the token bucket has been exhausted, further packets are discarded. Bursts up to the token bucket size are passed through the admission control filter; longer bursts are truncated at this size.

A variation of the admission control filter is to not immediately discard those packets that are outside the configured traffic profile, but to mark such packets in a unique fashion (see Figure 3.22). Such a mark could be a lower scheduling priority or an elevated discard preference. This form of traffic shaping attempts to provide a partial answer to the problem of adapting traffic to changing conditions. In this model, the burst is maintained through the rate-smoothing traffic filter, but only the smoothed component is marked as *in-profile*, while the burst is marked as *out-of-profile*. At a point when the network is under some form of load, the out-of-profile traffic component will be further delayed or discarded, while the network will attempt to pass the in-profile traffic that was within the configured profile. In this case, the decision to delete the traffic is deferred to the interior of the network. Only when there is congestion load within the network interior will an out-of-profile packet be discarded. If there is no congestion, there is no need to perform the discard. This function is identical to the Frame Relay burst capability, whereby out-of-CIR profile frames are marked with the Discard Eligible bit. Burst-rate admission

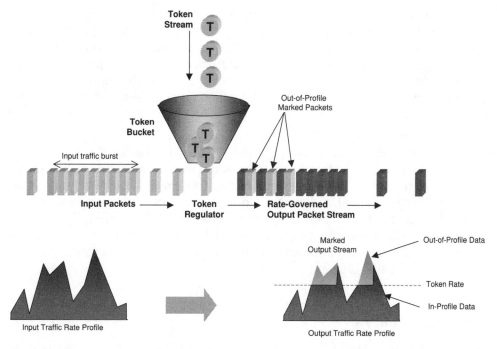

Figure 3.22 Traffic marking.

control filters can also use a marking action for excess burst rates, allowing the network to admit in-profile a certain burst size, and mark larger bursts as being an out-of-profile excess burst. We will look at such an approach in greater detail in Chapter 4, when we review the Assured Forwarding service defined within the Differentiated Services architecture.

With a sequence of admission control filters, multiple priorities, and multiple discard preference levels, the various traffic profile possibilities quickly balloon. A sequence of burst controllers enable the network manager to impose increasingly high discard preferences as a burst extends, while a sequence of peak rate controllers allows higher-intensity bursts to be marked with elevated discard priorities.

Network-Admission Policy

Network-admission policy is an especially troublesome topic, for which there is no clear approach throughout the networking industry. In fact, network-admission policy has various meanings, depending on the implementation and context. By and large, the most basic definition is: Admission to the network, or basic access to the network itself, is controlled by the imposition of a policy constraint. Some traffic (traffic that conforms to the policy) may be admitted to the network, and the remainder may not. Some traffic may be admitted under specific conditions, and when the conditions change, the traffic may be disallowed. Basically, this is the admission control function described in the previous section.

By contrast, admission may be based on identity through the use of an authentication scheme. A network admission system consists of the access control component to implement the admission control itself and an admission policy system that directs the control component's operation. *Access control* can be defined as allowing someone or something, generally a remote service or application, to gain access to a particular machine, device, or virtual service. By contrast, *admission policy* can be thought of as controlling the type of traffic that is allowed to enter or transit the network.

Admission Policy

Admission policy can be a very important area of networking, especially in the realm of providing QoS. Although being able to limit the sheer amount of traffic entering the network and the volume at which it enters is indeed important, it is much less important if the capability to distinguish services is unacceptable, because no real capability exists to treat one type of traffic differently from another.

One example of a simple method of providing admission policy is through the use of traffic filtering. Traffic filters can be configured on a device port, a destination IP host address, a source or destination TCP or UDP port address, and so on. Traffic that matches the filter is permitted or denied access to the network. These traffic filters could be used with a leaky- or token-bucket implementation to provide a more granular method of controlling traffic. The necessary level of granularity of isolating traffic elements may be lacking in this approach.

Admission policy is a crucial part of any QoS implementation. If you cannot control the traffic entering the network, you have no control over the introduction of congestion into the network system, and so you must rely on congestion control and avoidance mechanisms to maintain stability. This is an undesirable situation. If the traffic originators have the capability to induce severe congestion situations into the network, the network may be ill-conceived and improperly designed, or admission-control mechanisms may be inherently required to be implemented with prejudice. The prejudice factor can be determined by economics. Those who pay for a higher level of service, for example, get first shot at the available bandwidth, and those who do not get throttled, or dropped, when utilization rates reach a congestion state. In this fashion, admission control could feasibly be coupled with traffic shaping.

Admission policy, within the context of the IETF Integrated Services architecture, determines whether a network node has sufficiently available resources to supply the requested QoS. If the originator of the traffic flow requests QoS levels from the network, using parameters within the RSVP specification, the admission-control module in the RSVP process checks the requester's reservation to determine whether the resources are available along each node in the transit path to provide the requested resources. In this vein, the performance-related QoS parameters are defined by the Integrated Service architecture.

The main issue of an admission policy is one of effective feedback between the current state of the admission function and the current load state of the network. The Integrated Services architecture scopes this issue through the creation of a maintained *path state* within the network, where the admission policy matches the characteristics of the reserved resources along the path. When looking at aggregate services, as in the case of the Differentiated Services architecture, the challenge is to relate local dynamic network congestion conditions to the current admission function state across all ingress points. Exactly how such a feedback mechanism can be constructed, and how efficiently it can operate, is still a matter of active network research, not a case of standardizing existing technologies. To date, the pragmatic outcome is an environment where the admission functions are configured with a static policy that deliberately takes a very conservative view of available resources. This deliberately limits the amount of elevated service traffic admitted into the network, such that there is a high probability of its carriage through the network within the associated service parameters, regardless of the dynamic local load states of the network. Higher subscription rates of premium-level services, in relation to total available network resources with associated dynamically adjusting admission functions, must await the outcomes of further research.

The Internet Research Task Force (IRTF) is composed of a number of focused, long-term small research groups that work on topics related to Internet protocols, applications, architectures, and technologies. Participants are individual contributors instead of representatives of organizations. The IRTF's mission is to promote research of importance to the evolution of the Internet. The IRTF Research Groups guidelines and procedures are described more fully in RFC2104.

The IRTF is managed by the IRTF chair, in consultation with the Internet Research Steering Group (IRSG). The IRSG membership includes the IRTF chair, the chairs of the various research groups, and possibly other individuals ("members at large") from the research community. The IRTF chair is appointed by the Internet Architecture Board (IAB); the research group chairs are appointed as part of the formation of research groups; and the IRSG members at large are chosen by the IRTF chair in consultation with the rest of the IRSG and on approval of the IAB.

In addition to managing the research groups, the IRSG may from time to time hold topical workshops focusing on research areas of importance to the evolution of the Internet or more general workshops to discuss research priorities from an Internet perspective, for example.

You can find more information on the IRTF at the IRTF Web site, located at www.irtf.org.

Protocol Performance Revisited

One major drawback in any high-volume IP network is that when there are congestion hot spots, uncontrolled congestion can wreak havoc on the overall performance of the network to the point of collapse. When thousands of flows are active at the same time, and a congestion situation occurs within the network at a particular bottleneck, each flow could experience loss at approximately the same time, creating what is known as *global synchronization*. Global, in this case, has nothing to do with an all-encompassing planetary phenomenon; it refers to all TCP flows in a given network that traverse a common path. Global synchronization occurs when hundreds or thousands of flows back off and go into TCP slow start at roughly the same time. Each TCP sender detects loss and reacts accordingly, going into slow start, shrinking its window size, pausing for a moment, then attempting to retransmit the data again. If the congestion situation still exists, each TCP sender detects loss once again, and the process repeats itself over and over, resulting in network grid-lock [Zhang 1990].

A variation of this behavior can be seen when a high-volume, high-speed TCP transfer is superimposed on a mix of smaller TCP and UDP traffic flows. The high-volume transfer will tend to emit large-volume bursts of back-to-back packets at regular intervals that match its RTT. This "clumping" of data forces all other simultaneous flows to synchronize with this larger flow, imposing that same "burstiness" on all other active flows, thereby imposing its RTT-based timing on all other flows.

Uncontrolled service models and uncontrolled congestion are detrimental to the network systems; behavior becomes unpredictable, system buffers fill up, packets ultimately are dropped, and the result is a large number of retrans-mits and reduced transmission efficiency.

Random Early Detection (RED)

Jacobson discussed the basic methods of implementing congestion avoidance in TCP in 1988 [Jacobson 1988]. However, his approach was more suited for a small number of TCP flows, a problem space that is much less complex to manage. Another approach, documented a year later, used *random early drop*, where an arriving packet is discarded with a fixed probability once the queue length exceeds some trigger threshold [Hashem 1989]. A source that was send-ing at a rate greater than its fair share would be prone to a greater loss from the random drop filter than a source with a lower packet sending rate.

In 1993, Floyd and Jacobson documented the concept of *random early detection* (RED). Using an exponential average of the queue length and a ran-dom drop probability that was based on a linear function of average queue length, RED refined the random early drop algorithm. Small bursts can pass through such a filter without experiencing elevated drop probability, while larger overload conditions will trigger increasingly higher discard rates.

The RED mechanism for avoiding congestion collapse is to randomly drop packets from arbitrary flows in an effort to avoid the problem of global syn-chronization and, ultimately, congestion collapse [Floyd 1993, RFC2309]. The principal goal of RED is to avoid a situation in which all TCP flows experience congestion at the same time, and subsequent packet loss, thus avoiding global synchronization. RED monitors the queue depth, and as the queue begins to fill, RED begins to randomly select individual TCP flows from which to drop packets, in order to signal the receiver to slow down (see Figure 3.23). The

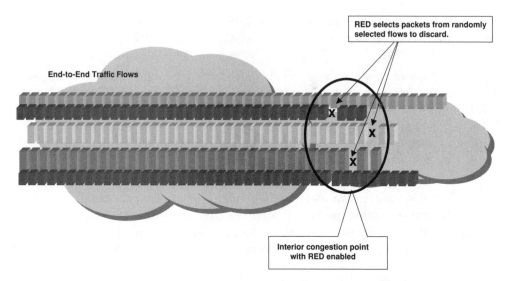

RED selects packets from randomly selected flows to discard.

End-to-End Traffic Flows

Interior congestion point with RED enabled

Figure 3.23 RED selects traffic from random flows to discard in an effort to avoid buffer overflow.

threshold at which RED begins to drop packets generally is configurable by the network administrator, as is the rate at which drops occur in relation to how quickly the queue fills. The more it fills, the greater the number of flows selected, and the greater the number of packets dropped (see Figure 3.24). This results in signaling a greater number of senders to slow down, thus preventing congestion avoidance.

The RED approach does not possess the same undesirable overhead characteristics as some of the non-FIFO queuing techniques discussed earlier. With RED, it is simply a matter of which packet gets into the queue in the first place; no packet reordering or queue management takes place. When packets are placed into the outbound queue, they are transmitted in the order in which they are entered. RED requires much less computational overhead than priority, class-based, and weighted-fair queuing that use packet reordering and queue management, but then again, RED performs a completely different function.

*Y*ou can find more detailed information on RED at www-nrg.ee.lbl.gov/floyd/red.html.

IP rate-adaptive signaling happens in intervals of the end-to-end round-trip time (RTT). When network congestion occurs, it can take some time to clear,

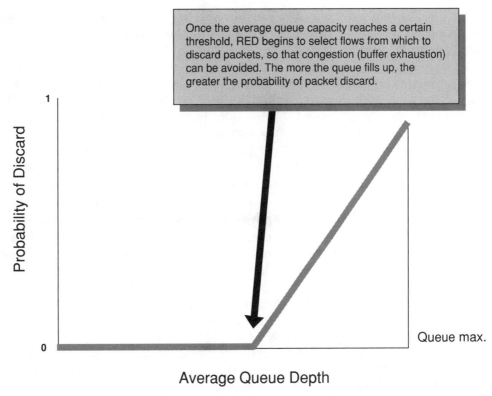

Once the average queue capacity reaches a certain threshold, RED begins to select flows from which to discard packets, so that congestion (buffer exhaustion) can be avoided. The more the queue fills up, the greater the probability of packet discard.

Figure 3.24 RED: The more the queue fills, the more traffic is discarded.

because transmission rates will not immediately back off in response to packet loss. The worst case is that packet loss will not be detected until the receiver's timer expires, and the transmitter will not see the signal until the receiver's NAK (negative acknowledgment) arrives back at the sender. Faster signaling is possible when the congestion event is a small-scale event, in which case trailing packets do make it through to the receiver. These out-of-sequence packets will cause duplicate ACK packets to be sent back to the sender, and these duplicate ACK packets will trigger a congestion response in the sender. In the best case, this takes slightly over half of the RTT interval; hence, when congestion occurs, it is not cleared quickly. The objective of RED is to start the congestion-signaling process slightly earlier than queue saturation. The use of random selection of flows to drop packets could be argued to favor dropping packets from flows in which the rate has opened up and flows are at their longest duration. These flows generally are considered to be the greatest contributor to the longevity of normal congestion situations.

The discard action can also be altered in RED to discard a packet selected at random from the queue, rather than discarding the arriving packet. The reasoning behind this alteration lies in the nature of the TCP flow-control algorithm, where three duplicate ACKs are used to trigger a fast retransmit. By selecting a packet within the queue, there is a probability of selecting a packet that already has subsequent packets within the queue. This in turn will reduce both the time it takes for the sender to receive the lost packet signal and for the sender to reduce its sending window. However, the benefits from such a refinement are somewhat marginal, and may not be adequate to compensate for the greater computational overhead of performing a queue removal operation at the head of the queue.

The critical issue in tuning RED systems is the selection of an algorithm for determining queue length. As we have already noted, IP traffic is self-similar in nature, so that the queue length is subject to high-frequency, short-term fluctuations. These short-lived queue variations should not be considered a trigger condition for packet drop. The task of the queue length algorithm is to pass the measurements of the queue length through a filter to remove any high-frequency component from the series of measurements. The resultant set of measurements can be used to form a weighted queue length measurement. It is this weighted queue length that should be used as the trigger value for RED actions. It is unlikely that there is a single robust weighting algorithm for queue-length calculation. The longer the average RTT of flows passing through the queue, the higher the damping factor required from the weighting function. Correspondingly, with shorter average RTT of flows, the queue-length calculation should be able to exhibit higher short-term variation.

Introducing Unfairness

RED can be said to be fair: It chooses random flows from which to discard traffic in an effort to avoid global synchronization and congestion collapse, as well as to maintain equity in traffic discard. Fairness is all well and good, but what is really needed here is a tool that can *induce* unfairness, meaning a tool that can allow the network administrator to predetermine which traffic to drop first or last when RED starts to select flows from which to discard packets. You can't differentiate services with fairness.

One approach uses a combination of the technologies already discussed to include multiple token buckets for traffic shaping. Exceeded token-bucket thresholds also provide a mechanism for marking a packet with a discard precedence. A precedence is set or policed when traffic enters the network (at ingress); a weighted congestion-avoidance mechanism implemented in the core routers can determine which traffic should be dropped first when congestion is anticipated due to queue-depth capacity. The higher the precedence indicated in a packet, the lower the probability of drop; the lower the precedence, the higher the probability of drop. When congestion avoidance is not actively discarding packets, all traffic is forwarded with equity.

Such marking of packets with a discard precedence is one way of achieving a differentiated service response from within the network, as you will learn when we examine the Differentiated Services architecture in Chapter 4.

Weighted Random Early Detection

For this type of operation to work properly, an intelligent congestion-control mechanism must be implemented on each router in the transit path. At least one mechanism is available that provides an unfair or weighted behavior for RED. This revision of RED, called *weighted random early detection* (WRED), yields the desired result for differentiated traffic discard in times of congestion (see Figure 3.25). A similar scheme, called *enhanced RED*, is documented in a paper by Feng, Kandlur, Saha, and Shin [Feng 1997].

The weighting of packets can be undertaken by a local traffic classification function, where local policy rules impose a weighted discard probability on each packet, depending on the classification function. An alternative, which again can be integrated into the Differentiated Services architecture, is to perform the packet classification function once upon ingress to the network, and mark the packet with a service precedence. In this case, the weighted discard probabilities can be determined by the packet's relative precedence level.

Active and Passive Admission Control

The network could simply rely on the end systems setting the IP precedence themselves. Alternatively, the ingress router (router A) could check and force the traffic to conform to administrative policies. The former is called *passive admission control*. The latter is a case of admission policing through an *active admission control* mechanism. Policing in this case is a good idea, as it is often unwise to explicitly trust all downstream users and applications to have complete knowledge of their application and the prevailing network conditions in order to determine how and when their traffic should be marked at a particular precedence. The danger always exists that users will try to attain a higher service classification than to which they are entitled.

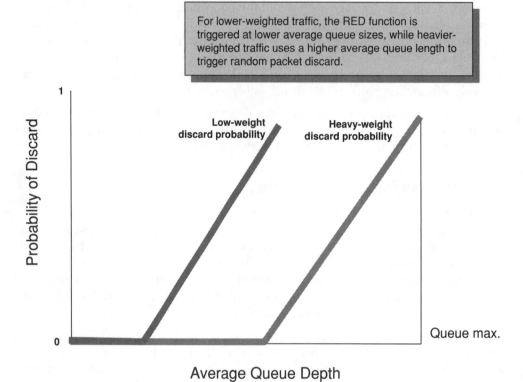

For lower-weighted traffic, the RED function is triggered at lower average queue sizes, while heavier-weighted traffic uses a higher average queue length to trigger random packet discard.

Figure 3.25 Weighted RED.

This is a policy issue, not a technical one, and network administrators need to determine whether they will passively admit traffic with precedence already set and simply bill accordingly for the service, or actively police traffic as it enters their network. The mention of billing for any service assumes that the necessary tools are in place to adequately measure traffic volumes generated by each downstream subscriber. Moreover, policing assumes that the network operator determines the mechanisms for allocation of network resources to competing clients; and, presumably, the policy enforced by the policing function yields a beneficial economic outcome to the operator, and meets the service requirements of the client base.

The case can be made, however, that the decision as to which flow is of greatest economic value to the customer can be made only by the customer and that the IP precedence level should be a setting that can be passed only into the network by the customer, not set (or reset) by the network. Obviously, in this scenario, the transportation of a packet in which precedence is set would attract a higher transport fee than if the precedence were not set, regardless of whether the network is in a state of congestion. It is not immediately apparent which approach yields the greatest financial return for the network operator. On the one hand, the policing function creates a network that exhibits congestion loading in a relatively predictable manner, as determined by the policies of the network administrator. The other approach allows the customer to identify the traffic flows of greatest need for priority handling. Here, network manages congestion in a way that attempts to impose degradation on precedence traffic to the smallest extent possible.

Explicit Congestion Notification

Behind a QoS scheme such as RED is the belief that TCP sets very few assumptions about the networks over which it must operate, and that it cannot count on any consistent performance feedback signal being generated by the network. As a minimal approach, TCP uses packet loss as its performance signal, interpreting small-scale packet loss events as peak load congestion events and extended packet loss events as a sign of more critical congestion load.

It is not necessary for RED to discard the randomly selected packet. The intent of RED is to signal the sender that there is the potential for queue exhaustion, and that the sender should adapt to this condition. An alternative mechanism is for the router experiencing the load to mark packets with an explicit Congestion Experienced (CE) bit flag, on the assumption that the sender will see and react to this flag setting in a manner comparable to its response to single packet drop [Floyd 1994, RFC2481].

To date, the RFC documents an experimental mechanism, and ECN does not form part of the current IP standards set. The ECN proposal uses a 2-bit scheme, claiming bits 6 and 7 of the IP version 4 TOS field (or the two Currently Unused (CU) bits of the IP Differentiated Services fields). Bit 6 is set by the sender to indicate that it is an ECN-Capable Transport system (the ECT bit). Bit 7 is the CE bit, and is set by a router once the average queue length exceeds configured threshold levels.

The proposed algorithm is that an active router will perform random early detection, as described. Once a packet has been selected, the router may mark the packet's CE bit if the ECT bit is set; otherwise, it will discard the selected packet (see Figure 3.26).

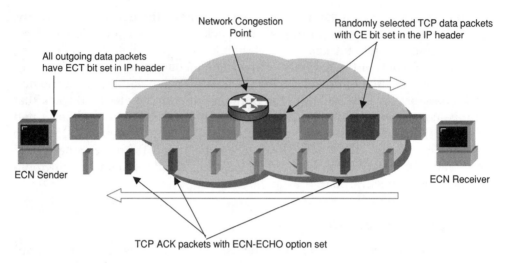

All outgoing data packets
have ECT bit set in IP header

Network Congestion
Point

Randomly selected TCP data packets
with CE bit set in the IP header

ECN Sender

ECN Receiver

TCP ACK packets with ECN-ECHO option set

Figure 3.26 Explicit congestion notification.

The TCP interaction is slightly more involved. The initial TCP SYN handshake includes the addition of ECN-echo capability and Congestion Window Reduced capability flags to allow each system to negotiate with its peer as to whether it will properly handle packets with the CE bit set during the data transfer. The sender sets the ECT bit in all packets sent. If the sender receives a TCP packet with the ECN-Echo flag set in the TCP header, the sender will adjust its congestion window (cwnd) as if it had undergone fast recovery from a single lost packet. The next sent packet will set the TCP Congestion Window Reduced flag (CWR), to indicate to the receiver that it has reacted to the congestion. The additional caveat is that the sender will react in this way at most once every RTT interval. Further, TCP packets with the ECN-Echo flag set will have no further effect on the sender within the same RTT interval. The receiver will set the ECN-Echo flag in all packets once it receives a packet with the CE bit set. This will continue until it receives a packet with the CWR bit set, indicating that the sender has reacted to the congestion. The ECT flag is set only in packets that contain a data payload. TCP ACK packets that contain no data payload should be sent with the ECT bit clear.

*Y*ou can find more detailed information on ECN at
www-nrg.ee.lbl.gov/floyd/ecn.html.

The connection does not have to await the reception of three duplicate ACKs—or the triggering of a retransmit timer—to detect the loss of a single packet, and the implicit congestion signal associated with such a loss. Instead, the receiver is notified of the incipient congestion condition through the explicit setting of a notification bit, which is in turn echoed back to the sender in the corresponding ACK. Simulations of ECN using a RED marking function indicate slightly superior throughput in comparison to RED as a packet discard function. However, widespread deployment of ECN is not considered highly likely in the near future, at least in the context of version 4 of IP. At this stage, there has been no explicit standardization of the field within the IPv4 header to carry this information, and the deployment base of IP is now so wide that any modifications to the semantics of fields in the IPv4 header would need to be very carefully considered to ensure that the changed field interpretation did not exercise some malformed behavior in older versions of the TCP stack or in older router software implementations.

ECN provides some level of performance improvement over a packet drop RED scheme. With large bulk data transfers, the improvement is moderate, based on the difference between the packet retransmission and congestion window adjustment of RED and the congestion window adjustment of ECN. The most notable improvements indicated in ECN simulation experiments occur with short TCP transactions (commonly seen in Web transactions), where a RED packet drop of the initial data packet may cause a six-second retransmit delay. Comparatively, the ECN approach allows the transfer to proceed without the lengthy delay.

The major issue with ECN is the need to change the operation of both the routers and the TCP software stacks to accommodate the operation of ECN. While the ECN proposal is carefully constructed to allow an essentially uncoordinated introduction into the Internet without negative side effects, the effectiveness of ECN in improving overall network throughput will be apparent only after this approach has been widely adopted. As the Internet grows, its inertial mass generates a natural resistance to further technological change; therefore, it may be some years before ECN is widely adopted in both host software and Internet routing systems. RED, on the other hand, has had a more rapid introduction to the Internet, as it requires only a local modification to router behavior, and relies on existing TCP behavior to react to the packet drop.

Threshold Triggering

The interesting aspect of the threshold-triggering approach is in the traffic-shaping mechanisms and the associated thresholds, which can be used to provide a method to mark a packet's relative precedence. As mentioned previously, precedence can be set either by the originating host or by the ingress router that polices incoming traffic. Using the latter, you can establish token-bucket thresholds to set precedence. You can implement a token bucket to define a particular bit-rate threshold, for example, and when this threshold is exceeded, mark packets with a lower-precedence value. Traffic transmitted within the threshold can be marked with a higher precedence to allow traffic that conforms to the specified bit rate to be marked as a *lower probability of discard* in times of congestion. This also allows traffic in excess of the configured bit-rate threshold to burst up to the port speed, but with a higher probability of discard than traffic that conforms to the threshold.

You can realize additional flexibility by adding multiple token buckets, each with similar or dissimilar thresholds, for various types of traffic flows. Suppose that you have an interface connected to a 45Mbps circuit. You could configure three token buckets: one for FTP, one for HTTP, and one for all other types of traffic. If, as a network administrator, you want to provide a better level of service for HTTP, a lesser quality of service for FTP, and a yet lesser quality for all other types of traffic, you could configure each token bucket independently. You also could select precedence values based on what you believe to be reasonable service levels or based on agreements between you and your customers. For example, you could strike a service level agreement that states that all FTP traffic up to 10Mbps is reasonably important, but that all traffic (with the exception of HTTP) in excess of 10Mbps simply should be marked as best effort. In times of congestion, this traffic is discarded first.

This approach gives a great deal of flexibility in determining the value of the traffic as well as deterministic properties in times of congestion. Again, this approach does not require a great deal of computational overhead, as do the more involved queuing mechanisms discussed earlier, because RED is still the underlying congestion-avoidance mechanism, and it does not have to perform packet reordering or complicated queue-management functions.

Varying Levels of Best Effort

The approaches just discussed use extant mechanisms in the TCP/IP suite to provide varying degrees of best-effort delivery. In effect, arguably, there is no way to guarantee traffic delivery in a Layer 3 IP environment. Delivering differentiated classes of service becomes as simple as ensuring that best-effort traffic has a higher probability of being dropped than premium traffic during times of congestion. (The differentiation is made by setting the IP service precedence bits.) If there is no congestion, everyone's traffic is delivered (hopefully, in a reasonable amount of time and with negligible latency) and everyone is happy. This is a very straightforward and simple approach to delivering differentiated service. You can apply these same principles to IP unicast or multicast traffic.

QoS Routing

So far in this chapter, we have examined in some detail the operation of various queuing disciplines and the associated packet discard function within the router to learn how these functions can be configured to deliver differentiated network service responses. Forwarding is another basic router function that can be manipulated to achieve some performance-related outcome.

QoS routing (QoSR) does not exist in production networks today; no standardized routing technology supports it, although there have been a number of attempts to add QoS capabilities into routing protocols in the past, and it remains the subject of study [ID-qosr-ospf]. In this section, we will look at these earlier efforts to implement QoS routing and the current state of QoS routing.

QoSR has the potential to be the missing piece of the puzzle in the effort to deliver real quality of service in IP networks. However, it must be remembered that QoSR still has not been deployed in an environment that remains keenly interested in QoS solutions. An effective QoSR environment presents considerable challenges and complexities in both concept and implementation. Despite interest from an eager market, a functional and operationally stable approach to QoSR is still the subject of study.

What Is QoS Routing?

To support local hop-by-hop forwarding decisions, a routing protocol is used to synchronize forwarding table information across all routers within the network's domain. The default routing behavior is based on destination-address path convergence, where each router takes the same forwarding decision for every packet destined to the same IP address. This leads to a set of paths defined within the network; at any location, there is a single path to the destination. This path minimizes the adopted additive hop-by-hop link-cost metric (this is a definition of *shortest path routing*). Thus, within this operational model, the path taken to reach a destination is not based on some match between the service characteristics of the path and the packet's requested service characteristics, nor is it based on the current congestion load within that path. The path taken to reach a metric is chosen to minimize the total *path metric*, and there is only one path selected by the routing protocol.

QoSR attempts to address this through the adoption of forwarding decisions that are sensitive to the packet's requested service profile and the characteristics of the various potential paths to the destination.

The most basic approach to QoSR is to expand the link metric to encompass a number of performance and service metrics. A reasonably robust QoSR mechanism must provide a method to calculate and select the most appropriate path based on this collection of metrics. These metrics should include information about the bandwidth resources available along each segment in the available path, end-to-end delay information, and resource availability and forwarding characteristics of each node in the end-to-end path. To explain this in some detail it is useful to start by examining the performance aspects of routing metrics.

Routing Protocol Link Metrics

Whether the network engineer chooses a distance vector protocol or a link-state routing protocol to manage interior routes, the goal is to build a structure for network metrics that generates traffic flows that match the network topology and capacity availability. Consistently, Internet routing protocols use network path metrics that are constructed as the sum of individual link metrics, and the protocol's path selection algorithm is based on the goal of minimizing the path metric. In this sense, a metric can be regarded as a cost function, and the path selection function one of cost minimization.

Many networks are configured with administratively determined link metrics for the operation of the routing protocol. The objective of a link metric is to manage the major traffic flows within the network with the greatest possible performance and efficiency and at the lowest cost to the network provider. In general, this implies that metrics should be set to prefer paths with higher capacity in which large-volume flows have to be supported and to prefer paths in which the path delay is minimized. Link metrics should be aligned to ensure that traffic flows pass over paths that have maximal capacity and minimal delay. Thus, short-delay, high-bandwidth links should be assigned a low link-cost metric, and longer-delay paths and lower-capacity links should be assigned correspondingly higher link-cost metrics. The routing path selection process will then converge on paths that have high bandwidth and short delay (see Figure 3.27). Given a base topology that provides adequate capacity between the major traffic concentration points within a network, the process

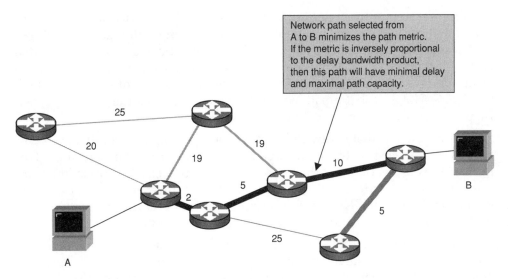

Network path selected from A to B minimizes the path metric. If the metric is inversely proportional to the delay bandwidth product, then this path will have minimal delay and maximal path capacity.

Figure 3.27 Link metrics and path selection.

of determining a set of metrics that make efficient use of the underlying topology is relatively straightforward. When the base topology is not well attuned to the traffic flows that are imposed on the network, there is little hope of being able to determine a set of path metrics that will alleviate this basic mismatch between installed capacity and traffic flow.

The alternative approach to manually defined metrics is to use some algorithm to determine the metric for each link. The simplest automated metric is to assign a value of 1 to every link. Under such a scheme, the outcome of the routing protocol will be to select paths that minimize the number of links, regardless of the performance characteristics of each of the links within the path. This is the default behavior of the RIP routing protocol, which by default uses such a unitary link metric [RFC1058]. This approach may be adequate for some simple networks, but in larger, diversely constructed networks, the most appropriate path may be one that maximizes the available bandwidth, rather than one that simply looks for a minimal link count.

An automated metric that can capture bandwidth information is based on the inverse of the link bandwidth. Here, the outcome of the routing search for lowest path metric will attempt to maximize the bandwidth of the path. However, the routing algorithm behavior is actually maximizing the sum of the individual link bandwidths as its path selection process. The true available bandwidth of a path is the minimum bandwidth of all links on the path. This difference can lead to anomalous path selection in certain cases, where the minimal sum of these per-hop inverse bandwidth metrics actually selects a path that has a low path bandwidth. The configuration in Figure 3.28 shows such a situation, in which the link metric is derived using 2048Kbps divided by the link bandwidth. Such inverse bandwidth metrics imply that the path from A to D is selected to transit B and C, with a path cost of 14, and the path (A, E, F, G, H, I, J, D) has a path cost of 17, despite the fact that the larger metric path has three times the path bandwidth.

The issue here is that bandwidth is not an additive metric; routing protocols that use a link-cost metric choose paths with a minimum metric sum. This leads to the observation that the more versatile approach for various link characteristic metrics is to define both the metric and the path calculation algorithm for the metric. The practical limitation today is that current routing protocols use a simple addition algorithm for the path metric.

Another candidate link metric is link delay. A delay measurement can include both link bandwidth and link distance. Bandwidth is relevant to delay, given the observation that a packet will take a minimum of *Delay* seconds to be transmitted, where:

$$Delay = \text{Link propagation delay} + (\text{Packet size} / \text{Bandwidth})$$

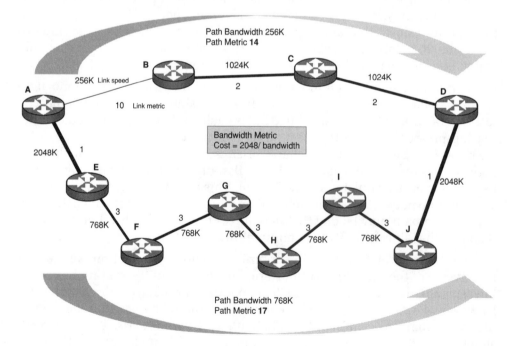

Figure 3.28 Bandwidth link-cost metric.

The action of a routing protocol that uses an additive metric will attempt to choose paths that minimize total path delay. This metric is very relevant to service performance, given that protocols work most efficiently when end-to-end delay is minimized, so that network efficiency is promoted within such a metric. Actual network financial cost normally has a significant distance component, and selecting paths that minimize network distance promote cost-efficient network paths.

Of course, the issue with any of these three metrics (hop count, inverse bandwidth, and delay), or indeed the issue with any statically fixed metric, is that the resulting network metrics are insensitive to dynamic network load. The question is whether a load-sensitive dynamically adjustable metric would allow the network to adjust path selection to enable the network to adapt to changing flow patterns.

Dynamic Link Metrics

Attempts to use a dynamically calculated link metric that is sensitive to current network load are by no means recent. The Hello protocol, documented in December 1983, used a link polling mechanism to measure the actual delay of each link and to generate a metric derived from these measurements [RFC891]. As the traffic load increases on the link, the router queue depth increases, which in turn increases the measured delay. As this delay increases, the routing protocol may locate a better path, in which case the protocol will converge on the better path. Subsequently, this will result in a new set of forwarding tables, shifting traffic loads toward more lightly loaded links where feasible (see Figure 3.29).

Using a dynamically calculated metric that is load-sensitive effectively creates a *negative feedback* path back from the outcome of the routing protocol to the input to the protocol. The term negative feedback is used because the feedback loop does not reinforce the outcomes of the routing protocol, but works in the opposite sense. As traffic levels increase across the paths selected by the routing process, the dynamic delay path metrics increase, which in turn may force traffic to use a different path. The same situation may occur on this new path, leading to the situation of routing oscillation and other forms of path instability. Dampening this feedback mechanism is an essential component of any dynamically calculated path metric, so that the metric reacts in very small metric units to traffic shifts.

The fundamental conclusion is that any load-sensitive dynamically calculated link metric requires some form of feedback control mechanism, where the control is intended to dampen high-frequency oscillations of the link metrics. But this dampening is not an optimal solution either, given that routing

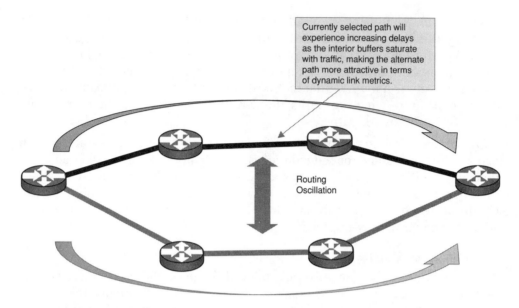

Figure 3.29 Route oscillation.

protocols effectively have a fixed set of stable states. When the dynamic metrics vary by degrees, no immediate shift occurs in selected paths. The change happens when the metric value exceeds the allowable variance in the current state; then the routing protocol converges to a new state. If this causes a reverse traffic movement, the network may immediately revert to its original state upon the next quantum change in link metrics, and so on. An alternative dampening mechanism is to increase the time interval over which the metric is held constant. This too may not produce a desirable outcome given that, at the expiration of every interval, the network may shift to a new state, with the consequent disruption to traffic flows that such a shift entails.

As a general observation on dynamically calculated metrics, it is perhaps overloading the function of the routing protocol to expect the protocol to undertake dynamic load management tasks as well as topology management and optimal path selection.

QoS Link Metrics

The approach used by these conventional routing protocols is to compute single paths, using a single metric. Multiple aspects of a link, such as a combination of delay, bandwidth, link error rate, and current link load, need to be taken into account for some form of QoS metric. In a single-path routing system, it is necessary to define the relative weightings to give to each component metric, and then to create a composite metric using a weighted addition of the component values.

A different approach is for the routing protocol to use a number of discrete metrics simultaneously, and compute a set of paths, where each selected path has the minimal path metric for a particular metric. Within each router there would exist a set of forwarding tables, with each table the outcome of the routing algorithm applied to a particular metric. To forward a packet, the router would first consult the requested service profile in the packet header and use it to select a corresponding forwarding table that would then be used to determine the next hop choice. For example, a set of metrics may consist of bandwidth, delay and delay jitter. The routing protocol would create three forwarding tables in each router corresponding to maximizing path bandwidth, minimizing delay, and minimizing delay jitter. When a router forwards a packet that has a service selector corresponding to jitter sensitivity, the router would use the delay jitter forwarding table.

QoSR extends the single-metric, shortest-path routing paradigm in a number of ways. The routing system must be able to identify multiple paths to each destination. It must also be able to maintain a set of alternative metrics describing each path, expressing path metrics such as minimal hops, minimal delay, or available bandwidth. Path service quality metrics that are based on current network conditions require additional management of the feedback loop between the network and the routing system. This represents one kind of approach to QoSR, where the routing protocol uses a vector of metrics to describe a corresponding set of QoS classes, and forwarding decisions are made based on the service selector field in the packet header, determining the use of a specific forwarding table.

In maintaining dynamic link metrics, QoSR attempts to use a feedback mechanism between the current state of the network and the routing system. QoSR encompasses the functionality of "a routing mechanism within which paths for flows are determined based on some knowledge of resource availability in the network, as well as the QoS requirements of flows" [RFC2386].

Equipping OSPF for QoSR

The OSPF routing protocol (successively refined in the protocol specifications RFC1131, RFC1247, RFC1583, RFC2178, and RFC2328) has been a consistent choice to carry QoS metrics. OSPF is a standardized intradomain routing protocol that provides a fast recalculation mechanism to determine changes in topology state and quickly notify neighboring nodes of link-state changes. The recalculation mechanism is called shortest-path first (SPF), based on the Dijkstra algorithm. SPF provides a mechanism to efficiently determine the shortest paths between one point and all other points, given a description of the network topology and associated link metrics. The OSPF routing protocol uses a mechanism to flood link attributes through the network, ensuring that each router has a complete, timely, and accurate topology description of the network, including link metrics. From this information, the router's OSPF process can use a local SPF calculation to construct a local forwarding table [Moy 1998].

Earlier specifications of the OSPF allowed for it to construct a distinct routing topology for each IP Type of Service (TOS) value. It was intended to allow for a form of QoS routing by explicit matching of the TOS field of the IPv4 packet header with the TOS-defined forwarding table. Accordingly, the TOS field is represented in OSPF in exactly the same format as in the IPv4 packet header. When a packet is forwarded by a TOS-capable router, the forwarding table used to make the forwarding decision is determined by the packet's TOS value.

To achieve this, each link had to be configured with TOS-defined metrics, in addition to the default (TOS = 0) link metric. The OSPF link state advertisements (LSAs) would then pass around a vector of link metrics for each link, pairing each metric with its associated TOS value. Each router used the SPF algorithm to construct a separate forwarding table for each distinct TOS value defined in the LSA. When forwarding a packet, the router performed the usual longest match on the destination address, then performed a forwarding table selection based on the packet's TOS field. If no TOS match was possible, the default (TOS = 0) forwarding table was used.

This approach demanded a very well-structured approach to QoS routing, as loops and routing black holes (dead ends) readily formed if some routers didn't support all the defined TOS values. The implication of the lack of neighbor TOS consistency checking within the protocol was that this task was intended to be passed to the network administrator to perform manually.

The major issue with this approach is that the IP TOS field, as defined in RFC1349 has not enjoyed widespread deployment in either end host IP software systems or within IP service networks. The emerging understanding of QoS now includes an appreciation that robust QoS relies as much on admission control, policing, and resource management as it requires signaling mechanisms within the packet and within routing protocols.

The most recent Internet Standard protocol specification of OSPF has removed the TOS routing option from OSPF [RFC2328]. This action was mandated by the lack of implementation experience with OSPF TOS routing.

Protocol standards specifications explicitly require that implementation experience be part of the standardization process. While this form of TOS routing is no longer part of the OSPF specification, for reasons of backward compatibility, the formats of OSPF's various LSAs remain part of the standard, maintaining the capability to specify TOS metrics in LSAs.

No QoS routing tools are widely deployed in the Internet environment today. As noted, QoS routing is an area of active research, which we will now address.

QoS Routing Requirements

Current routing technologies support best-effort services by computing lowest-cost paths through the network for each destination, where cost is defined on a single administratively assigned link metric. QoSR will need to extend this approach to support additional paths, by calculating additional paths through the network with each set of paths based on different link metrics.

Even within the scope of a single metric, QoSR will need to support the model of alternative path selection, as the interaction between the application and the routing system changes for QoSR. Under best-effort services, the interaction can be paraphrased by the request "please establish the best possible path to destination A." In QoSR the interaction can be paraphrased by the request "please establish a path to destination A that, at a minimum, meets the following service profile."

QoSR will also have to use dynamic metrics, where the metric value will fluctuate in response to changes in the service characteristic of the path. In the best-effort routing model, the routing process is opportunistic, and traffic flows will be shifted whenever a better path is located. In a model that uses service information as a feedback from the state of the network to the state of the routing protocol, such opportunism can lead to excessive instability of the network itself. QoSR will have to support some form of *path pinning*, so that paths will not oscillate in a highly unstable fashion.

QoSR and resource reservation protocols, such as RSVP, perform complementary functions. RSVP does not attempt to discover which path has sufficient available resources to accommodate the service requirements; and while QoSR can compute a path with certain service characteristics, it cannot reserve some component of the resource for use by a particular application. It could be argued that this division of function into two distinct protocols is artificial and unworkable, and that a truly functional QoS routing system would incorporate both resource management and resource reservation. On the other hand, part of the issue here is identifying the problem that QoSR is attempting to solve. There are two specific problems that QoSR could address. The first is to establish the current load status of the network and to identify those paths where network resources are available. The second task is to establish the current reservable status of the network and to identify the amount of resource available for further reservation.

Unfortunately, the complexities of the QoSR problem do not stop at just this level. The second level of functionality must also take into account any prioritization of traffic that occurs within the network. Traffic of an elevated precedence will displace some amount of lower-precedence traffic. The query that QoSR is intended to answer is: If there were to be a traffic flow of a certain service profile at a certain level of precedence, would sufficient traffic at lower priorities be displaced to support this additional load? Note that this is no longer a case of measuring various operational characteristics of each link and then applying a distribution protocol to determine paths of various characteristics. The calculation should, in theory, be able to include the displacement capability, and be able to compute paths that are feasible at various levels of traffic precedence. In other words, the view that QoSR is about measuring available resource within the network is constrained in that it implicitly assumes no differentiation of traffic. Active differentiation of traffic is one of the most immediate responses to provision of differentiated services. QoSR must be able to provide useful information to active differentiation systems. No longer is the issue one of tracking available capacity, rather it is tracking whether additional traffic corresponding to a certain service profile can be loaded into the network.

Objectives of a QoS Routing System

Obviously, QoS routing must be able to select a series of local forwarding decisions so that the cumulative transit meets the service requirements of each individual packet. On further reflection, though, perhaps this is not such an obvious objective, as the unit of service may be flows of packets, as generated by a real-time application. In this case, the service requirements that are passed to the QoS routing system are not based on per-packet attributes, but on the service requirements of the entire flow.

The second requirement is to make efficient use of the underlying network resource. If it were feasible to engineer the network so that there was an abundance of network resources, then there would be no need for the deployment of QoS routing. Deployment of QoS routing makes engineering and business sense only when the network resource is under some level of contention, when it is incumbent on the QoS system to resolve this resource contention with as little waste as possible. This efficiency is ideally stated as cost efficiency, where the desired outcome is to resolve the contention to the resource in a manner that maximizes total benefit—or, in financial terms, maximizes service revenue.

The third requirement is performance stability. Path oscillation should be damped, and continual path changes for a traffic flow should be avoided. This does not imply that all attempts to generate a change on a QoS-computed path should be damped. In the event of failure in the underlying transmission system, damping of route changes stands the risk of delaying the routing response to work around the problem. Selective damping of route updates is the more accurate statement of the objective.

QoSR Link Metrics

As the degree of granularity with which a routing decision can be made increases, so does the cost in performance. Although traditional destination-based routing protocols do not provide robust methods of selecting the most appropriate paths based on available path and intermediate node resources, they do impact less on forwarding performance, impose the least amount of computational overhead, and consume less memory than more elaborate schemes. As the routing scheme increases in complexity and granularity, so does the amount of resource expenditure. Resource consumption notwithstanding, path computation that is based on certain combinations of dynamic metrics, which include delay and jitter, raises issues of stability of routing information and impacts the stability of QoS-sensitive traffic flows.

Therefore, it is advisable to make certain concessions with regard to optimum path computation in exchange for decreased computational complexity. Some service-quality metrics may be considered of primary importance to the path calculation, while other metrics have a lesser service implication. By implementing a series of ordered, sequential computational comparisons, it may be sufficient to compare a small set of primary link performance metrics, instead of a computationally intensive comparison of all available metrics.

Administrative Controls

The principal objective of QoSR is that, given traffic with a clearly defined set of QoS parameters, routers must calculate an appropriate path based on available QoS path metrics. Therefore, administrative controls should be implemented to control the amount of traffic for which these types of path computations must be made. Of course, this leaves room for creative interpretation. Various administrative mechanisms that provide admission control, prejudicial packet drop with IP precedence, and partitioning of available bandwidth may be candidates.

Feedback Controls

Within QoSR, an immediate feedback loop exists between path selection and traffic flows. If a path is selected because of available resources, and traffic then is passed on this path, the available resource level of the path deteriorates. This action can cause the next iteration of the QoSR protocol to select an alternate path, which may in turn act as an attractor to the traffic flows. As established in the ARPAnet many years ago, where the routing protocol used measured RTT as the link metric, routing algorithms with feedback loops from traffic flows to the routing metric often suffer from the inability to converge to a stable state. Feedback loop management is a common requirement of all QoSR protocols.

Scaling

QoSR should be scalable, and the side effect of having to reduce the number of peers in a routing area or subdomain is somewhat disconcerting. Therefore, any QoSR approach also should provide a method for hierarchical aggregation, so that scaling properties are not significantly diminished by constraints imposed in smaller subdomains. Introducing aggregation into this equation may, however, cause a problem in maintaining accuracy in the path state information when senders and receivers reside in different subdomains.

Intradomain QoSR Requirements

An *intradomain* QoSR scheme must satisfy a set of minimum requirements [RFC2386]. The QoSR protocol must route a flow along a specific path that can accommodate its QoS requirements, or it must provide a mechanism to indicate why the flow has not been admitted. The protocol also must indicate disturbances in the route of a flow caused by topological changes. The QoSR protocol must accommodate best-effort traffic without requiring a resource reservation. The routing protocol should support QoS-based multicast traffic with receiver heterogeneity and shared reservation styles (when a resource reservation protocol is used). It also is recommended that QoSR make a reasonable effort to minimize computational and resource overhead and to provide some mechanisms to allow for administrative admission control.

Depending on the mechanism used for path selection, there are two possibilities for route processing at a *first-hop router* (the first router in the traffic path). A first-hop router simply may forward traffic to the most appropriate next-hop router, as is traditionally done in hop-by-hop routing schemes, or it may select each and every intermediate node through which the traffic will traverse in the end-to-end path to its destination, called *explicit routing*.

It also is appropriate to define the concept of *route pinning*, which loosely means that a particular hop-by-hop route is held in place for the flow duration so that changes in the routing topology or network load do not cause consistent rerouting of the flow. In addition, QoSR may need to use *path pinning* instead of route pinning, as what is being damped is the variability of the path for a given flow—not a route computed by the routing protocol. One way to achieve path pinning is to tie the path to an RSVP soft state that relies on RSVP message refreshes and time-out mechanisms to "pin" and "unpin" paths selected by the routing protocol. The resulting mutual dependence of RSVP and QoSR supports the earlier observation that splitting the roles of QoS routing and resource reservation into distinct protocols may not produce fully functional outcomes.

Interdomain QoSR

Interdomain routing can be described as the exchange of routing information between two dissimilar administrative routing domains or autonomous systems (ASes). Likewise, the basic definition of interdomain QoSR factors in the exchange of routing information, which includes QoS metrics. The principle concern of interdomain routing of any sort is stability. This continues to be a crucial issue in the Internet, because stability is a key factor in scalability. Instability affects the service quality of subscriber traffic.

For this reason, a link-state routing protocol may not be an ideal property in an interdomain QoSR protocol. First, LSAs are very useful within a single routing domain to quickly communicate topology changes to other routers in the routing area. Flooding of LSAs into another, adjacent routing domain (AS) most likely will inject instability into routing exchanges between neighboring domains. By the same token, an AS may not want to advertise details of its interior topology to a neighboring AS.

Also, link-state routing is desirable within intradomain routing because of the speed at which topology changes are computed and the granularity of information that can be disseminated quickly to peers within the routing domain. However, the utility of this mechanism would be greatly diminished by the aggregation of state and flow information, which normally is done within interdomain routing schemes as the routing information is passed between ASes. It also would be prudent to assume that limiting the rate of information exchanged between ASes is a good thing, because interdomain routing scalability is directly related to the frequency at which information is exchanged between routing domains. Therefore, a mechanism used to provide interdomain QoSR should provide only infrequent, periodic updates of QoS routing information, instead of frequently flooding information needlessly into neighboring ASes.

It is unclear at this time what mechanism would provide dynamic, interdomain QoSR routing. Given the pervasive deployed base of the Border Gateway Protocol (BGP), however, it is not difficult to imagine that a defined set of BGP extensions certainly could be deployed to provide this functionality. By the same token, it may be unnecessary to run a dynamic QoS-based routing protocol between different routing domains. The requirement in the interdomain environment is much more one of alignment of policies than discovery of resources within the interdomain connection space. It is not obvious that such a policy negotiation and management task should be performed by an interdomain routing protocol.

Resource Discovery and Resource Reservation

The underlying issue of QoSR requirements is that the task of resource discovery performed by QoSR is tightly bound to the outcomes of resource reservation. In this section, we will look at this interdependency in further detail.

Because the current routing paradigm is destination-based, you can assume that any reasonably robust QoSR mechanism must provide routing information based on both source and destination, flow identification, and some form of flow profile. RSVP side-steps the destination-based routing problem by making resource reservations in one direction only. Although an RSVP host may be both a sender and receiver simultaneously, RSVP senders and receivers are logically discrete entities. Path state is established and maintained along a path from sender to receiver, and subsequently, reservation state is established and maintained in the reverse direction from the receiver to the sender. Once this tedious process is complete, data is transmitted from sender to receiver. The efficiency concerns surrounding RSVP are compounded by its dependency on the stability of the routing infrastructure.

When end-to-end paths change because of the underlying routing system, the path and reservation state must be refreshed and reestablished for any given number of flows. The result can be a horribly inefficient expenditure of time and resources. Despite this cost, there is much to be said for this simple approach in managing the routing environment in very large networks. Current routing architectures use a unicast routing paradigm in which unicast traffic flows in the opposite direction of routing information. To make a path symmetrical, the unicast routing information flow also must be symmetrical. If destination A is announced to source B via path P, for example, destination

B must be announced to source A via the same path. Given the proliferation of asymmetric routing policies within the global Internet, asymmetric paths are a common feature of long-haul paths. Accordingly, when examining the imposition of a generic routing overlay, which attempts to provide QoS support, it is essential for the technology to address the requirement to support path asymmetry.

In addition, QoSR can, optionally, extend the IP routing model to include an element of per-flow context, held as *maintained state* within the routers. When the state of the network changes, there is no strict requirement to immediately shift existing traffic flows to new paths, as this may introduce harmful instabilities into the network. Paths that meet the service requirements of a traffic flow need not be continuously recomputed each time the service load on the network changes, so individual flows may remain on chosen paths for the duration of the flow.

QoSR presents several interesting problems, the least of which is determining whether the QoS requirements of a flow can be accommodated on a particular link or along a particular end-to-end path. The relationship between current path characteristics and load is only partially captured in QoSR. While QoSR can determine the existence of a path that can accommodate the requested service, the tools required to pin down the path for the duration of the application flow and to reserve resources along the path to ensure that the path attributes are not subsequently compromised through overload are both missing from the routing specification. For this reason, QoSR is seen as a tool to use in conjunction with a resource allocation mechanism, such as the RSVP. Some might argue that this problem has been solved with the tools provided by the IETF Integrated Services architecture and RSVP. But no integral relationship exists between the routing system and RSVP, a shortcoming in the Integrated Services architecture that QoSR is intended to address.

Link Fragmentation and Packet Compression

Regardless of the queuing discipline chosen and the routing system used, packet-switched networks have a residual induced jitter component caused by the use of data packets. No matter what the relative priorities of the packets may be, if a packet has been partially transmitted, any subsequent packets must await completion of the packet transmission. Such a delay causes jitter in the traffic flow.

For high-speed links, this is not a significant factor. A maximally sized 64Kbyte IP packet will take some 3 milliseconds to enter a 155Mbps transmission system. A more common 576-byte packet will take some 30 microseconds to enter such a circuit. For slower speeds, this delay is more critical. For a 64Kbps circuit, the same 576-byte packet will take 72 milliseconds to enter the circuit, while a maximally sized 64Kbyte packet will take 8 seconds. Practically, this means that if two 576-byte packets collide for the use of a 64Kbps circuit, one packet will be delayed for slightly less that one-tenth of a second while waiting for the first packet to enter the circuit. We can describe this delay as a *packet quantization jitter*, as the delay variation is caused by the network handling the data in minimum units of packets rather than bits or bytes. This delay is cumulative, and across a 20-hop network of 64K channels, the worst-case delay attributable to packet quantization alone is some 1.5 seconds. This is in addition to the normal propagation delay for the bits to travel from one end of the circuit to the other. This combination of large packet sizes and low transmission speeds is more prevalent at the periphery of a network. Such delays are in addition to queuing delays, and one large packet can effect a series of successive packets within the queue.

One approach to lessen the packet quantization jitter component of a network is to reduce a link's maximum transfer unit down to a level that allows high-priority real-time traffic to access the output driver within an acceptable delay. Another approach uses the multilink extensions to the PPP, and invokes explicit interleaving of high-priority packets within the fragments of a large packet. In both cases, the intention is identical: to address the jitter induced by packet delay by setting a ceiling on the maximum packet size carried across the link, and to use interleaving of higher-priority packets within the fragments of the larger packet (see Figure 3.30).

Of course, such fragmentation should not be undertaken lightly, as packet fragmentation can have a severe negative impact on the performance of reliable data transfer protocols. A careful design choice must be made, one that compromises between the desire to eliminate unnecessary jitter from real-time data flows and allows reliable data transfer protocols to operate efficiently.

Another issue relating to packet sizes and service performance is protocol layering. For example, an 8Kbps compressed audio encoding stream, transmitting at a rate of one packet every tenth of a second will carry 100 octets of payload in each packet. If this is transmitted across a serial line, then each packet will contain a Level 2 PPP frame header, an IP header, a UDP header, an RTP header, then the audio data in the RTP payload field. The overhead of these four levels is minimally some 41 octets, or a 30 percent protocol overhead. If a higher-frequency signal sampling time is used for the signal, and the RTP packets are increased in frequency, this protocol overhead will pose

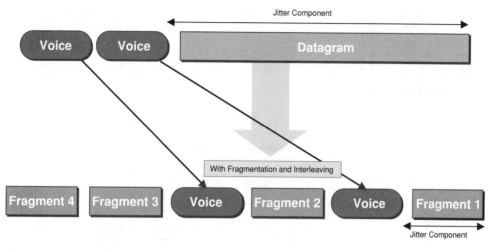

Figure 3.30 Packet fragmentation and interleaving.

severe performance issues for low-speed links. For example, increasing the packet rate to 100 packets per second, each containing a 10-octet payload will result in a 77 percent protocol overhead, where only 10 octets of the 53-octet packet contains the audio payload. This becomes a significant performance issue for modem connections and low-speed wireless connections, such as packet radio, where the available bandwidth is very limited.

Given that such bandwidth limitation issues are found more commonly at the edge of the network, where the number of concurrent data flows is limited, it is possible for the two end points of the flow to maintain some per-flow state information, and compress the packet header to transmit only those elements of the packet that have changed in unpredictable ways (see Figure 3.31).

The compression of headers is based on the observation that many header fields within the same data flow remain fixed. If both sides of the link are aware of the set of fixed values, these need not be retransmitted for every packet. Similarly, fields that change by known amounts, such as sequence counters that change by the payload length, need not be transmitted. Some fields change by a constant amount for each packet. For such value sequences, where the second-order derivative is zero, these fields also need not be transmitted. Of the remaining fields, there are a number where the change values over successive packets is small. For such fields, the value can be replaced by sending the difference between the previous value and the current value. The techniques for compressing TCP/IP headers was first described in RFC1144. A

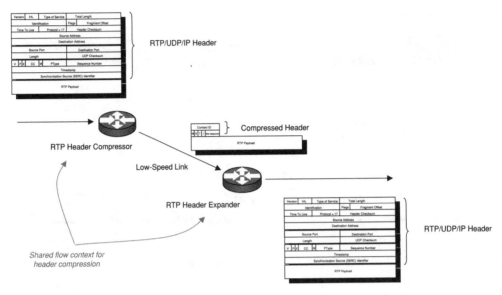

Figure 3.31 Packet header compression.

refinement of this approach, extending it for UDP/IP and RTP/UDP/IP headers is described in RFC2508.

The best-case compression outcome is compressing the 40 octets of IP/UDP/RTP header into 2 octets. In the previous example of the 10-octet payload, the protocol overhead is reduced from 77 percent to 16 percent, or 2 octets of compressed header plus the 10-octet payload.

Active Networks

Active Network technology is included in this chapter because it is logically a performance tool for networks. However, like QoS routing, this topic is the subject of current research, and no production tools exist for support of Active Networking in today's networks.

Within the conventional model of packet networks, the packet's header contains a number of fields intended to trigger predictable behaviors from the router. The Destination Address field is to be used to make a forwarding decision. The Service Class field is to control the queuing and discard behavior of the router, and the Time To Live field is to act as a discard trigger if its value ever falls to zero. Active Networking uses a different approach, where each data packet header contains a micro program for a router, or additional data for a router-resident program, such that the router's behavior can be directly driven by the packet. In this model, the packet is changed from a passive object handled by the active router elements to an active element that uses the router's mechanisms in a manner determined by the packet itself [Tennenhouse 1996].

The relevance of Active Networking to performance management is based on the observation that feedback-based congestion control systems do not scale well within a large-scale network. The longer the delay between the congestion point and the two end points of the data stream, the longer it takes for information relating to the congestion event to reach the sending system. During this period, the sending system will continue to send data into the congestion point, making the condition worse. The higher the bandwidth-delay product of the network, the longer the duration of local congestion condition if feedback is the means of control of senders [Bolot 1990]. As networks grow in bandwidth and reach further in span, this delay-bandwidth product continues to grow. Active Networking may provide a mechanism of addressing this feedback control structure.

One area of study has been in Active Network Congestion Control (ACC) [Faber 1998]. In this model, the initial packet of a data stream programs the routers along the path on how to react to congestion that may be experienced by the stream, in effect replicating the sending system's feedback-triggered congestion response in each of the routers along the network path. When a router along the path detects a local congestion condition, it can elect to trigger a congestion response from this data stream, immediately discarding packets from the data stream that the sender would not have sent were it immediately aware of the congestion condition. In addition, the triggered router can generate a message to the sending system, informing the end point to transition to a new congestion state to match the actions already taken by the router. Simulations of such an approach are certainly promising, with greater effective data throughput being measured due to shorter congestion events within the network.

Active Networks and performance management are areas of active research; and while some of the Active Network concepts being studied may not emerge within production network environments, Active Networking does offer considerable promise as a means of further increasing the efficiency of packet networks and of increasing the flexibility of packet networks in carrying a wider range of application service requirements.

Interoperability and Complexity

Implicit in this examination of classification functions, queuing disciplines, admission policies, routing capabilities, and fragmentation functions, has been the assumption that such disciplines and policies are implemented uniformly across all equipment deployed within a performance domain. Also assumed is that the configuration settings used to define specific performance and service policies are set to a consistent intended outcome.

Such assumptions are not well founded in practice. Performance domains may span multiple networks. Equipment may vary between networks or even within a single network. The implementations of the queuing disciplines and admission filters will vary in their behavior to some extent. Performance and service policies may not be uniform within the overall performance domain, and different queuing disciplines may be used within various network subdomains. Even where the same queuing discipline is used, the same packet may be classified into different service classes within various network subdomains, causing random end-to-end service outcomes.

The conclusion to draw is that the operational Internet environment tends to be far more chaotic than one would expect. Consequently, the task of creating a set of well-ordered performance and service profiles within a large heterogeneous network environment should not be treated lightly. Of added importance is that the operating cost of maintaining a distributed set of configurations all tuned to deliver a consistent set of service outcomes is considerable. The estimated cost-benefit outcome of performance-managed services is always a useful start to any form of performance engineering activity, as in many cases the value of the ultimate benefits are dwarfed by the implementation and operational costs.

It is important to remember that there is always an alternative engineering response to adding complexity into the switching environment. Complexity is often an outcome of the desire for greater control in managing contention for shared resources. In such cases, an alternative approach is to reduce the level and frequency of contention by increasing the amount of available resource. One answer to managing congestion and erratic performance within a network is the comprehensive deployment of a more complex weighted sharing queuing discipline. But an alternative approach often exists. It may be that a simple FIFO queuing discipline and admission controls, combined with ample bandwidth and switch capacity within the network, are necessary and sufficient measures.

References

[Bolot 1990] *Dynamical Behavior of Rate-Based Flow Control Systems*, J. Bolot, Computer Communications Review, vol. 20, no. 2, ACM SIGCOMM, April 1990.

A study of feedback-controlled flow systems that illustrates that the duration of local congestion events in a feedback-controlled network are directly related to the network's bandwidth-delay product.

[Demers 1990] *Analysis and Simulation of a Fair Queuing Algorithm*, A. Demers, S. Keshav and S. Shenker, Internetwork: Research and Experience, vol. 1, no. 1, pp. 3–26, September 1990.

The original paper describing the fair queuing technique.

[Faber 1998] *ACC: Using Active Networking to Enhance Feedback Congestion Control Mechanisms*, T. Faber, IEEE Network, vol. 12, no. 3, May 1998.

A paper describing the use of active networking as a means of providing faster reaction to local congestion events within a network.

[Feng 1997] *Techniques for Eliminating Packet Loss in Congested TCP/IP Networks,* W. Feng, D. Kandlur, D. Saha and K. Shin, Technical Note CSE-TR-349-97, Univ. Michigan, November 1997.

A description of enhanced random early detection that uses differentiated responses to congestion through variation of the probabilities of random packet discard.

[Floyd 1993] *Random Early Detection Gateways for Congestion Avoidance*, S. Floyd and V. Jacobson, IEEE/ACM Transactions on Networking, vol. 1, no. 4, August 1993.

A description of a congestion avoidance technique through active queue management. The mechanism described uses the technique of weighted queue length to drive the probability of random packet discard prior to queue exhaustion.

[Floyd 1994] *TCP and Explicit Congestion Notification*, S. Floyd, ACM Computer Communication Review, vol. 24, no. 5, October 1994.

The original proposal to use a field in the IP header as a means of explicit signaling of local congestion within the network, rather than relying on packet loss as the sole means of communicating local congestion events to the sending system.

[Hashem 1989] *Analysis of Random Drop for Gateway Congestion Control*, E. Hashem, Technical Report LCS/TR-465, Laboratory for Computer Science, Massachusetts Institute of Technology, Cambridge, MA, 1989.

An early analysis of the effects of active queue management as a means of congestion control.

[Jacobson 1988] *Congestion Avoidance and Control*, V. Jacobson, Computer Communication Review, vol. 18, no. 4, August 1988.

A landmark paper describing end system flow control mechanisms for TCP that are intended to avoid network congestion collapses.

[Keshav 1997] *An Engineering Approach to Computer Networking: ATM Networks, the Internet, and the Telephone Network*, S. Keshav, Addison-Wesley, 1997.

A comprehensive examination of the engineering design that supports the Public Switched Telephone Network (PSTN), Asynchronous Transfer Mode (ATM) networks, and the Internet.

[Kleinrock 1975] *Queuing Systems, Vol. 1: Theory*, L. Klenirock, John Wiley & Sons, Inc., 1975.

A comprehensive reference on the theory of queuing systems.

[Kleinrock 1976] *Queuing Systems, Vol. 2: Computer Applications*, L. Kleinrock, John Wiley & Sons, Inc., 1976.

A comprehensive reference on the application of queuing systems to computer applications and network switching environments.

[Leland 1993] *On the Self-Similar Nature of Ethernet Traffic*, W. Leland, W. Willinger, M. Taqqu, and D. Wilson, ACM SIGComm '93, San Francisco, 1993.

One of a number of studies on the nature of data traffic. The paper describes a self-similar traffic pattern that has a strong fractal nature, where bursting is visible over any chosen timebase. This traffic pattern has serious implications for the design, control, and analysis of high-speed data networks. A copy of this paper is available online at www.sobco.com/e.132/files/sigcomm93.ps.

[Moy 1998] *OSPF: Anatomy of an Internet Routing Protocol*, J. Moy, Addison-Wesley, 1998.

Complete coverage of the OSPF routing protocol, including its implementation, configuration, and management.

[Partridge 1994] *Gigabit Networking*, C. Partridge, Addison-Wesley, 1994.

A reference that surveys the issues and available technologies for very high-speed data networks.

[Tennenhouse 1996] *Towards an Active Network Architecture*, D. Tennenhouse, Computer Communications Review, vol. 26, no. 2, ACM SIGCOMM, April 1996.

An article that describes the Active Network architecture.

[Zhang 1990] *Oscillating Behavior of Network Traffic A Case Study Simulation*, L. Zhang, and D. Clark, Internetwork: Research and Experience, vol. 1, no. 2, John Wiley & Sons, Inc., September 1990.

A study of how multiple concurrent TCP sessions synchronize their dynamic behavior, causing the total network load to oscillate with a common period of oscillation.

RFCs

Request for Comments documents (RFCs) are published by the RFC editor. They are available online at www.rfc.editor.org.

[RFC891] *DCN Local-Network Protocols*, D. Mills, RFC891, December 1983.

The original specification of the HELLO routing protocol, which used dynamic feedback of network characteristics into the path metric calculation.

[RFC970] *On Packet Switches with Infinite Storage*, J. Nagle, RFC970, December 1985.

A description of a packet-switching technique that uses per-flow queuing inside the switch. This work was subsequently refined as the fair queuing technique.

[RFC1058] *Routing Information Protocol*, C. Hedrick, RFC 1058, Historic RFC, June 1988.

The original specification of the Routing Information Protocol (RIP). This specification has been obsoleted by RIP, version 2, standardized in RFC2453.

[RFC1131] *OSPF specification*, J. Moy, RFC1131, Proposed Standard, October 1989.

The original specification of the OSPF routing protocol, subsequently revised as part of the Internet Standards process as OSPF, version 2, in RFC2328.

[RFC1144] *TCP/IP Compression for Low-Speed Serial Links*, V. Jacobson, RFC1144, Proposed Standard, February 1990.

This RFC describes a method to improve TCP/IP performance over low-speed serial links by saving per-flow state information at each end of a link and sending only those fields that change from packet to packet within a TCP flow.

[RFC1247] *OSPF Version 2*, J. Moy, RFC1247, Draft Standard, July 1991.

The first submission of the specification of the OSPF routing protocol, submitted to the Internet Standards process. The final standard specification for OSPF, version 2 is RFC2328.

[RFC1349] *Type of Service in the Internet Protocol Suite*, P. Almquist, RFC1349, Proposed Standard RFC, July 1992.

This RFC defines some aspects of the semantics of the Type of Service octet in the Internet Protocol header. The handling of IP Type of Service by both hosts and routers is specified in some detail. This document is superseded by the IETF Differentiated Services DS field definition.

[RFC1583] *OSPF Version 2*, J. Moy. RFC 583, Draft Standard, March 1994.

An intermediate revision of the specification of the OSPF routing protocol, submitted to the Internet Standards process. The final standard specification for OSPF, version 2 is RFC2328.

[RFC2104] *IRTF Research Group Guidelines and Procedures*, A. Weinrib and J. Postel, RFC2104, Best Current Practice RFC, October 1996.

This document describes the guidelines and procedures for formation and operation of IRTF research groups.

[RFC2178] *OSPF Version 2*, J. Moy, RFC2178, Draft Standard, July 1997.

An intermediate revision of the specification of the OSPF routing protocol, submitted to the Internet Standards process. The final standard specification for OSPF, version 2 is RFC2328.

[RFC2309] *Recommendations on Queue Management and Congestion Avoidance in the Internet*, R. Braden, D. Clark, J. Crowcroft, B. Davie, S. Deering, D. Estrin, S. Floyd, V. Jacobson, G. Minshall, C. Partridge, L. Peterson, K. Ramakrishnan, S. Shenker, J. Wroclawski, and L. Zhang, RFC2309, Informational RFC, April 1998.

This informational document recommends the implementation of active queue management within the network through the adoption of random early detection (RED) or similar functioning mechanisms. It also urges a concerted effort of research, measurement, and ultimate deployment of router mechanisms to protect the Internet from flows that are not sufficiently responsive to congestion notification.

[RFC2328] *OSPF Version 2*, J. Moy, RFC2328, Internet Standard, April 1998.

This Internet Standard RFC document, version 2, of the OSPF protocol. OSPF is a link-state routing protocol designed to be run internal to a single autonomous system. Each OSPF router maintains an identical database describing the autonomous system's topology. From this database, a routing table is calculated by constructing a shortest-path tree.

[RFC2386] *A Framework for QoS-based Routing in the Internet*, E. Crawley, R. Nair, B. Rajagopalan, and H. Sandick, RFC2386, Informational RFC, August 1998.

This informational RFC describes some of the QoS-based routing issues and requirements, and proposes a framework for QoS-based routing in the Internet. This framework is based on extending the current Internet routing model of intra- and interdomain routing to support QoS.

[RFC2401] *Security Architecture for the Internet Protocol*, S. Kent and R. Atkinson, RFC2401 Proposed Standard, November 1998.

This proposed standard RFC specifies the base architecture for IPsec-compliant systems. The goal of the architecture is to provide various security services for traffic at the IP layer, in both the IPv4 and IPv6 environments. This document describes the goals of such systems, their components, and how they fit together and into the IP environment.

[RFC2475] *An Architecture for Differentiated Service*, S. Blake, D. Black, M. Carlson, E. Davies, Z. Wang, and W. Weiss, RFC2475, Proposed Standard, December 1998.

The description of the Differentiated Services architecture, which allows service responses to be generated from the network in a stateless fashion using packet marking and per-hop behaviors in the network's routers. The architecture is intended to offer a broad range of service responses with a very low protocol overhead.

[RFC2481] *A Proposal to Add Explicit Congestion Notification (ECN) to IP*, K. Ramakrishnan and S. Floyd, RFC2481, Experimental RFC, January 1999.

This RFC document describes a proposed addition of explicit congestion notification (ECN) to IP. The document describes TCP's use of packet drops as an indication of congestion. The document argues that, with the addition of active queue management (e.g., RED) to the Internet infrastructure, where routers detect congestion before the queue overflows, routers are no longer limited to packet drops as an indication of congestion. Routers can instead set a Congestion Experienced (CE) bit in the packet header of packets from ECN-capable transport protocols. It describes when the CE bit would be set in the routers and what modifications would be needed to TCP to make it ECN-capable.

[RFC2508] *Compressing IP/UDP/RTP Headers for Low-Speed Serial Links*, S. Casner, V. Jacobson, RFC 2508, Proposed Standard, February 1999.

This document describes a method for compressing the headers of IP/UDP/RTP datagrams to reduce overhead on low-speed serial links. In many cases, all three headers can be compressed to 2 to 4 bytes.

IETF Internet Drafts

IETF Internet drafts are work-in-progress documents. The documents are valid within the IETF process for a maximum period of six months from the date of submission. Internet drafts are not normally referenced, but those cited here are the best pointers to current research and developmental efforts, and so have relevance to this topic of quality of service and network performance. Typically, these documents follow the Internet Standards process and are published as RFCs sometime in the future. The current collection of Internet drafts, along with pointers to the RFC documents, can be found at www.ietf.org references.

[ID-qosr-ospf]   *QoS   Routing   Mechanisms   and   OSPF   Extensions*,   G. Apostolopoulos, R. Guerin, A. Amat, A. Orda, T. Przygienda, and D. Williams, IETF Internet draft, draft-guerin-qos-routing-ospf-05.txt, April 1998.

The focus of this document is on the algorithms used to compute QoS routes and on the necessary modifications to OSPF to support this function; for example, the information needed, its format, how it is distributed, and how it is used by the QoS path selection process. Aspects related to how QoS routes are established and managed are also briefly discussed. The goal of this document is to identify a framework and possible approaches to allow deployment of QoS routing capabilities with the minimum possible impact to the existing routing infrastructure.

chapter four

Quality of Service Architectures

Good, fast, cheap. Pick any two.

The preceding chapters examined the elements of Internet networks, in particular how each element contributes to the delivery of service performance to the network application. We also examined a number of performance tools, including classification functions, queuing disciplines, admission policies, and discard control and routing mechanisms, which are intended to allow the network to be configured to deliver various responses to support a range of service applications. In this chapter, we explore which of these tools can be applied in concert to create reliable performance outcomes.

There are two basic approaches to styling tools to deliver QoS services. One is to create a reservation state within the network that corresponds to a service request, and then maintain this state for the duration of the associated data flow. This approach is used within the Integrated Services architecture. The other approach is to dispense with reservation processes and use a combination of admission control and packet-based service responses to create service outcomes. This approach is the basis of the Differentiated Services architecture.

Why do we need a service delivery architecture at all? So far we have examined a set of performance tools that are contained within IP, within the underlying transmission systems, and within the routers. These tools range from what is the equivalent of a pair of tweezers to that of a heavy mallet. Priority queuing (PQ), for example, has very coarse service outcomes, while weighted random early detection (WRED) has relatively subtle outcomes for many traffic profiles. Certain combinations of tools will not work well together, and, at worst, may interact in wildly chaotic ways. Other combinations will interact in a way that is unproductive, reducing the usable capacity of the network without generating a particular service outcome. To create a predictable service outcome, it is necessary to assemble a set of component tools that are mutually consistent and to configure their operation in a way that works to produce the desired service outcomes. By doing so, the performance engineering activity follows a consistent theme. This is what is referred to as a service delivery architecture.

There is no single "correct" performance architecture, and opinions vary wildly in regard to how to provide differentiation of service outcomes and where such differentiation can be effected optimally within the network architecture.

The intent of each of the architectures examined here is broadly similar: to create a service environment that disassembles the single best-effort service model and replaces it with a set of service classes so that traffic within each class will experience a different service outcome. But though the broad outcome of service differentiation may be similar, each approach is quite distinct, and the fine detail of the service outcomes differ markedly. We will examine the general approaches used in each architecture, to gain an understanding of their respective strengths and weaknesses.

Why use IP to deliver QoS in the first place? Quality-agile transport protocols, such as ATM, are often presented as potential solutions for QoS-enabled networks. The relative merits of IP and ATM in supporting QoS have been the subject of many heated exchanges in both engineering and marketing forums, yet on at least one premise it is safe to assume there is sufficient agreement. That premise is that the most appropriate place to provide internet-based service delivery management mechanisms is at the level of the most common denominator (where common is defined in terms of the amount of end-to-end deployment in today's networks). In this vein, the issue becomes one of determining which network service component has the most prevalence in the end-to-end traffic path. In other words, the relevant question is: What is the common end-to-end bearer service? In a service internet platform, it is undeniable that the common bearer service is at the IP transport level, rather than at any particular data-link level, or at a level of virtual circuits. IP is indeed the common denominator.

This thought process leading to a choice of IP has several supporting lines of reason. The common denominator is chosen based on the hope of using the most pervasive and ubiquitous protocol in the network, whether it be at the level of individual link control protocols or at the level of the end-to-end signaling control protocol. Using the most pervasive protocol makes implementation, management, and troubleshooting much easier, and yields a greater possibility of successfully providing a QoS implementation. Also, this particular technology operates in an end-to-end fashion, using a signaling mechanism that spans the entire traversal of the network in a consistent fashion. IP is the end-to-end transportation service in most cases. Although it is possible to create QoS services in substrate layers of the protocol stack, such services only cover part of the end-to-end data path. Such partial measures often have their effects masked by those of the traffic distortion generated from the remainder of the end-to-end path in which they are absent. In these cases, the overall outcome of a partial QoS structure often is ineffectual, and sometimes may be quite negative. Therefore, it is generally accepted that the most appropriate place is indeed within the most common denominator, and that common denominator is in the context of the IP protocol level.

Service Delivery Using IP Type of Service

The Type of Service (TOS) field in the IP packet header is an 8-bit field. The original IP service definition used this field to encompass a 3-bit IP precedence value and the 4-bit service profile indicator. The original objective of this field was described as providing "an indication of the abstract parameters of the quality of service desired. These parameters are to be used to guide the selection of the actual service parameters when transmitting a datagram through a particular network" [RFC791]. What was envisaged was that this field would modify both per-hop queuing behaviors and per-hop forwarding behavior, allowing packets with different TOS fields to be managed with different service levels within the network, and, potentially, take different paths across the network. In this light, the TOS field was "expected to be used to control . . . routing and queuing algorithms" [RFC1122]. The precedence field in particular was seen as a "scheme for allocating resources in a network based on the importance of different traffic flows" [RFC1812]. Despite such grand expectations, the IP TOS field has not been widely used as a custom feature in internet service environments. We'll look at the reason for this in more detail in this section.

There are two subfields in the header, the Service Profile selector field and the Precedence field.

IP Service Profiles

The Service Profile selector field has always been open to variable interpretation, which has proved problematic, because an ambiguous feature specification can be a significant barrier to widespread adoption of the feature. Ambiguous features do not readily translate into production-mode service systems. The Service

Profile selector field was intended to trigger one or more underlying performance responses from the network that created an overall service "characteristic." Given that the original IP specification, RFC791, used the rather vague terms *low delay*, *high throughput*, and *high reliability* to describe the setting within this selector field, the precise configuration of the network's performance tools to elicit such responses were open to a very broad range of interpretations. In the spirit of "what you get is exactly what you ask for," a vague service specification can do no better than generate an equally vague response.

The refinement of the specification of the service profile field in RFC1349 did not improve this situation. The inclusion of an additional service type that requested the network service be provided with *minimized cost*, along with the rewording of *low* to *minimize* and *high* to *maximize* for the existing three service specifications, did little to improve the accuracy of the service description. Again, the only response that could be phrased to such vague service descriptions was an equally vague network service configuration.

Such service specifications do not lend themselves to a specific configuration of performance tools. For example, is a minimized delay service allowed to incur a higher packet discard probability as a result of using shorter service queues? Does a maximized throughput service allow the network to impose an increase in average delay and imposed jitter? The practical difficulty in implementing these types of parameters has proved to be more difficult than the protocol designers could have foreseen; and for the most part, these service profiles have never been deployed in large-scale operational networks in the Internet with any degree of success.

IP Precedence

The 3-bit IP Precedence field is somewhat easier to interpret. A packet with a higher precedence should arrive no later than a packet with a lower precedence, if both packets are passed into the network at the same time with the same destination and same IP options (see Figure 4.1). In addition, the IP Precedence field is related to discard probability, meaning that if a sequence of such packets is passed along such a path, the packet sequence with the elevated precedence should experience a lower discard probability. This can be translated into a set of performance tools, as indicated in the IETF *Requirements for IP Version 4 Routers* document [RFC1812]. This RFC document discusses:

> **Precedence-ordered queue service.** Provides a mechanism for a router to specifically queue traffic for the forwarding process based on highest precedence.

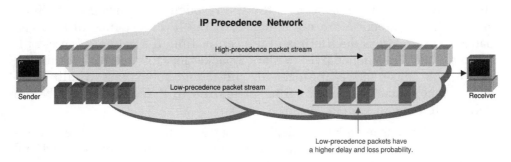

Figure 4.1 IP Precedence.

Precedence-based congestion control. Causes a router to drop packets based on precedence during periods of congestion.

Data-link-layer priority features. Cause a router to select service levels of the lower layers in an effort to provide preferential treatment.

Although RFC1812 does not explicitly describe how to perform these IP Precedence-related functions in detail, it does furnish references on previous related works and an overview of some possible implementations in which administration, management, and inspection of the IP Precedence field may be quite useful.

This is the benefit of the intended semantics of the IP Precedence field: Instead of the router attempting to perform an initial classification of a packet into one of potentially many thousands of active flows, and then applying a service rule that applies to that form of flow, you can use the IP Precedence field to reduce the scope of the task considerably. This approach is meant to mark the IP packet with the desired precedence indicator *before* the packet enters the network. On all subsequent interior routers, the required action is to look up the IP Precedence field and apply the associated service action to the packet. This approach can scale quickly and easily—assuming that the range of differentiated service actions is a finite number that does not grow with traffic volume. In contrast, per-flow service delivery includes a computational task that is related in processor load to traffic volumes.

You, should, however, exercise care when you undertake a network action that raises the precedence value of a TCP packet without the knowledge of the end systems. The original TCP specification, RFC793, allows for a TCP stack to reset a TCP connection if the IP Precedence settings used by both ends of the connection are not equal, or if the IP Precedence of received packets varies across the lifetime of the connection. If a network ingress filter sets the IP Precedence field to mark a packet for QoS delivery within the network, a conservative approach to end-to-end robustness would be to ensure that a matching egress filter clears the IP Precedence field before delivery to the end system.

IP Precedence is not a widely deployed performance feature of Internet service networks, for a number of reasons, including a poorly understood and undefined service interface to the client of such priority services, an uncertain tariff model that would apply to such services, and considerable confusion about how such a mechanism could extend beyond the administrative boundary of each network domain in a robust fashion. In addition, IP Precedence alone is insufficient to transform a mechanism into a service. An overloaded high-priority queue may offer worse service performance than a lightly loaded low-priority queue. Controlled service delivery requires not only internal service differentiation, but also some form of admission control function to ensure that the service channel is not saturated.

Universal support of the IP Precedence field was never going to happen, and perhaps the most reasonable expectation for its adoption was a heterogeneous service environment, with some network "islands" of IP precedence support with limited interconnection capability. This observation led to one proposal advocating using a dynamic discovery mechanism for end-to-end IP precedence, treating the precedence attribute of a path in the same vein as the maximum transmission unit [Huston 1996].

The IP Precedence field has now been subsumed into the Differentiated Services architecture, with the Class Selector Per Hop Behavior group providing a direct mapping of the IP Precedence field into a group of Differentiated Services code points. The critical change to the Differentiated Services architecture was the introduction of explicit admission controls that are intended to avoid overload of any priority service level. We will examine this in more detail when we look at the Differentiated Services architecture later in this chapter.

Precedence and 802.1p

An interesting set of enhancements to the IP Precedence field has been proposed by the IEEE 802.1 Internetworking Task Group. These enhancements provide a method of identifying 802-style frames based on a simple priority. A supplement to the original IEEE MAC Bridges standard [IEEE-802.1D], the proposed 802.1p specification [IEEE-802.1P] provides a method to allow preferential queuing and access to media resources by traffic class on the basis of a *user_priority* value signaled in the frame. The IEEE 802.1p specification, if adopted, will provide a way to transport this value (*user_priority*) across the subnetwork in a consistent method for Ethernet, Token Ring, or other MAC-layer media types using an extended frame format. Of course, this also implies that 802.1p-compliant hardware may have to be deployed to fully realize these capabilities.

The current 802.1p draft defines the *user_priority* field as a 3-bit value, resulting in a variable range of values between 0 and 7, with 7 indicating high priority and 0 indicating low priority. The IEEE 802 specifications do not make any suggestions on how the *user_priority* should be used in the end system by network elements. It only suggests that packets may be queued by LAN devices based on their *user_priority* values.

A proposal submitted in the Integrated Services over Specific Link Layers (ISSLL) Working Group of the IETF suggests how to use the IEEE 802.1p user_priority value with an Integrated Services class [ID-isl802-map]. Because no practical experience exists for mapping these parameters, the suggestions are somewhat arbitrary and provide only a framework for further study. As shown in Table 4.1, two of the user_priority values provide separate classifications for guaranteed services traffic with different delay requirements. The less-than-best effort category could be used by devices that tag packets that are not conformant to a traffic commitment and that may be dropped elsewhere in the network.

Because no explicit traffic class or *user_priority* field exists in Ethernet 802.3 packets [IEEE-802.3], the *user_priority* value must be regenerated at a

Table 4.1 IEEE 802.1p user_priority Mapping to Integrated Services Classes

User Priority	Service
0	Less than-best effort
1	Best effort
2	Reserved
3	Reserved
4	Controlled load
5	Guaranteed service, 100ms bound
6	Guaranteed service, 10ms bound
7	Reserved

downstream node or LAN switch by some predefined default criteria, or by looking further into the higher-layer protocol fields in the packet and matching some parameters to a predefined criteria. Another option is to use the IEEE 802.1Q encapsulation proposal tailored for VLANs (Virtual Local Area Networks), which may be used to provide an explicit traffic class field on top of the basic MAC format [IEEE-802.1Q]. As with IP Precedence, the relative priorities will only produce desired outcomes as long as the load in each individual service class is tightly constrained. It remains to be seen how the 802.1 effort will address this basic requirement for associated load control.

The Token Ring standard does provide a priority mechanism that can be used to control the queuing of packets and access to the shared media [IEEE-802.5]. This mechanism is implemented using bits from the Access Control (AC) and Frame Control (FC) fields of an LLC frame. The first 3 bits (the token priority bits) and the last 3 bits (the reservation bits) of the AC field dictate which stations get access to the ring. Theoretically, a Token Ring station is capable of separating traffic belonging to each of the eight levels of requested priority and transmitting frames in the order of indicated priority. The last 3 bits of the FC field (the user priority bits) are obtained from the higher layer in the *user_priority* parameter when it requests transmission of a packet. This parameter also establishes the access priority used by the MAC. This value usually is preserved as the frame is passed through Token Ring bridges; thus, the *user_priority* can be transported end-to-end unmolested. There are few, if any, users of the Token Ring priority settings.

IP Type of Service Observations

A service is predictable in terms of its performance only when the load imposed on a service mechanism remains within the network's design parameters. Once the load exceeds the service capability, the delivered service will exhibit delay and loss. Priority systems attempt to shift the side effects of such overload onto various classes of traffic. But if the elevated priority load exceeds the entire available capacity, than all other service classes are denied any service whatsoever. The conclusion: Service delivery architecture requires more than a simple precedence indicator in the packet header. In addition, some form of admission control is necessary to ensure that the imposed load remains within the service delivery capabilities for the network. As you will see later in this chapter, this important concept is the groundwork for the emergence of the Differentiated Services architecture.

We will look at each of these architectures in the approximate order they were developed; up first is the Integrated Services architecture.

Integrated Services Architecture

Anyone faced with the task of reviewing, analyzing, and comparing approaches to providing service delivery architectures understandably will feel overwhelmed by the complexities involved in the IETF Integrated Services architecture, described as an overview in RFC1633. The Integrated Services (IntServ) architecture was designed to provide a set of extensions to the best-effort traffic delivery model currently in place in the Internet. The framework was set up to enable special handling for certain types of traffic and to provide a mechanism for applications to choose between multiple levels of delivery services for its traffic. The driving motivation behind this architecture's design was the desire for multimedia support within the Internet architecture. As the Integrated Services overview points out:

> . . . an important technical element is still missing: real-time applications often do not work well across the Internet because of variable queuing delays and congestion losses. The Internet, as originally conceived, offers only a very simple quality of service (QoS), point-to-point best-effort data delivery. Before real-time applications such as remote video, multimedia conferencing, visualization, and virtual reality can be broadly used, the Internet infrastructure must be modified to support real-time QoS, which provides some control over end-to-end packet delays [RFC1633].

The direction proposed in the Integrated Services architecture is to use resource reservation to provide service guarantees to applications.

The Integrated Services architecture, is very similar in concept to the technical mechanics of ATM—namely, in its effort to provision guaranteed services, as well as differing levels of best-effort via a controlled-load mechanism. In fact, the Integrated Services architecture and ATM are somewhat analogous; Integrated Services provides end-to-end signaling for QoS parameters at Layer 3 in the Open Systems Interconnection (OSI) reference model, and ATM provides signaling for QoS parameters at Layer 2 of that model. The common architectural theme is that each reservation of resource requires a state to be maintained within the network. This state links a determined amount of resource with the resource load required to service those packets that fall within the reservation profile. In ATM, this state is maintained as a virtual circuit with a predetermined admission policy. Within Integrated Services, this state is maintained as a flow, with an associated admission policy and per-hop packet-handling characteristics.

The architectural model can be described as a host-centric model, where the end host is responsible for directing the network to allocate adequate resources to the end-to-end traffic flow (see Figure 4.2). The network can either accept or reject the allocation request, but it cannot negotiate with the host. If the allocation is accepted by the network, the commitment is intended to be nonnegotiable and indefinite. The host does not specify the duration of the resource allocation, and the network is expected to maintain the commitment for as long as the host application remains active. The network can renege on the entire allocation at any time, but it cannot negotiate a downgrade of the allocation quantity. In many ways, this model is like a telephone call. If the network and the remote party both accept the call, the network is expected to support the call's resource requirements indefinitely. This resource commitment extends until one of the parties terminates the call.

Integrated Services Framework: Background

The Integrated Services framework was built under the assumption that the basic underlying Internet architecture needs some level of modification in order to provide customized service support for different classes of applications. For example, service support for a voice-over-IP service implies some bound on delay, loss, and jitter, while support for consistent data transfer transactions implies some bound on available capacity. The Integrated Services architecture proposes a set of service extensions that provide services beyond the traditional best-effort offerings. The rational for this extension is that a single best-effort delivery service is simply inadequate for such

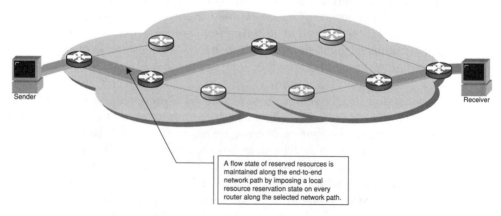

A flow state of reserved resources is maintained along the end-to-end network path by imposing a local resource reservation state on every router along the selected network path.

Figure 4.2 The Integrated services per-flow service model.

customized service responses. A number of aspects of networks underlie this perceived inadequacy:

The need for resource management within the network. In the immediate future, network capacity will not be in such overwhelming abundance everywhere such that every application will be able to secure the resources that it requires. As a consequence there will be resource contention within the network itself. There is no doubt that there is a rapid growth of available network capacity in many parts of today's networks. The capacity of transmission systems is growing at a rapid rate—the capacity of optical fibers continues to increase with the adoption of Dense Wave Division Multiplexing (DWDM), and this is coupled with the continued deployment of more optical fiber systems to service Internet demand. However, at the same time, the demand for capacity is also increasing, at a similar or even greater rate. Not only is this increase based on a growing client population for existing Internet services, but on an expanding service model where the Internet is used to deliver an ever broadening array of services. One outcome of the greater availability of network capacity is a continuing drop in the price of network capacity and a flow-on in terms of the continuing drop in the price of Internet services. As the network capacity increases and the unit price of capacity use decreases, additional communications applications become economically and technically viable at various price points. To date, the demand model for Internet-based services has far outstripped available supply of network capacity. In addition, the nature of the demand is such that it is unreasonable to expect a sudden reversal in the relationship between demand and supply. This is the basis of the case for a continuing need for some form of resource management within the network.

Abundance is not uniform. Even in a world of abundant network capacity, there is no assurance that such capacity will be well in excess of demand everywhere. Probably, some level of contention for network resources within a network is an enduring feature, and, therefore, some form of prioritization of the various resource demands is necessary.

Simple priority is not enough. Prioritization of packet service by any means is, in itself, inadequate as a resource management strategy. As we noted when examining the shortfalls of the IP Precedence field in IP Type of Service, prioritization of traffic alone will not yield predictable service outcomes. A saturated high-priority queue provides worse service levels than an unloaded best-effort service. Any form of comprehensive resource management should include control of the load placed on the resource. It is not enough to simply allocate a certain amount of resource, as the load must also be constrained so that it does not overwhelm the allocated resource.

Inelastic applications. Some applications cannot adapt arbitrarily to the prevailing network conditions. Within a multiservice environment, there will be data flows that use external real-time data clocks. Today, such flows typically carry real-time interactive voice and video streams, although no doubt the family of such applications will grow over time. Such applications require some level of consistency of response from the network. The level of tolerance to variance in this service response is often quite small.

To address these inadequacies, the Integrated Services architecture adds the concept of *resource reservation* into the Internet service model, so that a *flow* of data can be handled by each router in the path using the reserved resources to manage the flow. This is coupled with some form of *admission policy* to ensure that the flow does not saturate the reserved resource.

In terms of the Internet architecture, this is a fundamental step away from the end-to-end model of Internet service [Clark 1988]. One of the major reasons for the robustness and cost-effectiveness of the Internet is that all flow-related state is contained in the end systems, allowing the interior of the network to operate without the maintenance of any form of remembered state or any associated resource limitation. The Integrated Services architecture moves away from this base model into one that implies a greater overhead within the network. But the Integrated Services model also admits some increased vulnerability of the robustness of the network service, and such a service-enabled network may not operate at the same level of cost-efficiency as that of a best-effort service benchmark level. The proponents of this architectural approach of course argue that the negative impacts on the Internet service model are more than offset by the gains. They describe the positive aspects as being able to offer guaranteed service levels to a range of real-time network applications, and they anticipate that the outcome will be a single multiservice network platform supporting both real-time and nonreal-time traffic that will operate far more reliably and cost-efficiently than a collection of parallel service networks, each attuned for a particular service.

Contextual QoS Definitions

Before we can discuss the underlying mechanisms in the Integrated Services model, we need to review a number of definitions within the context of the Integrated Services architecture.

Quality of service. In the context of the Integrated Services framework, refers to the nature of the packet delivery service provided by the network, characterized by parameters that include available bandwidth, packet delay, and packet loss rates.

Network node. Any component of the network that handles data packets and is capable of imposing QoS control over data flowing through it. Nodes include routers, subnets (the underlying link-layer transport technologies), and end systems.

QoS-capable or Integrated Services (IS)–capable node. A network node that can provide one or more of the services defined in the Integrated Services model.

QoS-aware or IS-aware node. A network node that supports the specific interfaces required by the Integrated Services service definitions but that cannot provide the requested service. Although a QoS-aware node may not be able to provide any of the QoS services themselves, it can understand the service request parameters and deny QoS service requests accordingly.

Service or QoS control service. Refers to a coordinated set of QoS control capabilities provided by a single node. The definition of a service includes a specification of the functions to be performed by the node, the information required by the node to perform these functions, and the information made available by a specific node to other nodes in the network.

Integrated Services Model Components

First and foremost, the IntServ architecture assumes that resource use in the network must be controlled in order to deliver managed services. It is a fundamental building block of the architecture that traffic managed under this model must be subject to admission-control mechanisms.

In addition to admission control, the IntServ architecture makes provisions for a resource-reservation mechanism, which is equally important in providing differentiated services. Integrated Services proponents argue that real-time applications, such as real-time video services, cannot be accommodated without resource guarantees, and that resource guarantees cannot be realized without resource reservations and admission controls. Normally, *guarantee* has a very precise meaning, that of an absolute service level, but in this context, the term must be afforded some latitude of interpretation; the service guarantees may be approximate and imprecise. The Integrated Services architecture defines guaranteed service as *predictable* service, within some bounds of approximation, that a user can request from the network for the duration of a particular session.

There are four major components in the reference implementation framework for an IntServ router: the resource reservation controller, the admission control routine, the classifier, and the packet scheduler [RFC1633]. The reference model is shown in Figure 4.3.

Flow resource reservation. The basic approach adopted by the IntServ architecture is to deliver service quality through a process of reservation of resources. This implies the requirement for a reservation setup protocol and associated mechanisms to allow each router to create and maintain a reservation state for each reserved traffic flow. There are a number of components to a reservation, including the desired scheduling characteristics, the resource profile of the flow, and the classification criteria for the flow and the next-hop forwarding decision, as well as a reservation identifier. All of these components are summarized as a *flow specification*, or *flowspec*. The information in the flowspec is necessary for the router to construct and maintain an internal state to support the reservation, including classification, forwarding and scheduling actions taken in the data path. We will look at this area in further detail as we examine the resource reservation protocol (RSVP).

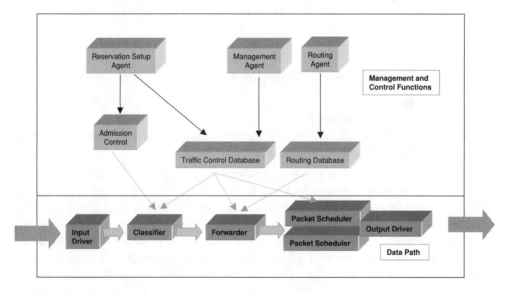

Figure 4.3 The Integrated Services reference model.

Flow admission control. Admission control in this context is the decision process used to grant or deny a reservation request. Each node along a proposed end-to-end path must make a local decision as to whether the node has sufficient reservable resources remaining to admit the reservation's traffic profile. The admission control function may include authentication of the request and some local prioritization function to ensure that local resources are allocated in conformance to some defined policy framework.

Packet classifier. While resource reservation and admission control are part of the router's control functions, the packet classifier is part of the packet processing function. Each arriving packet must be mapped to a local service class. Every packet within a service class is treated identically, and may include such actions as preferential queuing, lower drop probability through buffering into extended queues, or a low jitter service using shorter local queues. The classifier does not necessarily have to classify packets so that each flow reservation is visible, but if it doesn't, then a separate network function has to police the traffic profile of each reservation flowspec upon entry to the network, and the forwarding state of the reservation flow may be lost. A more robust approach is for the admission classifier to perform a classification into a known flowspec, pass the flow through the associated traffic filter, make a forwarding decision based on the flow path, then select a scheduler action when queuing the packet to the selected output driver.

Packet scheduler. The final component is the packet scheduler, which manages service delivery through management of the output queuing function. To implement a range of service classes, there may be a number of distinct queue sets, each with an associated queuing discipline. As we noted in Chapter 3, "Performance Tuning Techniques," a real-time low jitter service uses a queuing discipline with a short queue and an elevated scheduling priority. A high-capacity, high-reliability service may use a queuing discipline with larger queues and some form of active congestion management, such as RED or ECN. Enforcing relative bandwidth allocations to each of these service classes can be achieved with a WFQ discipline as the mixer.

QoS Requirements

The Integrated Services architecture is concerned primarily with traffic *time-of-delivery*; therefore, per-packet delay is the central theme in determining service commitments. Understanding the characterization of real-time and nonreal-time, or elastic, applications and how they behave in the network is an important aspect of the IntServ environment.

The fundamental difference between real-time and elastic data-streaming models lies in flow-control and error-handling mechanisms. In terms of flow control, real-time traffic is intended to be self-paced, whereby the sending data rate is determined only by the characteristics of the sending application. The task of the network is to attempt to impose minimal distortion on this sender-defined pacing of the data stream, so that flow control is nonadaptive and is based on some condition imposed by the sender. Elastic applications typically use an adaptive flow-control algorithm in which the data flow rate is based on the sender's view of available network capacity and available space at the receiver's end, instead of on some external condition.

With respect to error handling, real-time traffic may have a timeliness factor associated with it; if the packet cannot be delivered within a certain elapsed time frame and within the sender-sequencing format, it is irrelevant to playback, and can be discarded. Packets that are in error, are delivered out of sequence, or arrive outside of the delivery time frame, must be discarded. Elastic applications do not have a timeliness factor, and accordingly, there is no time at which the packet is deemed to be irrelevant. The relevance and importance of delivery of the packet depend on other factors within such applications. In general, elastic applications use error-detection and retransmission-recovery mechanisms at the application layer to recover cleanly from data-transmission errors (such as TCP).

The effect of real-time traffic flows is to lock the sender and receiver into a common clocking regime, where the only major difference is the offset of the receiver's clock, as determined by the network's propagation delay. This state of end-to-end integrity of data clocking imposes some constraints on the network service profile. For elastic applications, the sender attempts to adapt the data clocking rate to the current state of the network.

IntServ predominantly focuses on real-time classes of application traffic. Real-time applications generally can be defined as those with playback characteristics; in other words, a data stream that is packetized at the source and transported through the network to its destination, where it is depacketized and played back by the receiving application. The reassembly and playback of the real-time data must reproduce the original timing of the data. To achieve this functionality for real-time applications, the packet-switched Internet network has to emulate a synchronously clocked data circuit with constant end-to-end delay. In this way, packets can be presented to the decoder with the same relative timing as they were generated by the encoder, reproducing the original signal. Data that is lost within the network or delayed beyond the limits of the reassembly process is discarded, as the timing of the data is of greater importance than integrity of the data itself when reassembling the entire original signal. The other class of application traffic is elastic, where there is no strict requirement to preserve the original timing of the data stream. Here the data has no time dimension, and the objective of the transfer is to reproduce precisely the entire data set, creating an exact replica of the original data at the receiver.

As the data is transported along its way through the network, latency is inevitably introduced at each point in the transit path. The degree of latency introduced is variable, because latency is the sum of the per-hop transmission delays and router queuing hold times, and the queuing hold times can be highly variable. This variation in latency, or jitter, in the real-time signal is what must be smoothed by the playback. The receiver compensates for this jitter by buffering the received data for a period of time (an offset delay) before playing back the data stream, in an attempt to negate the effects of the jitter introduced by the network (see Figure 4.4). The trick is in calculating the optimum offset delay, because having an offset delay that is too short for the current level of jitter effectively renders the original real-time signal worthless, while a lengthy offset delay consumes host resources and is a significant impediment to interactive applications.

Jitter-Tolerant and -Intolerant Real-Time Applications

The ideal scenario is to have a mechanism that can dynamically calculate and adjust the offset delay in response to fluctuations in the average jitter induced by the network. An application that can adjust its offset delay is called an *adaptive playback application*. Such adaptive playback applications are slightly more tolerant of induced jitter. To assist in the reproduction of the original timing of the data at playback, the data stream of an adaptive real-time application may be encoded using the Real-Time Protocol (RTP) [RFC1889]. This protocol adds relative timing information to packets, allowing the receiver to play back the payload of a sequence of packets in the correct order and with the correct interpacket delays. RTP is used by SIP and H.323 multimedia and telephony applications.

RTP allows the classification of real-time applications to be further broken down into two subcategories: those that are *tolerant* and those that are *intol-*

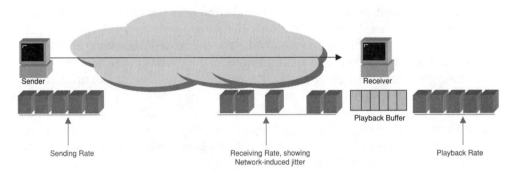

Figure 4.4 Jitter compensation for real-time flows.

erant of induced jitter. Jitter-tolerant applications can be characterized as those that can function in the face of nominal induced jitter and still produce an acceptable signal quality when played back. Examples of such tolerant real-time applications are various packetized audio-streaming voice-over-IP applications that use RTP to time the data, and may use the silent periods to adjust the playback buffer size to the prevailing jitter bounds. Jitter-intolerant applications can be characterized as those in which induced jitter and packet delay result in enough introduced distortion to effectively render the playback signal quality unacceptable or the application nonfunctional. Examples of more intolerant applications are two-way telephony applications (interactive voice, as opposed to noninteractive audio playback), high-quality real-time video flows, and circuit-emulation services.

For jitter-tolerant applications, IntServ recommends the use of a *predictive* service, otherwise known as a *controlled-load* service. For jitter-intolerant applications, IntServ recommends a *guaranteed* service model. The fundamental difference in these two models is that one provides a reliable upper bound on delay (guaranteed) and the other (controlled-load) provides a less-than-reliable delay bound.

Integrated Services Control and Characterization

The Integrated Services architecture uses a set of general and service-specific control parameters to characterize the QoS service requested by the end system's application. The general parameters appear in all QoS service-control services. Service-specific parameters are used only with the controlled-load or guaranteed service classes—for example, a node that may need to occasionally export a service-specific value that differs from the default. In the case of the guaranteed service class, if a node is restricted by some sort of platform- or processing-related overhead, it may have to export a service-specific value (such as smaller maximum packet size) that is different from the default value, so that applications using the guaranteed service will function properly. These parameters are used to characterize the QoS capabilities of nodes in the path of a packet flow, and are described in detail in RFC2215. A brief overview of these control parameters is presented here so that you can begin to understand how the Integrated Services model fits together.

NON_IS_HOP

The NON_IS_HOP parameter provides information about the presence of nodes that do not implement QoS control services along the data path. In this vein, the *IS* portion of this and other parameters also stands for *Integrated Services-aware;* an IS-aware element is one that conforms to the requirements specified in the Integrated Services architecture. A flag is set in this object if a node does not implement the relevant QoS control service or if it knows that there is a break in the traffic path of nodes that implement the service. This also is called a *break bit*, because it represents a break in the chain of network elements required to provide an end-to-end traffic path for the specified QoS service class.

NUMBER_OF_IS_HOPS

The NUMBER_OF_IS_HOPS parameter is represented by a counter that is a total incremented by 1 at each IS-aware hop. This parameter is used to inform the flow end points of the number of IS-aware nodes that lie in the data path. Valid values for this parameter range from 1 to 255; in practice, the value is limited by the bound on the IP hop count.

AVAILABLE_PATH_BANDWIDTH

The AVAILABLE_PATH_BANDWIDTH parameter provides information about the available bandwidth along the path followed by a data flow. This is a local parameter, and provides an estimate of the bandwidth available for traffic following the path. Values for this parameter are measured in bytes per second and range in value from 1 byte per second to 40 terabytes per second (which is believed to be of the same order of magnitude as the theoretical maximum bandwidth of a single strand of fiber).

MINIMUM_PATH_LATENCY

The MINIMUM_PATH_LATENCY local parameter is a representation of the latency in the forwarding process associated with the node, where the latency is defined to be the smallest possible packet delay added by the node itself. This delay results from speed-of-light propagation delay, packet-processing limitations, or both. It does not include any variable queuing delay that may be introduced. The purpose of this parameter is to provide a baseline minimum path latency figure to be used with services that establish estimates or bounds on additional path delay, such as the guaranteed service class. Together with the queuing delay bound offered by the guaranteed service class, this parameter gives the application a priori knowledge of both the minimum and maximum packet-delivery delay. Knowing both minimum and maximum latencies experienced by traffic allows the receiving application to attempt to accurately compute buffer requirements to remove network-induced jitter.

PATH_MTU

The PATH_MTU parameter is a representation of the maximum transmission unit (MTU) for packets traversing the data path, measured in bytes. This parameter informs the end point of the packet MTU size that can traverse the data path without being fragmented. A correct and valid value for this parameter must be specified by all IS-aware nodes. This value is required to invoke QoS control services that require the IP packet size to be strictly limited to a specific MTU. Existing MTU discovery mechanisms cannot be used, because they provide information only to the sender; they do not directly allow for QoS control services to specify MTUs smaller than the physical MTU. The local parameter is the IP MTU, where the MTU of the node is defined as the maximum size the node can transmit without fragmentation, including upper-layer and IP headers but excluding link-layer headers. Remember that each node must be able to classify a packet into either a reserved flow or into the default best-effort service class. Fragments of an IP packet contain no UDP packet header, so that the additional port information used in such classification filters is unavailable. To ensure the integrity of an IntServ flow, packet fragmentation must be avoided.

TOKEN_BUCKET_TSPEC

The TOKEN_BUCKET_TSPEC parameter (TSpec is used in this context as an abbreviation of traffic specification) describes traffic parameters using a simple token-bucket filter, and is used by data senders to characterize the traffic it expects to generate. This parameter also is used by QoS control services to describe the parameters of traffic for which the subsequent reservation should apply. This parameter takes the form of a token-bucket specification plus a peak rate, a minimum policed unit, and a maximum packet size. The token-bucket specification itself includes an average token rate and a bucket depth. The token-bucket parameters refer to the token filter described in Figure 4.5.

The token rate, r, is measured in bytes of IP datagrams per second; it may range in value from 1 byte per second to 40 terabytes per second, the average rate to be sustained through the token filter. The token-bucket depth, b, is measured in bytes; its value ranges from 1 byte to 250 gigabytes, the maximum burst volume that can be accommodated by the token filter. The peak traffic rate, p, is measured in bytes of IP datagrams per second and may range in value from 1 byte per second to 40 terabytes per second, the maximal rate used for burst traffic passing through the token filter. The minimum policed unit, m, is an integer measured in bytes, where any packet less than the minimum unit of bytes is treated for the purposes of the token filter as being the minimal unit of bytes in size. The purpose of this parameter is to allow a reasonable estimate of the per-packet resources needed to process a flow's packets; the maximum packet rate can be computed from the values expressed in b and m, where the maximum packet rate is b divided by m. The packet size includes the application data and all associated protocol headers at or above the IP layer. It does not include the link-layer headers, because these may change in size as a packet traverses different portions of a network. The maximum packet size, M, is the largest packet that will conform to the traffic specification, also measured in bytes. Packets transmitted that are larger than M

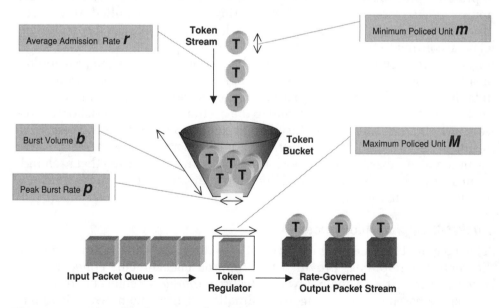

Figure 4.5 Token-bucket specification.

may not receive QoS-controlled service, because they are considered to be nonconformant with the traffic specification.

The range of values that can be specified in these parameters has been designed to be large enough to allow for future network technologies; an IntServ node is not expected to support the full range of values.

Resource-Sharing Requirements

The allocation of network resources is accomplished on a flow-by-flow basis, and although each flow is subject to admission-control criteria, many flows share the available resources on the network, which is described as *link-sharing*. With link-sharing, the aggregate bandwidth in the network is shared by various types of traffic, which generally can be different network protocols (for example, IP, IPX, SNA), different services within the same protocol suite (for example, telnet, FTP, SMTP), or simply different traffic flows that are segregated and classified by sender. An important criterion within the IntServ architecture is that different traffic types do not unfairly utilize more than their fair share of network resources; such a situation could result in a disruption of other traffic. The IntServ architecture also focuses on link-sharing by aggregate flows and link-sharing with an additional admission-control function.

There are two problems to be addressed within the associated queuing discipline adopted to support IntServ. The first is to segregate traffic into the various service classes, effectively separating IntServ-controlled flows from best-effort traffic. The second is to schedule traffic within the IntServ service class according to the service objectives. Priority queuing can be used to segregate the IntServ traffic class from the best-effort traffic. As the priority load has been conditioned by admission filters and the router has accepted the total IntServ load via RSVP signaling, it is expected that the total IntServ traffic load will not cause service denial to the best-effort traffic class. For this reason, priority queuing can be used with some confidence that there is no resultant service denial side effect. Within each service class, a WFQ mechanism can be used as the resource-sharing mechanism, ensuring that each individual flow receives a resource allocation that meets its requested level of resource (see Figure 4.6).

Packet-Dropping Allowances

The Integrated Services architecture outlines different scenarios in which traffic control is implicitly provided by dropping packets. One concept is that some packets within a given flow may be *preemptable*, or subject to drop. This concept is based on local congestion situations where the network is in dan-

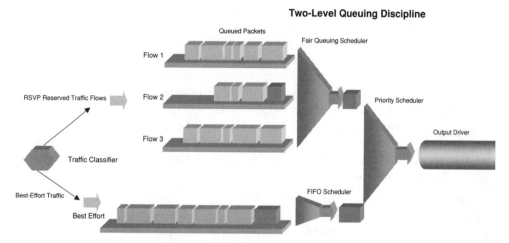

Figure 4.6 Integrated Services support queuing model.

ger of reneging on established service commitments. A router simply could discard traffic by acting on a particular packet's *preemptability option* to avoid disrupting established commitments. Another approach classifies packets that are not subject to admission-control mechanisms as being the first priority for packet discard.

Controlled Load Service Class

The Integrated Services definition for *controlled-load service* attempts to provide end-to-end traffic behavior that closely approximates traditional best-effort services within the environmental parameters of unloaded or lightly utilized network conditions [RFC2211]. In other words, controlled load service corresponds to a *better-than-best-effort* delivery; that is, the service attempts to provide traffic delivery within the same bounds as an unloaded network in the same situation. Assuming that the network is functioning correctly (routing and forwarding), applications may assume that a very high percentage of transmitted packets will be delivered successfully and that any latency introduced into the network will not greatly exceed the minimum delay experienced by any successfully transmitted packet.

To ensure that these conditions are met, the application requesting the controlled-load service provides the network with an estimation of the traffic it will generate—the TSpec. The controlled-load service uses the TOKEN_ BUCKET_TSPEC structure to describe a data flow's traffic parameters. In turn, each node handling the controlled-load service request ensures that sufficient resources are available to accommodate the request. The degree of accuracy to which the TSpec matches available resources in the network does not have to be high. If the requested resources fall outside the bounds of what is available, the traffic flow may experience some small additional amount of induced delay, or possibly a slightly higher rate of dropped packets because of congestion situations. However, the degree at which traffic may be dropped or delayed should be low enough for the adaptive real-time applications to function without noticeable degradation.

The controlled-load service does not accept or use specific values for control parameters that include information about delay or loss. Acceptance of a controlled-load request implies a commitment to provide a better-than-best-effort service that approximates network behavior under nominal network-utilization conditions.

The method a node uses to determine whether adequate resources are available to accommodate a service request is purely a local matter, and may be implementation-dependent; only the control parameters and message formats are required to be interoperable.

Links on which the controlled-load service is run are not allowed to fragment packets. Packets larger that the MTU of the link must be treated as nonconformant with the Tspec. The controlled-load service is provided to a flow when traffic conforms to the TSpec given at the time of flow setup. When nonconformant packets are presented within a controlled-load flow, the node must ensure that these three things happen:

◆ The node must ensure that it continues to provide the contracted QoS to those controlled-load flows that are conformant.

◆ The node must prevent nonconformant traffic in a controlled-load flow from unfairly impacting other conformant controlled-load flows.

◆ If sufficient resources are available, the node must attempt to forward nonconformant traffic on a best-effort basis. Nodes should not assume that nonconformant traffic is indicative of an error, because large numbers of packets may be nonconformant as a matter of course. This nonconformancy occurs because some downstream nodes may not police extended bursts of traffic to conform with the specified TSpec, and in fact will borrow available bandwidth resources to clear traffic bursts that have queued up. If a flow obtains its exact fixed-token rate in the presence of an extended burst, for example, there is a danger that the queue will fill up to the point of packet discard. To prevent this situation, the controlled-load node may allow the flow to exceed its token rate in an effort to reduce the queue buildup. Thus, nodes should be prepared to accommodate bursts larger than the advertised TSpec.

Guaranteed QoS Service Class

The guaranteed service class, described in RFC2212, provides a framework for delivering traffic for applications that require a guarantee on both available bandwidth and an upper delay bound from the network. This service is intended for delay-sensitive real-time applications, such as high-quality interactive audio and video streams. The guaranteed service does not attempt to place a bound on jitter, but adopts the approach of placing a bound on the maximal queuing delay in the end-to-end traffic path; the fixed delay in the traffic path is introduced by factors other than queuing, such as speed-of-light propagation, and setup mechanisms used to negotiate end-to-end traffic parameters. The guaranteed service framework mathematically asserts that the queuing delay is a function of two factors: the token-bucket depth, b, and the data rate, r, of the TSpec. Because the application controls these values, it has an a priori knowledge of the queuing delay provided by the guaranteed service.

The guaranteed service profile commits the network to ensure that packets will arrive within a certain delivery time and will not be discarded because of queue overflows, provided that the flow's traffic stays within the bounds of its specified traffic parameters. The guaranteed service does not control the minimal or average delay of traffic, and it doesn't attempt to control or minimize jitter—it only controls the maximum queuing delay.

The guaranteed service is invoked by a sender specifying the flow's traffic parameters (via the TSpec) and the receiver subsequently requesting a desired service level (the corresponding RSpec, or reservation specification). The guaranteed service specification uses the TOKEN_BUCKET_TSPEC parameter as the TSpec. The RSpec consists of a data rate, R and a slack term, S, where R must be greater than or equal to the token-bucket data rate r. The rate R is measured in bytes of IP datagrams per second and has a value range of between 1 byte per second to 40 terabytes per second. The slack term S is measured in microseconds. The RSpec rate can be larger than the TSpec rate, because higher rates are assumed to reduce queuing delay. The slack term represents the difference between the desired delay and the delay obtained by using a reservation level of R. The slack term also can be used by the network to reduce its resource reservation for the flow.

Because of the end-to-end and hop-by-hop calculation of two service error metrics, every node in the data path must implement the guaranteed service for this service class to function. The first error metric, C provides a cumulative representation of the delay a packet might experience because of rate parameters of a flow, also referred to as *packet serialization*. The error term C is measured in bytes. The second error metric, D, is a rate-independent, per-element representation of delay imposed by time spent waiting for transmission through a node. The error metric D is measured in units of microseconds. The cumulative end-to-end calculation of these error metrics, $Ctot$ and $Dtot$, represent a flow's deviation from the *fluid model*. The *fluid model* states that service flows within the available total service model can operate independently of each other.

As with the controlled-load service, links on which the guaranteed service is run are not allowed to fragment packets. Packets larger than the MTU of the link must be treated as nonconformant with the TSpec.

Two types of traffic policing are associated with the guaranteed service: simple policing and reshaping. Policing is done at the edges of the network, and reshaping is done at intermediate nodes within the network. *Simple policing* compares traffic in a flow against the TSpec for conformance. *Reshaping* consists of an attempt to restore the flow's traffic characteristics to conform to the TSpec. Reshaping causes delay in forwarding of datagrams until they are in conformance with the TSpec. Reshaping is done by combining a token bucket with a peak-rate regulator and buffering a flow's traffic until it can be forwarded in conformance with the token-bucket r and peak-rate p parameters. Such reshaping may be necessary because of small levels of distortion introduced by the packet-level use of any transmission path. This packet-level *quantization* of flows is what is addressed by reshaping. In general, reshaping adds a small amount to the total delay, while reducing the overall imposed jitter of the flow.

The mathematical computation and supporting calculations for implementing the guaranteed service mechanisms are documented in RFC2212, and thus, mercifully, are not duplicated here.

Provisions for Usage Accounting

Although it is commonly recognized that usage accounting is a necessary part of any potential IntServ deployment plan, the IntServ description does not go into a great deal of detail on this topic. In fact, RFC1633 says little about it at all, other than that usage feedback appears to be a highly contentious issue. The probable reason is that the primary need for accounting data is for subscriber billing, which is a matter of institutional business models and becomes a local policy issue. Although perhaps technical documentation eventually will emerge that produces mechanisms that provide accounting data that could be used for these purposes, it is the technical mechanisms themselves that have received the detailed examination to date—not the policy that drives the use of such mechanisms.

Arguably, the omission of accounting for usage of resources is a weakness of the IntServ architecture. A refinement of the best-effort service model that includes the preemption of resources to meet service guarantees will inevitably include some form of usage accounting and an associated premium on the service price. If there is a commonly specified usage accounting structure within the service architecture, the task of supporting such services across a multiprovider service platform will be made slightly easier. Without such a specification, each provider is able to make a local interpretation of the parameters of measuring the use of such a service, and local interpretations of the semantics of the IntServ architecture are a barrier to multiprovider IntServ interoperability.

There are, however, some aspects of this architecture that assist in the accounting task. A service initiator requests a certain service from the network, and upon confirmation of the request the initiator then passes traffic into the network along the reserved service path. Within this model is an identifiable network transaction, a transaction initiator, and an associated transaction duration, traffic volume, and service characteristics. It is possible to associate a per-transaction accounting function with each IntServ service request. When each request is confirmed, an accounting function can be triggered at the network admission control point. This accounting function can measure the transaction characteristics for the duration of the transaction, then pass the accumulated data to an accounting database, similar to the manner of the Radius accounting model [RFC2139]. The weakness of this accounting model is that it does not measure what the network actually delivers, but what is *anticipated* to be delivered.

An alternative approach is to undertake egress accounting, measuring what is passed out of the network as the delivered service. For an IntServ flow that spans multiple service providers, it may be appropriate to extend this model even further, placing an egress accounting function on the egress of each service provider's network, forming the accounting basis of an interprovider financial settlement model.

The complete accounting model is more detailed than either ingress or egress accounting, at least from the customer's perspective. The accounting model should include the service profile being requested, the profile of the admitted traffic, and the profile of the traffic passed out of the network, measuring average data rate, peak burst rates, packet loss rates, and average interpacket delay and delay variance. This would allow the ingress traffic profile to be matched against the egress traffic profile, creating a more accurate picture of the network's delivered service quality. However, given the complexities associated with such an approach, egress accounting appears to offer the next best model in terms of measuring the delivered service quality for an IntServ flow.

RSVP: Resource Reservation Protocol

As described earlier, the Integrated Services architecture provides a framework for applications to choose between multiple controlled levels of delivery services for their traffic flows. Two basic requirements exist to support this framework. The first is for the nodes in the traffic path to support the QoS control mechanisms that are consistent with support of the controlled-load and guaranteed services. The second requirement is for a mechanism by which the applications can communicate their QoS requirements to the nodes along the transit path, and for the network nodes to communicate the QoS requirements that must be provided for the particular traffic flows between one another. This functionality can be provided by a resource reservation setup protocol called RSVP.

The information presented here is intended to provide a synopsis of the internal mechanics of RSVP; it is not intended to be a complete detailed description of the internal semantics, object formats, or syntactical construct of the protocol fields. You can find a more in-depth description in the RSVP version 1 protocol specification, RFC2205, and an excellent practical description of RSVP and its application in the recently published Inside the Internet's Resource reSerVation Protocol *[Durham 1999].*

There is a logical separation between the Integrated Services QoS control services and RSVP. RSVP, as a path maintenance signaling protocol, can be used to support a variety of services, and, by the same token, the QoS control services are designed to be managed by a variety of control protocols.

The design of RSVP is sufficiently general that it can maintain edge-to-edge paths across an internet network, where the path possesses a set of attributes. In addition to supporting QoS paths within the IntServ environment, RSVP is currently pivotal in setting up and maintaining Label Switched Paths with a MPLS switched network.

RSVP does not define the internal format of the protocol objects related to characterizing QoS control services; it treats these objects as opaque. In other words, RSVP is simply the signaling mechanism, and the QoS control information is the signal content. RSVP is not a routing protocol, but is designed to interoperate with existing unicast and multicast IP routing protocols. RSVP uses the local routing table in routers to determine routes to the appropriate destinations. For multicast, a host sends Internet Group Management Protocol (IGMP) messages to join a multicast group and then sends RSVP messages to reserve resources along the delivery path(s) of that group.

In general terms, RSVP is used to provide QoS requests to all router nodes along the transit path of the traffic flow, and to maintain the state necessary in the router required to actually provide the requested services. RSVP requests generally result in resources being reserved in each router in the transit path for each flow. RSVP establishes and maintains a *soft state* in nodes along the transit path of a reservation data path. A *hard state* is what other technologies provide when setting up virtual circuits for the duration of a data-transfer session; the connection is torn down after the transfer is completed. A *soft state* is maintained by periodic refresh messages sent along the data path to maintain the reservation and path state. In the absence of these periodic messages, which typically are sent every 30 seconds, the state is deleted as it times out. This soft state is necessary, because RSVP is essentially a QoS reservation protocol and does not associate the reservation with a specific static path through the network. Therefore, it is entirely possible that the path will change, so that the reservation state must be refreshed periodically.

RSVP also provides dynamic QoS; the resources requested may be changed at any given time for a number of reasons:

♦   An RSVP receiver may modify its requested QoS parameters at any time.

◆ An RSVP sender may modify its traffic-characterization parameters, defined by its Sender TSpec, and cause the receiver to modify its reservation request.

◆ A new sender can start sending to a multicast group with a larger traffic specification than existing senders, thereby causing larger reservations to be requested by the appropriate receivers.

◆ A new receiver in a multicast group may make a reservation request that is larger than existing reservations.

The last two reasons are related inextricably to how reservations are merged in a multicast tree. Figure 4.7 shows a simplified version of RSVP in hosts and routers.

Method of Operation

RSVP requires the receiver, instead of the sender, to be responsible for requesting specific QoS services. This was intentionally designed as part of RSVP in an attempt to efficiently accommodate large groups (generally, for multicast traffic), dynamic group membership (also for multicast), and diverse receiver requirements.

An RSVP sender sends Path messages *downstream* toward an RSVP receiver destination (see Figure 4.8). Path messages are used to store path information

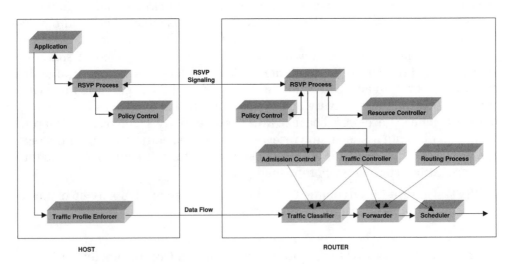

Figure 4.7 RSVP in hosts and routers.

in each node in the traffic path. Each node maintains this path state characterization for the specified sender's flow, as indicated in the sender's TSpec parameter. After receiving the Path message, the RSVP receiver sends Resv (reservation request) messages back *upstream* to the sender (see Figure 4.9) along the same hop-by-hop traffic path the Path messages traversed when traveling toward the receiver. Because the receiver is responsible for requesting the desired QoS services, the Resv messages specify the desired QoS and set up the reservation state in each node in the traffic path. After successfully receiving the Resv message, the sender begins sending its data. If RSVP is being used in conjunction with multicast, the receiver first joins the appropriate multicast group, using IGMP, prior to the initiation of this process.

RSVP requests only unidirectional resources—resource reservation requests are made in one direction only. Although an application can act as a sender and receiver at the same time, RSVP treats the sender and receiver as logically distinct functions.

RSVP Reservation Styles

A Reservation (Resv) request contains a set of options, which collectively are called the *reservation style*, characterizing how reservations should be treated in relationship to the sender(s). These reservation styles are particularly relevant in a multicast environment.

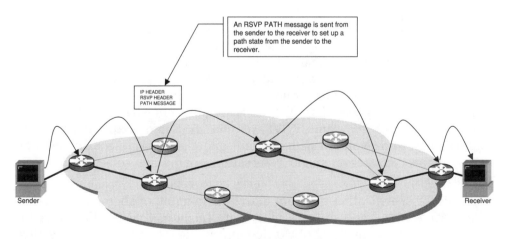

Figure 4.8 Traffic flow of the RSVP Path message.

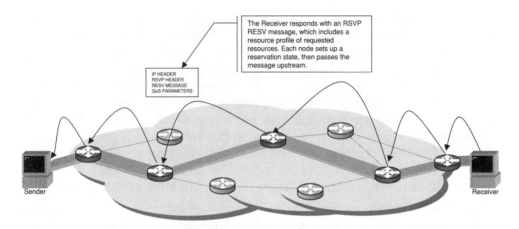

Figure 4.9 Traffic flow of the RSVP Resv message.

One option concerns the treatment of reservations for different senders within the same RSVP session. The option has two modes: establish a *distinct* reservation for each upstream sender or establish a *shared* reservation used for all packets of specified senders.

Another option controls the selection of the senders. This option also has two modes: an *explicit* list of all selected senders or a *wildcard* specification that implicitly selects all senders for the session. In an explicit sender-selection reservation, each Filter Spec must match exactly one sender. In a wildcard sender selection, no Filter Spec is needed.

As shown in Figure 4.10, the Wildcard-Filter (WF) style implies a shared reservation and a wildcard sender selection. A WF style reservation request creates a single reservation shared by all flows from all upstream senders. The Fixed-Filter (FF) style implies distinct reservations with explicit sender selection. The FF style reservation request creates a distinct reservation for a traffic flow from a specific sender; the reservation is not shared with another sender's traffic for the same RSVP session. A Shared-Explicit (SE) style implies a shared reservation with explicit sender selection. An SE style reservation request creates a single reservation shared by selected upstream senders.

The RSVP specification does not allow the merging of shared and distinct style reservations, because these modes are incompatible. Neither does the specification allow merging of explicit and wildcard style sender selection, because doing so most likely would produce unpredictable results for a

Sender Selection	Reservations	
	Distinct	Shared
Explicit	Fixed-Filter (FF) Style	Shared-Explicit (SE) Style
Wildcard	(No Style Defined)	Wildcard-Filter (WF) Style

Figure 4.10 Reservation attributes and styles.

receiver that may specify an explicit style sender selection. As a result, the WF, FF, and SE styles are all incompatible with one another.

RSVP Messages

An RSVP message contains a Message-Type field in the header that indicates the function of the message. Although seven types of RSVP messages exist, two are fundamental RSVP message types: the Resv (reservation) and Path messages, which provide for the basic operation of RSVP. As mentioned earlier, an RSVP sender transmits Path messages downstream along the traffic path provided by a discrete routing protocol (for example, OSPF). Path messages store path information in each node in the traffic path, which includes at a minimum the IP address of each previous hop (PHOP) in the traffic path. The IP address of the previous hop is used to determine the path on which the subsequent Resv messages will be forwarded. The Resv message is generated by the receiver and is transported back upstream toward the sender, creating and maintaining reservation state in each node along the traffic path, and following the reverse path on which Path messages previously were sent and the same path data packets will subsequently use.

RSVP control messages are sent as raw IP datagrams using protocol number 46. Although raw IP datagrams are intended to be used between all end systems and their next-hop intermediate node router, the RSVP specification allows for end systems that cannot accommodate raw network I/O services to encapsulate RSVP messages in User Datagram Protocol (UDP) packets.

Path, PathTear, and ResvConf messages must be sent with the Router Alert option set in their IP headers [RFC2113]. The Router Alert option signals routers on the arrival of IP datagrams that need special processing. Therefore, routers that implement high-performance forwarding designs can maximize forwarding rates in the face of normal traffic, and be alerted to situations where they may have to interrupt this high-performance forwarding mode to process special packets.

PATH MESSAGES

The RSVP Path message contains information in addition to the PHOP address, which characterizes the sender's traffic. These additional information elements are called the *Sender Template*, the *Sender TSpec*, and the *Adspec*.

The Path message is required to carry a Sender Template, which describes the format of data traffic the sender will originate. The Sender Template contains information called a Filter Spec (filter specification), which uniquely identifies the sender's flow from other flows present in the same RSVP session on the same link. The Path message must contain a Sender TSpec, which characterizes the traffic flow the sender will generate. The TSpec parameter characterizes the traffic the sender expects to generate; it is transported along the intermediate network nodes and received by the intended receiver(s). The Sender TSpec is not modified by the intermediate nodes.

Path messages may contain additional fragments of information contained in an Adspec. When an Adspec is received in a Path message by a node, it is passed to the local traffic-control process, which updates the Adspec with resource information and then passes it back to the RSVP process to be forwarded to the next downstream hop. The Adspec contains information required by receivers that allow them to choose a QoS control service and determine the appropriate reservation parameters. The Adspec also allows the receiver to determine whether a non-RSVP-capable node (a router) lies in the transit path or a specific QoS control service is available at each router in the transit path. The Adspec also provides default or service-specific information for the characterization parameters for the guaranteed service class.

Information also can be generated or modified within the network and used by the receivers to make reservation decisions. This information may include specifics on available resources, delay and bandwidth estimates, and various parameters used by specific QoS control services. This information also is carried in the Adspec object and is collected from the various nodes as it makes its way toward the receiver(s). The information in the Adspec represents a cumulative summary, computed and updated each time the Adspec passes through a node. The RSVP sender also generates an initial Adspec object that characterizes its QoS control capabilities. This forms the starting point for the accumulation of the path properties; the Adspec is added to the RSVP Path message created and transmitted by the sender.

As mentioned earlier, the information contained in the Adspec is divided into fragments; each fragment is associated with a specific control service. This allows the Adspec to carry information about multiple services and allows the addition of new service classes in the future without modification to the mechanisms used to transport them. The size of the Adspec depends on the number and size of individual per-service fragments included, as well as the presence of non-default parameters.

At each node, the Adspec is passed from the RSVP process to the traffic-control module. The traffic-control process updates the Adspec by identifying the services specified in the Adspec and calling each process to update its respective portion of the Adspec as necessary. If the traffic-control process discovers a QoS service specified in the Adspec that is unsupported by the node, a flag is set to report this to the receiver. The updated Adspec then is passed from the traffic-control process back to the RSVP process for delivery to the next node in the traffic path. After the RSVP Path message is received by the receiver, the Sender TSpec and the Adspec are passed up to the RSVP Application Programming Interface (RAPI).

The Adspec carries flag bits that indicate that a non-IS-aware (or non-RSVP-aware) router lies in the traffic path between the sender and receiver. These bits are called break bits and correspond to the NON_IS_HOP characterization parameter described earlier. A set break bit indicates that at least one node in the traffic path did not fully process the Adspec, so the remainder of the information in the Adspec is considered unreliable.

RESV MESSAGES

The Resv message contains information about the reservation style, the appropriate Flowspec object, and the Filter Spec that identify the sender(s). The pairing of the Flowspec and the Filter Spec is referred to as the *Flow Descriptor*. The Flowspec is used to set parameters in a node's packet-scheduling process, and the Filter Spec is used to set parameters in the packet-classifier process. Data that does not match any of the Filter Specs is treated as best-effort traffic.

Resv messages are sent periodically to maintain the reservation state along a particular traffic path. This is referred to as *soft state*, because the reservation state is maintained by using these periodic refresh messages.

Various bits of information must be communicated between the receiver(s) and intermediate nodes to appropriately invoke QoS control services. Among the data types that need to be communicated between applications and nodes is the information generated by each receiver that describes the QoS control service desired, a description of the traffic flow to which the resource reservation should apply (*Receiver TSpec*), and the necessary parameters required to invoke the QoS service (*Receiver RSpec*). This information is contained in the Flowspec (flow specification) object carried in the Resv messages. The information contained in the Flowspec object may be modified at any intermediate node in the traffic path because of reservation merging and other factors.

The format of the Flowspec depends on whether the sender is requesting controlled-load or guaranteed service. When a receiver requests controlled-load service, only a TSpec is contained in the Flowspec. When requesting guaranteed service, both a TSpec and an RSpec are contained in the Flowspec object. (The RSpec element was described earlier in relation to the guaranteed service QoS class.)

In RSVP version 1, all receivers in a particular RSVP session are required to choose the same QoS control service. This restriction is due to the difficulty of merging reservations that request different QoS control services and the lack of a service-replacement mechanism. This restriction may be removed in future revisions of the RSVP specification.

At each RSVP-capable router in the transit path, the Sender TSpecs arriving in Path messages and the Flowspecs arriving in Resv messages are used to request the appropriate resources from the appropriate QoS control service. State merging, message forwarding, and error handling proceed according to the rules defined in the RSVP specification. Also, the merged Flowspec objects arriving at each RSVP sender are delivered to the application, informing the sender of the merged reservation request and the properties of the data path.

OTHER MESSAGE TYPES

Aside from the Resv and Path messages, the remaining RSVP message types concern path and reservation errors (*PathErr* and *ResvErr*), path and reservation teardown (*PathTear* and *ResvTear*), and confirmation for a requested reservation (*ResvConf*).

The PathErr and ResvErr messages are simply sent upstream to the agent that created the request; they do not modify the path state in the nodes through which they pass. A PathErr message indicates an error in the processing of Path messages. The PathErr message is directed to the flow sender who requested the path. A ResvErr message indicates an error in the processing of the Resv message. The ResvErr message is directed to the flow receiver who requested the reservation.

RSVP teardown messages remove path or reservation state from nodes as soon as they are received. It is not always necessary to explicitly tear down an old reservation, however, because the reservation eventually times out if periodic refresh messages are not received. PathTear messages are generated explicitly by senders or by the time-out of path state in any node along the traffic path, and are sent to all receivers. An explicit PathTear message is forwarded downstream from the node that generated it; this message deletes path state and reservation state that may rely on it in each node in the traffic path. A ResvTear message is generated explicitly by receivers or any node in which the reservation state has timed out, and is sent to all pertinent senders. Basically, a ResvTear message has the opposite effect of a Resv message.

A ResvConf message is sent by each node in the transit path that receives a Resv message containing a reservation confirmation object. When a receiver wants to obtain a confirmation for its reservation request, it can include a confirmation request (RESV_CONFIRM) object in a Resv message. A reservation request with a Flowspec larger than any already in place for a session normally results in a ResvErr or a ResvConf message being generated and sent back to the receiver. Thus, the ResvConf message acts as an end-to-end reservation confirmation.

Merging

The concept of *merging* is essential to the interaction of multicast traffic and RSVP. Merging of RSVP reservations is required because of the method multicast uses for delivering packets: replicating packets that must be delivered to different next-hop nodes. At each replication point, RSVP must merge reservation requests and compute the maximum of their Flowspecs.

Flowspecs are merged when Resv messages, each originating from different RSVP receivers and initially traversing diverse traffic paths, converge at a merge point node and are merged prior to being forwarded to the next RSVP node in the traffic path (see Figure 4.11). The largest Flowspec from all merged Flowspecs—the one that requests the most stringent QoS reservation state—is used to define the single merged Flowspec, which is forwarded to the next-hop node. Because to RSVP Flowspecs are opaque data elements, the methods for comparing them are defined outside of the base RSVP specification.

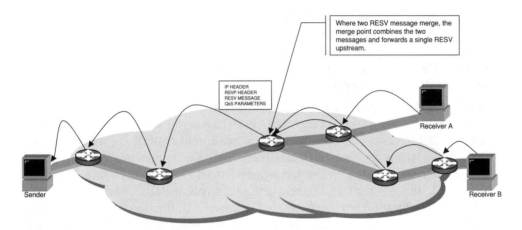

Figure 4.11 Flowspec merging.

As mentioned earlier, different reservation styles cannot be merged, because they are fundamentally incompatible. You can find specific ordering and merging guidelines for message parameters within the scope of the controlled-load and guaranteed service classes in RFC2211 and RFC2212, respectively.

Non-RSVP Clouds

RSVP can function across intermediate nodes that are not RSVP-capable. End-to-end resource reservations cannot be made, however, because non-RSVP-capable devices in the traffic path cannot maintain reservation or path state in response to the appropriate RSVP messages. Although intermediate nodes that do not run RSVP cannot provide these functions, they may have sufficient capacity to be useful in accommodating tolerant real-time applications.

Because RSVP relies on a discrete routing infrastructure to forward RSVP messages between nodes, the forwarding of Path messages by non-RSVP-capable intermediate nodes is unaffected. Recall that the Path message carries the IP address of the previous hop (PHOP) RSVP-capable node as it travels toward the receiver. As the Path message arrives at the next RSVP-capable node after traversing an arbitrary non-RSVP cloud, it carries with it the IP address of the previous RSVP-capable node. Therefore, the Resv message can be forwarded directly back to the next RSVP-capable node in the path.

Although RSVP functions across non RSVP-capable nodes, its use in such contexts may severely distort the QoS request by the receiver.

Low-Speed Links

Low-speed links, such as analog telephone lines, ISDN connections, and wireless systems present unique problems with regard to providing QoS, especially when multiple flows are present. It is problematic for a user to receive consistent performance, for example, when different applications are active at the same time, such as a Web browser, an FTP (File Transfer Protocol) transfer, and a streaming-audio application. Although the Integrated Services model is designed implicitly for situations in which some network traffic can be treated preferentially, it does not provide tailored service for low-speed links such as those described earlier.

One proposal submitted to the IETF's Integrated Services over Specific Link Layers (ISSLL) working group suggests the combination of enhanced, compressed, real-time transport protocol encapsulation, optimized header compression, and extensions to the Point-to-Point Protocol (PPP) to permit fragmentation and a method to suspend the transfer of large packets in favor of packets belonging to flows that require QoS services [ID-isl-slow]. The relationship of this proposal and the IETF Integrated Services model is outlined in a related proposal [ID-isl-svcmap].

Integrated Services and RSVP-over-ATM

There are functional disconnects between IP and ATM services, the least of which are their principal modes of operation: IP is nonconnection-oriented and ATM is connection-oriented. An obvious contrast is how each delivers traffic. In the base mode of operation IP is a best-effort datagram delivery service, whereas ATM has underlying technical mechanisms to provide differentiated levels of QoS for traffic on virtual connections. ATM uses point-to-point and point-to-multipoint VCs. Point-to-multipoint VCs allow nodes to be added and removed from VCs, providing a mechanism for supporting IP multicast.

---

*T*he issues discussed in this section are based on information from a collection of documents resulting from study conducted under the auspices of the IETF. These documents are referenced individually throughout this section. The discussion in this section is based on the ATM Forum Traffic Management Specification Version 4.0 [AF-tm].

Although several models exist for running IP-over-ATM networks, any one of these methods will function, as long as RSVP control messages (IP protocol 46) and data packets follow the same data path through the network [RFC1932]. The RSVP Path messages must follow the same path as data traffic so that path state may be installed and maintained along the appropriate traffic path. With ATM, this means that the ingress and egress points in the network must be the same in both directions (remember that RSVP is unidirectional only) for RSVP control messages and data.

Supporting documentation that addresses the interaction of ATM and upper-layer protocols can be complex and confusing, to say the least. And when ATM support for Integrated Services and RSVP are factored in, the result can be overwhelming.

Background

The technical specifications for running so-called Classical IP-over-ATM is detailed in RFC2225. Classical IP-over-ATM is based on the concept of a Logical IP Subnetwork (LIS), where hosts within an LIS communicate via the ATM network, and communication with hosts that reside outside the LIS must be through an intermediate router. Classical IP-over-ATM provides a method, called an ATM ARP (Address Resolution Protocol) server, for resolving IP host addresses to native ATM addresses. The ATM Forum offers similar methods for supporting IP-over-ATM in its Multi-Protocol-Over-ATM (MPOA) and LAN Emulation (LANE) specifications [AF-mpoa, AF-lane]. By the same token, IP multicast traffic and ATM interaction can be accommodated by a Multicast Address Resolution Server (MARS) [RFC2022].

The technical specifications for LANE, Classical IP, and Next-Hop Resolution Protocol (NHRP) discuss methods of mapping best-effort IP traffic onto ATM switched virtual connections (SVCs) [RFC2332]. However, when QoS requirements are introduced, the mapping of IP traffic becomes complicated by the need to differentiate the mapping into ATM virtual circuits by the requested IP packet's quality profile as well as the packet's route through the network. Ongoing research is necessary in this area to enable the complete integration of RSVP and ATM.

Using RSVP-over-ATM PVCs can be rather straightforward. ATM PVCs emulate dedicated point-to-point circuits in a network, so the operation of RSVP is no different than when it is implemented on any point-to-point network model using leased lines. Here the ATM PVC potentially supports a number of RSVP flows, where the characteristics of the individual flows are consistent with the service profile of the ATM PVC. The QoS of the PVCs, however, must be consistent with the Integrated Services classes being implemented, to ensure that RSVP reservations are handled appropriately in the

ATM network. Therefore, today, there is no apparent reason why RSVP cannot be successfully implemented in an ATM network that uses only PVCs.

Using SVCs in the ATM network is more problematic in general. (The exception is where SVCs are used to do auto-setup of PVCs, where the first packet sets up the SVC, and the circuit stays up indefinitely.) The complexity, cost, and efficiency of setting up SVCs can impact their benefit when used in conjunction with RSVP. Scaling issues may also be introduced when a single VC is used for each RSVP flow. The number of VCs in any ATM network is limited, so the number of RSVP flows that can be accommodated by any one device is depends entirely on the number of VCs available to a device.

The IP and ATM interworking requirements are compounded by VC management issues introduced in multicast environments. A primary concern in this regard is how to integrate the many-to-many connectionless features of IP multicast and RSVP to the one-to-many, point-to-multipoint, connection-oriented realm of ATM.

ATM point-to-multipoint VCs provide an adequate mechanism for dealing with multicast traffic. The ATM Forum 4.0 introduced a new concept called *Leaf Initiated Join* (LIJ), which allows an ATM end system to join an existing point-to-multipoint VC without necessarily contacting the source of the VC. This reduces the resource burden on the ATM source when setting up new branches, and it more closely resembles the receiver-based model of RSVP and IP multicast. Still, several scaling issues exist, and new branches added to an existing point-to-multipoint VC will end up using the extant QoS parameters as the existing branches, posing yet another problem. Consequently, a method must be defined to provide better handling of heterogeneous RSVP and multicast receivers with ATM SVCs.

There is a major difference in how ATM and RSVP QoS accomplish negotiation. ATM is sender-oriented and RSVP is receiver-oriented. At first glance, this might appear to be a major discrepancy. In practice, RSVP receivers actually determine the QoS required by the parameters included in the sender's TSpec, which is included in received Path messages. Thus, while the resources in the network are reserved in response to receiver-generated Resv messages, the resource reservations actually are initiated by the sender. This means that senders will establish ATM QoS VCs, and receivers must accept incoming ATM QoS VCs. This is consistent with how RSVP operates, and allows senders to use different RSVP flow-to-VC mappings for initiating RSVP sessions.

Efforts have been made to provide QoS by using *traditional* IP (an IP network's undifferentiated best-effort service response), plus underlying ATM mechanics to provide specific service responses. Some of these same efforts have driven study at the IETF and elsewhere to develop methods for using RSVP and Integrated Services with ATM to augment the existing service model. The integration of ATM and the Integrated Services model is important in two primary areas: QoS translation, or mapping between Integrated Services and ATM, and VC management, which deals with establishing VCs and the traffic flows that are forwarded over the VCs. An ATM edge device (router) in an IP network must provide IP and ATM interworking functions, servicing the requirements of each network (see Figure 4.12). In the case of RSVP, it must be able to process RSVP control messages, reserve resources, maintain soft state, and provide packet scheduling and classification services. It also must be able to initiate, accept, or refuse ATM connections via UNI signaling. Combining these capabilities, the edge device also must translate RSVP reservation semantics to the appropriate ATM VC establishment parameters.

The task of providing a translation mechanism between the Integrated Services controlled-load and guaranteed services and appropriate ATM QoS parameters is a complex problem with many facets, as stated in RFC2381. This document provides a proposal for a mapping of both the controlled-load and guaranteed service classes, as well as best-effort traffic to the appropriate ATM QoS services.

As shown in Figure 4.13, the mapping of the Integrated Services QoS classes to ATM service categories appears to be straightforward. The ATM CBR and rt-VBR service categories possess characteristics that make them a reasonable match for guaranteed service, whereas the nrt-VBR and ABR

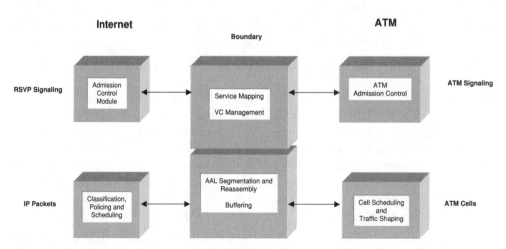

Figure 4.12 ATM edge function mappings.

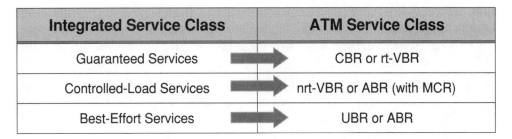

Integrated Service Class	ATM Service Class
Guaranteed Services	CBR or rt-VBR
Controlled-Load Services	nrt-VBR or ABR (with MCR)
Best-Effort Services	UBR or ABR

Figure 4.13 Integrated Services and ATM QoS mapping.

(albeit with an MCR) service categories have characteristics that are the most compatible with the controlled-load service. Best-effort traffic fits well into the UBR service class.

The practice of tagging nonconformant cells with CLP = 1, which designates cells as lower priority, can have a special use with ATM and RSVP. As outlined previously, you can determine whether cells are tagged as conformant by using a Generic Cell Rate Algorithm (GCRA), a leaky-bucket algorithm. Also recall that traffic in excess of controlled-load or guaranteed service specifications must be transported as best-effort traffic. Therefore, dropping cells with the CLP bit set should be implemented with excess guaranteed service or controlled-load traffic. Of course, this is an additional nuance you should consider in an ATM/RSVP interworking implementation.

Several ATM QoS parameters exist for which there are no IP layer equivalents. Therefore, these parameters must be configured manually as a matter of local policy. Among these parameters are Cell Loss Ratio (CLR), Cell Delay Variation (CDV), Severely Errored Cell Block Ratio (SECBR), and Cell Transfer Delay (CTD).

The following section briefly outlines the mapping of the Integrated Services classes to ATM service categories, and it is discussed in more detail in RFC2381.

Guaranteed Service and ATM Interaction

The Guaranteed Service class requires reliable delivery of traffic with as little delay as possible. Thus, an ATM service category that accommodates applications with end-to-end time synchronization also should accommodate guaranteed service traffic, because the end-to-end delay should be closely monitored and controlled. The nrt-VBR, ABR, and UBR service categories, therefore, are not good candidates for guaranteed service traffic. The CBR service category provides ideal characteristics in this regard, but it does not adapt to changing data rates, and can leave large portions of bandwidth under-utilized. In most scenarios, CBR proves to be a highly inefficient use of available network resources.

Apparently, rt-VBR is the most appropriate ATM service category for guaranteed service traffic because of its inherent adaptive characteristics. Choosing this service category requires that two specified rates be quantified: the sustained cell rate (SCR) and peak cell rate (PCR). These two parameters provide a burst and average tolerance profile for traffic with bursty characteristics. The rt-VBR service also should specify a low enough CLR for guaranteed service traffic so that cell loss is minimized.

Earlier iterations of RFC2381 suggested that the ATM traffic descriptor values for PCR, SCR, and MBS should be set within the following bounds when mapping guaranteed service onto an rt-VBR VC:

$R <= PCR <=$ minimum p or minimum line rate

$r <= SCR <= PCR$

$0 <= MBS <= b$

where

$R = RSpec$

$p = $ peak rate

$r = $ Receiver TSpec

$b = $ bucket depth

In other words, the RSpec should be less than or equal to the PCR, which in turn should be less than or equal to the minimum peak rate, or alternatively, the minimum line rate. The Receiver TSpec (assumed here to be identical to the Sender TSpec) should be less than or equal to the SCR, which in turn should be less than or equal to the PCR. The MBS should be greater than or equal to zero (generally, greater than zero, of course) but less than or equal to the leaky-bucket depth defined for traffic shaping.

Controlled-Load Service and ATM Interaction

Of the three remaining ATM service categories (nrt-VBR, ABR, and UBR), only nrt-VBR and ABR are viable candidates for the controlled-load service. The UBR service category does not possess strict enough traffic capabilities for the controlled-load service. UBR does not provide any mechanism to allocate network resources, which is the goal of the controlled-load service class. Recall that traffic appropriate for the controlled-load service is characterized as tolerant real-time applications that are somewhat forgiving of packet loss but still require reasonably efficient performance.

The ABR service category aligns reasonably well with the model for controlled-load service, which is characterized as a variable service with base minimum service requirements. Therefore, if the ABR service class is used for controlled-load traffic, it requires that a Minimum Cell Rate (MCR) be specified to provide a lower bound for the data rate. The TSpec rate should be used to determine the MCR. The interaction of the inner ABR control loop and the outer TCP control loop make the use of ABR in a network with TCP traffic somewhat inefficient. This has led to consideration of CBR and nrt-VBR as better service substrates for controlled-load network services.

The nrt-VBR service category is potentially a better match for controlled-load traffic. However, the Maximum Cell Transfer Delay (maxCTD) and Cell Delay Variation (CDV) parameters must be chosen for the edge ATM device. This is done manually as a matter of policy.

Earlier iterations of RFC2381 suggested that the ATM traffic descriptor values for PCR and MBS should be set within the following bounds when mapping controlled-load service onto an nrt-VBR VC:

$r <= SCR <= PCR <=$ minimum p or minimum line rate

$0 <= MBS <= b$

where

p = peak rate

r = Receiver TSpec

b = bucket depth

Integrated Services, Multicast, and ATM

For RSVP to be useful in multicast environments, flows must be distributed appropriately to multiple destinations from a given source. Both the Integrated Services model and RSVP support the idea of heterogeneous receivers. Conceptually, not all receivers of a particular multicast flow are required to ask for the same QoS parameters from the network. The following example explains why this concept is important. When multiple senders exist on a shared-reservation flow, enough resources can be reserved by a single receiver to accommodate all senders in the shared flow. This is problematic in conjunction with ATM. If a single maximum QoS is defined for a point-to-multipoint VC, network resources could be wasted on links where the reassembled packets eventually will be dropped. Furthermore, a maximum QoS may impose a degradation in the service provided to the best-effort branches. In ATM networks, additional end points of a point-to-multipoint VC must be set up explicitly. Because ATM does not currently provide support for this situation, any RSVP-over-ATM implementations must make special provisions to handle heterogeneous receivers.

It has been suggested that RSVP heterogeneity can be supported over ATM by mapping RSVP reservations onto ATM VCs by using one of four methods proposed in RFC2382. This proposal is discussed briefly here.

In the *full heterogeneity* model, a separate VC is provided for each distinct multicast QoS level requested, including requests for best-effort traffic.

In the *limited heterogeneity* model, each ATM device participating in an RSVP session would require two VCs: one point-to-multipoint VC for best-effort traffic and one point-to-multipoint QoS VC for RSVP reservations. Both these approaches require what could be considered inefficient quantities of network resources. Though the full heterogeneity model can provide users with the QoS they require, it makes the most inefficient use of available network resources. The limited heterogeneity model requires substantially fewer network resources, but it is still somewhat inefficient, because packets must be duplicated at the network layer and sent on two VCs.

The third model is a *modified homogeneous* model. In a homogeneous model, all receivers—including best-effort receivers—on a multicast session use a single QoS VC that provides a maximum QoS service that can accommodate all RSVP requests. This model most closely matches the method by which the RSVP specification handles heterogeneous requests for resources. Although this method is the simplest to implement, it may introduce a couple of problems. One is that users who expect to make a small or no reservation may end up not receiving any data at all, to include best-effort traffic; their request may be rejected because of insufficient resources in the network. The modified homogeneous model proposes that special handling be added to address the situation in which a best-effort receiver cannot be added to the QoS VC, by generating an error condition that triggers a request to establish a best-effort VC for the appropriate receivers.

The fourth model, the *aggregation* model, proposes that a single, large point-to-multipoint VC be used for multiple RSVP reservations. This model is attractive for a number of reasons, primarily because it solves the inefficiency problems associated with full heterogeneity, because concerns about induced latency imposed by setting up an individual VC for each flow are negated. The primary problem with the aggregation model is that it may be difficult to determine the maximum QoS for the aggregate VC.

The term *variegated VCs* has been coined to describe point-to-multipoint VCs that allow a different QoS on each branch [RFC2382]. However, cell-drop mechanisms require further research to retain the best-effort delivery characterization for nonconformant packets that traverse certain branch topologies. Implementations of Early Packet Discard (EPD) should be deployed in these situations so that all cells belonging to the same packet can be discarded— instead of only a few arbitrary cells from several packets, making them useless to their receivers.

Another issue of concern is that IP-over-ATM currently uses a multicast server or reflector that can accept calls from multiple senders and redirect them to a set of senders through the use of point-to-multipoint VCs. This moves the scaling issue from the ATM network to the multicast server. However, the multicast server needs to know how to interpret RSVP messages to enable VC establishment with the appropriate QoS parameters.

Integrated Services over Local Area Network Media

The Integrated Services discussions up to this point have focused on two basic network entities: the host and the intermediate nodes or routers. There may be many cases in which an intermediate router does not lie in the end-to-end path of an RSVP sender and receiver; or perhaps several link-layer bridges or switches may lie in the data path between the RSVP sender or receiver and the first intermediate IS-aware router. Because LAN technologies, such as Ethernet and Token Ring, typically constitute the last-hop link-layer media between the host and the wide area network, it is interesting to note that, currently, no standard mechanisms exist for providing service guarantees on any of these LAN technologies. Given this consideration, link-layer devices such as bridges and LAN switches may need a mechanism that provides customized admission-control services to provide support for traffic that requests Integrated Services QoS services. Otherwise, delay bounds specified by guaranteed services end systems may be impacted by intermediate link-layer devices that are not IS-aware, and controlled-load services may not prove to be better-than-best-effort.

The concept of a subnetwork *Bandwidth Manager* first was described in [ID-isl802-frame] and provides a mechanism to accomplish several things on a LAN subnet that otherwise would be unavailable. Among these are admission control, traffic policing, flow segregation, packet scheduling, and the capability to reserve resources (to include maintaining soft state) on the subnet.

The model of operation is fairly straightforward. The Bandwidth Manager model consists of two major components: a requester module (RM) and a bandwidth allocator (BA), as illustrated in Figure 4.14. The RM resides in every end system that resides on the subnetwork, and provides an interface between the higher-layer application (which can be assumed to be RSVP) and the Bandwidth Manager. For the end system to initiate a resource reservation, the RM is provided with the service desired (guaranteed or controlled-load); the traffic descriptors in the TSpec and the RSpec define the amount of resources requested. This information is extracted from RSVP Path and Resv messages. The Subnet Bandwidth Manager concept is expanded and described in more detail in the following section.

The Subnet Bandwidth Manager

The concept of the Subnet Bandwidth Manager (SBM) is defined further in [ID-isl802-sbm], which is a proposal for a standardized signaling protocol for LAN-based admission control for RSVP flows on IEEE 802-style LANs. The SBM proposal suggests that this mechanism, when combined with per-flow policing on the end systems and traffic control and priority queuing at the link layer, will provide a close approximation of the controlled-load and guaranteed services. In the absence of any link-layer traffic controls or priority-queuing mechanisms in the LAN infrastructure (for example, on a shared media LAN), the SBM mechanism limits only the total amount of traffic load

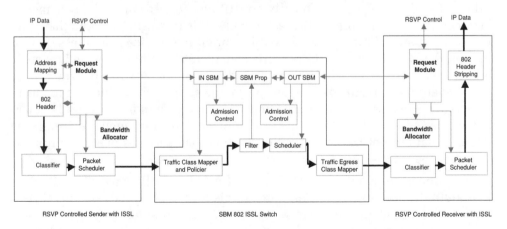

Figure 4.14 Model of operation of Integrated Services over a LAN.

imposed by RSVP-associated flows. In environments of this nature, no mechanism is available to separate RSVP flows from best-effort traffic. This brings into question the usefulness of using the SBM model in a LAN infrastructure that does not support the capability to forward packets tagged with an IEEE 802.1p priority level.

In each SBM-managed segment, a single SBM is designated to be the Designated Subnet Bandwidth Manager (DSBM) for the managed LAN segment. The DSBM is configured with information about the maximum bandwidth that can be reserved on each managed segment under its control. Although this information most likely will be statically configured, future methods may dynamically discover this information. When the DSBM clients come online, they attempt to discover whether a DSBM exists on each of the segments to which they may be attached. This is done through a dynamic DSBM discovery-and-election algorithm, which is described in Appendix A of [ID-isl802-sbm]. If the client itself is capable of serving as a DSBM, it may choose to participate in the election process.

When a DSBM client sends or forwards a Path message over an interface attached to a managed segment, it sends the message to its DSBM instead of to the RSVP session destination address, as is done in conventional RSVP message processing. After processing, and possibly updating the Adspec, the DSBM forwards the Path message to its destination address. As part of its processing, the DSBM builds and maintains a Path state for the session and notes the previous hop (PHOP) of the node that sent the message. When a DSBM client wants to make a reservation for an RSVP session, it follows the standard RSVP message-processing rules and sends an RSVP Resv message to the corresponding PHOP address specified in a received Path message. The DSBM processes received Resv messages based on the bandwidth available and returns ResvErr messages to the requester if the request cannot be granted. If sufficient resources are available, and the reservation request is granted, the DSBM forwards the Resv message to the PHOP based on the local Path state for the session. The DSBM also merges and orders reservation requests in accordance with traditional RSVP message-processing rules.

In the example in Figure 4.15, an "intelligent" LAN switch is designated as the DSBM for the managed segment. The "intelligence" is only abstract, because all that is required is that the switch implement the SBM mechanics as defined in [ID-isl802-sbm]. As the DSBM client, host A, sends a Path message upstream, it is forwarded to the DSBM, which in turn forwards it to its destination session address, which lies outside the link-layer domain of the managed segment. The Resv message processing occurs in exactly the same

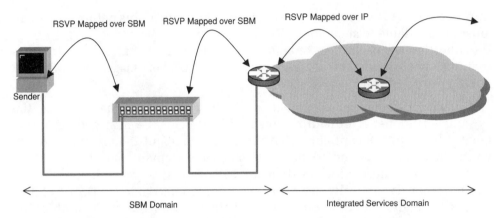

Figure 4.15 SBM-managed LAN segment: forwarding RSVP messages.

order, following the Path state constructed by the previously processed Path message.

The inclusion of the DSBM for admission control for managed segments results in some additions to the RSVP message-processing rules at a DSBM client. Where a DSBM needs to forward a Path message to an egress router for further processing, the DSBM may not have the Layer 3 routing information available to make the necessary forwarding decision when multiple egress routers exist on the same segment. Therefore, specific RSVP objects have been proposed, called LAN_NHOP (LAN Next Hop) objects, to keep track of the Layer 3 hop as the Path message traverses a Layer 2 domain between two Layer 3 devices.

When a DSBM client sends out a Path message to its DSBM, it must include LAN_NHOP information. In the case of unicast traffic, the LAN_HOP address indicates the destination address or the IP address of the next-hop router in the path to the destination. As a result, when a DSBM receives a Path message, it can look at the address specified in the LAN_NHOP object and forward the message to the appropriate egress router. However, because the link-layer devices (LAN switches) must act as DSBMs, the level of intelligence of these devices may not include an ARP capability that enables it to resolve media access control (MAC) addresses to IP addresses. For this reason, [ID-isl802-sbm] requires that LAN_HOP information contain both the IP address (LAN_NHOP_L3) and corresponding MAC address (LAN_NHOP_L2) for the next Layer 3 device.

Because the DSBM may not be able to resolve IP addresses to MAC addresses, a mechanism is needed to dispense with this translation requirement when processing Resv messages. Therefore, the RSVP_HOP_L2 object is used to indicate the Layer 2 MAC address of the previous hop. This provides a mechanism for SBM-capable devices to maintain the Path state necessary to accommodate forwarding Resv messages along link-layer paths that cannot provide IP-address-to-MAC-address resolution.

At least one new RSVP object has been proposed, called the TCLASS (Traffic Class) object. The TCLASS object is used with IEEE 802.1p *user-priority* values, which can be used by Layer 2 devices to discriminate traffic based on these priority values. These values are discussed in more detail in the following section. The priority value assigned to each packet is carried in the new extended frame format defined by IEEE 802.1Q [IEEE-802.1Q]. As an SBM Layer 2 switch, which also functions as an 802.1p device, receives a Path message, it inserts a TCLASS object. When a Layer 3 device (a router) receives Path messages, it retrieves and stores the TCLASS object as part of the process of building Path state for the session. When the same Layer 3 device needs to forward a Resv message back toward the sender, it must include the TCLASS object in the Resv message.

The Integrated Services model is implemented via an SBM client in the sender, as shown in Figure 4.14. This figure also shows the SBM implementation in a LAN switch. The components of this model are defined in the following summary [ID-isl802-map]:

Local admission control. One local admission control module on each switch port manages available bandwidth on the link attached to that port. For half-duplex links, this involves accounting for resources allocated to both transmit and receive flows.

Input SBM module. One instance per port. This module performs the *network* portion of the client-network peering relationship. This module also contains information about the mapping of Integrated Service classes to IEEE 802.1p *user_priority*, if applicable.

SBM propagation. Relays requests that have passed admission control at the input port to the relevant output port's SBM module(s). This requires access to the switch's forwarding table and port spanning-tree states.

Output SBM module. Forwards messages to the next Layer 2 or Layer 3 network hop.

Classifier, queuing, and scheduler. The classifier function identifies the relevant QoS information from incoming packets and, with information contained in the normal bridge forwarding database, uses this to determine to which queue of the appropriate output port to direct the packet for transmission. The queuing and scheduling functions manage the output queues and provide the algorithmic calculation for servicing the queues to provide the promised service (controlled-load or guaranteed).

Ingress traffic class mapper and policing. This optional module may check whether the data in the traffic classes conforms to specified behavior. The switch may police this traffic and remap to another class or discard the traffic altogether. The default behavior should be to allow traffic through unmodified.

Egress traffic class mapper. This optional module may apply remapping of traffic classes on a per-output port basis. The default behavior should be to allow traffic through unmodified.

Observations on the Integrated Services Architecture

It has been suggested that the Integrated Services architecture and RSVP are excessively complex and possess poor scaling properties. This criticism undoubtedly has been prompted by the underlying complexity of the signaling requirements. Conversely, it can be said that RSVP is no more complex than some of the more advanced routing protocols, such as BGP. An alternative viewpoint might be that the underlying complexity is required because of the inherent difficulty in establishing and maintaining path and reservation state information along the transit path of data traffic. The observation that RSVP has poor scaling properties does, however, deserve further examination, because deployment of RSVP has not been widespread enough to determine the scope of this assumption. There is some interest in exploring what can be removed—in the style of an "RSVP-lite"—from RSVP, such as RSVP merge states, to address some of these issues.

As discussed in RFC2208, there are several areas of concern about the wide-scale deployment of RSVP. With regard to RSVP scalability, the resource requirements (computational processing and memory consumption) for running RSVP on routers increase in direct proportion to the number of separate RSVP reservations, or sessions, accommodated. Therefore, supporting a large number of RSVP reservations could introduce a significant negative impact on router performance. Router-forwarding performance may also be impacted adversely by the packet-classification and scheduling mechanisms intended to provide differentiated services for reserved flows. IntServ's guaranteed services pose some challenges to the queuing mechanisms, where the requirement to allocate absolute levels of egress bandwidth to individual flows, while still supporting an unmanaged low priority best-effort traffic class, call for layered queuing disciplines to achieve the correct packet scheduling behavior. These scaling concerns tend to suggest that organizations with large, high-speed networks will be reluctant to deploy RSVP in the foreseeable future, at least until these concerns are addressed. The underlying implication of these concerns is that, without deployment by Internet service providers, who own and maintain the high-speed backbone networks in the Internet, deployment of pervasive RSVP services in the Internet will not be forthcoming.

Several interesting proposals have been submitted to the IETF for further study that suggest a rationale for grouping similar guaranteed service flows to reduce the bandwidth requirements that each flow might consume individually. These proposals do not necessarily suggest an explicit implementation method to provide this grouping, but instead offer the reasons for identifying identical guaranteed service flows in an effort to group them. Each of these proposals suggests, offhand, that some sort of tunneling mechanism could be used to transport flow groups from one intermediate node to another, which could conceivably reduce the amount of bandwidth required in the nodes through which a flow group tunnel passes. Although well-intentioned, the obvious flaw in this proposal is that it only partially addresses the scaling problems introduced by the Integrated Services model. The flow group still must be policed at each intermediate node to provide traffic-conformance monitoring, and path and reservation state still must be maintained at each intermediate node for individual flows.

On a slightly more contentious note, the benefits of resource reservation may not outweigh the associated operational cost of supporting such a service. The distinguishing factor of this form of soft state service platforms, when compared to a stricter form of time division multiplexing (TDM), is the assumption of the presence of elastic flows that can make use of any available transmission resource, as a low-priority *filler* flow set. The thinking is that by admitting a greater total payload volume into the network through such mixing of elastic and inelastic flows, the network could operate within an improved level of cost efficiency. Within the Internet, such elastic applications are typically managed through TCP.

As examined in Chapter 2, "The Performance Toolkit," TCP uses a flow control algorithm whereby the flow rate is varied using a timebase of the end-to-end round-trip time. If the availability timescale of resources within the idle space of the inelastic real-time applications is less than some multiple of the average TCP round-trip times, then TCP will be relatively ineffectual in improving the total traffic throughput of the multiservice network (see Figure 4.16). The next question to ask is whether the deployment of more conventional TDM-based systems may provide a more efficient outcome. This remains an open issue for many Internet service providers. For those who already operate a TDM-based network, such as a telephony service provider, the advantages of moving the entire set of TDM-based applications to an IntServ platform may be questionable. The optimal approach may be to shift only voice-signaling systems over to the IP platform, while retaining the circuit-switched platform to carry the real-time data flows within a strictly time-synchronized carriage environment. For the more recent entrants to this industry, the initial capital costs of establishing a TDM-based platform may be prohibitive. For them, the question is whether it is necessary to use ATM as an intermediate switching level, above the basic transmission plant, in order to provide an

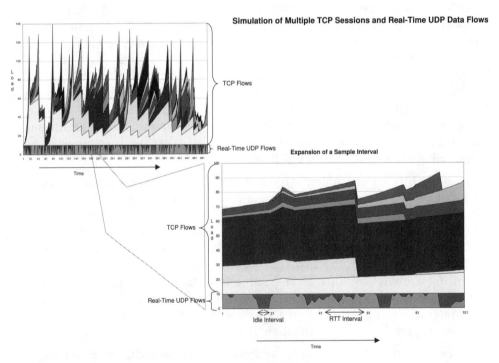

Figure 4.16 Mixing real-time and TCP data flows.

approximation of TDM channels, or whether the IntServ approach can provide sufficient functionality and reliability to allow the elimination of a second switching level within the carriage platform.

Another important concern expressed in RFC2208 deals with policy-control issues and RSVP. Policy control addresses the question who is authorized to make reservations and provisions to support access control and accounting. Although the current RSVP specification defines a mechanism for transporting policy information, it does not define the policies themselves, because the policy object is treated as an opaque element. Some vendors have indicated that they will use this policy object to provide proprietary mechanisms for policy control. At the time of this writing, the IETF RSVP working group has been chartered to develop a simple policy-control mechanism to be used in conjunction with RSVP. Ongoing work on this issue continues at the IETF. Several mechanisms already have been proposed to deal with policy issues, in addition to the aforementioned vendor-proprietary policy-control mechanisms. In fact, a working group (called the RSVP Admission Policy—RAP group) has been formed at the IETF to address policy issues in conjunction with RSVP.

The key recommendation contained in RFC2208 is that, given the current form of the RSVP specification, multimedia applications within smaller, private networks are the most likely to benefit from the deployment of RSVP. The inadequacies of RSVP scaling and lack of policy control may be more manageable within the confines of the smaller, more controlled network environment than in the expanse of the global Internet. It certainly is likely that RSVP will provide genuine value and find legitimate deployment uses in smaller networks, both in the peripheral Internet networks and in the private arena, where these issues of scale are far less important. Therein lies the key to successfully delivering quality of service using RSVP. After all, the purpose of the Integrated Services architecture and RSVP is to provide a method to offer quality of service, not to degrade the service quality.

The Differentiated Services Architecture

In mid-1997, upon the completion of the RSVP protocol specification, many Internet service providers expressed concern to the IETF over their difficulty in effectively, and pragmatically, deploying QoS services based on the Integrated Services model. These concerns revolved around many issues discussed in RFC2208. The consensus was that large-scale deployment of RSVP in the global Internet was impractical, and that a simpler and more scalable solution was needed—one that provided simple differentiation of traffic, while not requiring per-flow resource reservation state, which consumes critical network processing resources.

In the early stages of this discussion, two papers were published that laid the groundwork for later differentiated services specifications [Nichols 1997, Clark 1997]. Both of these documents began to characterize the environments in which differentiated services might be deployed, based on the premise that IP packets could be individually "marked" as to their relative priority, such that each packet could be afforded differentiated service on a hop-by-hop basis. This concept proves to possess significant scaling properties; in contrast, maintaining individual end-to-end flow and path state at each intermediate hop sets scaling restrictions and raises operational issues for large service networks.

It is said that there is nothing new under the sun, and the reader may wonder if this concept of per-packet marking with intended service behaviors is the same as that of the per-packet marking defined in the original intended use of the Type of Service field RFC791, RFC1349]. In this aspect, the mechanics of the Differentiated Services architecture (DiffServ) are little different from the mechanics of the Type of Service field. However, the DiffServ effort has undertaken significant development in the area of admission control mechanisms, and this is where the real value of the DiffServ approach lies. An overloaded service class will always yield poor service outcomes. Service quality is the result of both the definition of various service classes and the associated management of load to ensure that the imposed load lies within the capability of the service class. It is on this foundation that DiffServ has been constructed.

The approach adopted by the Differentiated Services architecture (DiffServ) is to divide the problem into a number of component tasks, so that each task can be addressed relatively independently of the others. We have already seen this approach used to decouple the task of packet forwarding from the actions of routing protocols. In the DiffServ architecture, the per-hop service behaviors are decoupled from the admission control and policy administration tasks.

The objective of the DiffServ architecture is quite straightforward: to define simple and coarse methods of providing differentiated classes of service for Internet traffic. There is no intent to make DiffServ networks respond to individual application sessions (or *micro-flows*, to use the terminology adopted by DiffServ). Instead, the DiffServ architecture is designed to offer service to *aggregate* service classes, where a number of individual session flows are grouped and treated consistently by the network. The DiffServ architecture is also designed to support various types of applications and specific business requirements for network administrators and Internet service providers who want to differentiate traffic based on locally generated traffic priorities. The DiffServ approach to providing QoS in networks employs a small, well-defined set of building blocks from which a variety of services may be built.

Traffic entering a DiffServ network is first *classified*, then passed through some form of *admission filter*, intended to shape (or *condition*) the traffic to meet the policy requirements associated with the classification. The conditioned traffic stream is then assigned to a particular *behavior aggregate*, by marking the IP header DS field of the component packets with the appropriate *Differentiated Services Code Point* (DSCP). When passed through the DiffServ network, this DSSP value triggers a selected per-hop behavior (PHB) from all interior active elements of the network. This architecture attempts to push the more intensive processing load of traffic classification and profiling to the

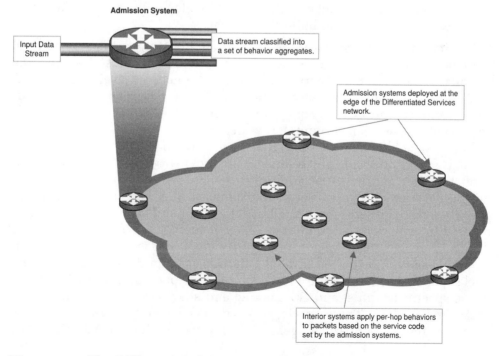

Figure 4.17 The Differentiated Services architecture.

edge of the network, and to perform this function upon ingress and potentially at egress. Within the interior of the network, all active network elements have to respond to the service marking contained in the packet's DiffServ header field, and to apply a service response that is consistent with this marking. No further classification or profiling is performed in the interior of the network (see Figure 4.17).

Three important components comprise this architecture: the initial classification and conditioning of traffic entering the network, the assignment of a DSCP value to the IP header DS field, and the subsequent PHBs that are triggered within the network by the DS field value as the packet is passed through the network.

Not only is it important to distinguish between a packet's DSCP value and the network's PHBs, it is also important to distinguish between *services* and PHBs. The Differentiated Services architecture document defines a service as "the overall treatment of a defined subset of a customer's traffic within a DS-domain, or end-to-end" [RFC2475]. In other words, the service is some quantitative treatment in the end-to-end network path between the sender and receiver of traffic. The DiffServ architecture is adamant about not defining the services; the service operators are responsible for doing so. What the architecture does define is the syntax of the DS field codepoints and a collection of PHBs, which can be used to build these services.

Another important aspect of the DiffServ architecture is the service level agreement, the SLA, which is negotiated between the subscriber and the provider. The broader category of SLAs include network availability assurances, frame (Frame-Relay services) or cell (ATM services) delivery assurances, and so forth. The notion of an SLA is not new; indeed, these agreements have been around for years. However, without an SLA that describes which DiffServ services the service provider will be delivering (or perhaps more important, which services the provider will not be delivering or cannot guarantee), the expectation of differentiated services may be sorely misplaced. Therefore, a subset of the standard SLA has been defined that describes the DiffServ components of the standard SLA. This subset is referred to as the *Traffic Conditioning Agreement* (TCA). The TCA specifies detailed service parameters for each service level, and they may include (but are not limited to) expected throughput, drop probability, or latency, token-bucket parameters that define a service profile, disposition of traffic in excess of the specified profile, and DS marking and shaping services preferred for each service class. These service classes may be defined by the service provider, the subscriber, or, generally, via negotiation of the SLA/TCA between both parties.

The DiffServ architecture is designed to provide a framework where network designers and Internet Service Providers can offer each subscriber a range of network services differentiated on the basis of performance, in addition to subscriber preferences and pricing tiers that may have been previously used. Subscribers can request a specific performance level on a packet-by-packet basis by marking the DS field of each packet with a specific value. These values specify the PHB to be designated to each packet within the service provider's network. Typically, the subscriber and the ISP negotiate some sort of policing profile that describes the rate at which traffic can be accepted by the network within each service level (packets that are transmitted in excess of this profile may not be allocated the service level requested by the sender. A principal design goal of DiffServ is its scalability, which allows it to be deployed in very large networks. This scalability is achieved by forcing as much complexity out of the core of the network and into edge devices, which process lower volumes of traffic and fewer numbers of flows.

The DiffServ framework and DiffServ architecture are both described in further detail in the documents [ID-diffserv-frame] and RFC2475, respectively.

DiffServ Network Elements

We have already discussed the concepts of traffic classification and traffic conditioning in some amount of detail. These principles are extraordinarily important in a DiffServ context, since packets must be classified, metered, policed (conditioned), marked, and forwarded at the network ingress point according to the TCA negotiated between the network service provider and the network subscriber. That said, however, one or more of these functions may be done at other nodes in the network, such as at interior DiffServ nodes in the network path, or at the network egress point, when crossing network domain boundaries. As we describe the DiffServ elements, you will see why these functions are critical to the success of differentiated services in the network.

The DS Field

The header field used in the DiffServ architecture is based on reclaiming the Type of Service (TOS) 8-bit field within the IPv4 header or the Traffic Class field in the IPv6 header. The value in this field is interpreted by DiffServ routers within the network to indicate a particular forwarding treatment, or per-hop behavior (PHB), at each network node in the traffic path. This field is known as the DS field, DS referring to the role as DiffServ indicator, as illustrated in Figure 4.18. The definition and format of the DS field is described in RFC2474.

IP Header

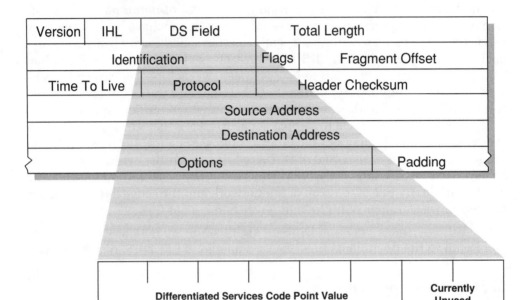

Figure 4.18 The DS field.

The first 6 bits of the DS field contain the Differentiated Services codepoint (DSCP), which is mapped to the PHB that a packet will receive as it transits each node. The mapping between DSCPs and PHBs is not necessarily one to one. A service provider may use a three-level priority PHB set, and map the eight Class Selector DSCPs into these three PHBs. The last 2 bits in this field are currently unused (CU), and must be ignored by Differentiated Services network elements (the Explicit Congestion Notification, or ECN, proposal has a potential use for these 2 bits: as an ECN indicator and ECN congestion-experienced flag). It is also important to note that the structure of the DS field is incompatible with the existing definition of the IPv4 TOS field; although the functionality is similar, the detailed bit boundaries and semantics are quite different.

CODEPOINTS AND PER-HOP BEHAVIORS

There is an important distinction between the PHBs themselves and the corresponding packet *codepoints*, or rather, the values contained with the DSCP field. The PHB is the DiffServ behavior expected at each node—the treatment afforded to packets at each intermediate device as they pass through the network. The packet DSCP, on the other hand, is simply the value contained within the first 6 bits of the DS byte; it indicates a particular PHB that should be given to the corresponding packet. Therefore, packet codepoints are the markings a packet will receive to indicate how it will be treated by each DiffServ-aware node in the traffic path. The PHB is fairly self-explanatory; it is the behavior given to a particular *DS-coded* packet (based on the packet DSCP), and can be based on any number of criteria (which we will discuss at length). A default PHB, correlating to standard best-effort forwarding, must be supported by all DS-capable devices. The DSCP that maps to the default PHB is all zeros in the DSCP field. At present, there are three sets of PHBs specified by the IETF DiffServ working group (aside from the default PHB): the Class Selector PHB (CS) [RFC2474], the Expedited Forwarding (EF) PHB [RFC2598], and the Assured Forwarding (AF) PHB [RFC2597]. We will examine all three in more detail. By saying there are three does not mean that there will not be other PHBs specified; it only means that, currently, three PHBs have been specified at the time of this writing.

At this juncture, it is important to point out two critical changes in the definition of the DS field: Firstly, whereas all other IP header fields have universal semantics (meaning that a value in the field is interpreted uniformly by an Internet network element), the DS field has values that are local to each network service provider. One service provider may elect to associate a given DSCP value with a low-delay PHB, while another provider may elect to use the same DSCP to select a low-drop precedence PHB. Each service provider can select up to 64 PHBs to map to DSCP values. (But note, there may well be many more potential PHBs than the 64 mappable PHBs. The full encoding of PHBs is described using a 32-bit value for each.) Each service provider may use a number of *default* PHB-to-DSCP mappings, and augment them with a number of individually selected PHBs with a custom mapping to DSCPs.

Secondly, this field is not end-to-end transparent, nor is it symmetric. This implies that the DSCP value set by the sender may not be the same as the received DSCP value, nor will any response packets have a corresponding DSCP value. The network is allowed to alter the DSCP value in accordance with the network's resource management policies.

The DiffServ architecture is quite explicit on where the DS field is marked: upon ingress to the network, by the network. This is a significant shift in the model from the earlier interpretation of the role of the IP TOS header, which was to be set by the end host, and carried, without alteration, by the network on an end-to-end basis. Within the DiffServ architecture, the DS field may be re-marked each time the packet passes from one network domain to the next. In terms of the end-to-end model, a sender cannot assume that the DSCP value placed in a packet will be the same as that received at the destination. There is no end-to-end transparency of the original value of the DS field. Even allowing for various re-marking of the DS field values, it could be assumed that the PHB selected by the original DSCP value will be carried end to end. Again, this is not the case. At a network boundary, the admission classification of the next network in the end-to-end path may have admission policies that rewrite the DSCP value to select a new PHB. As an example, a host application may elect to use a DSCP to select expedited forwarding. Upon ingress to the first network, the first network's admission policies may preserve the expedited forwarding PHB as a service selection, but rewrite the DS field to the local mapping of this PHB to a DSCP value. The packet's destination may determine that the packet then be passed to a second network provider that has no expedited forwarding TCA with the first network provider. Upon ingress at the second network, the admission system may rewrite the packet's DSCP to the local best-effort PHB value, clearing the service selection. As this simple example demonstrates, end-to-end DiffServ responses require careful coordination at both a technical and a policy level in the interprovider domain.

The DiffServ Traffic Classifier

All traffic entering a DiffServ network is passed through a *traffic classifier*. The classifier selects packets based on a match between a configured admission policy and the header fields of the packet. The DiffServ architecture defines two forms of classification:

One based on the DSCP field alone. This is termed a *behavior aggregate classifier* (BA). It selects packets according to their DSCP value on ingress.

Another based on a more general set of classification conditions. These conditions may include any of the IP, TCP, or UDP packet header field values, such as the source address, destination address, DSCP, protocol value or TCP or UDP port address; they may also include this ingress interface or similar local information. This classifier is termed the *multifield classifier* (MF). The classified packet streams are then passed to a traffic profile meter.

It should be noted that the same issues relating to IP fragmentation and the IntServ classifiers also apply to DiffServ MF classifiers. Within the trailing fragments there is insufficient packet header information to allow the packet to be correctly classified. This implies that a conservative sender should undertake some form of path MTU discovery in a DiffServ-enabled environment in order to avoid network level packet fragmentation.

The DiffServ Traffic Profile Meter

A DiffServ traffic profiler examines each aggregate traffic stream, as selected by the classifier, and applies profile rules to determine whether the packet is in or out of the associated profile as defined by the TCA. The logical view of a packet classifier and traffic conditioner is shown in Figure 4.19.

Traffic profiles can be described as specific properties of a traffic stream that allow packets to be classified as either in-profile or out-of-profile by the traffic profiler. For example, a specific profile at an ingress node may be based on a token-bucket implementation and may classify packets with a specific DSCP against a specific burst rate:

DSCP = X, use token-bucket (r, b)

This simple example of a profile indicates that all packets received with a DSCP of X be measured against a token-bucket meter with a rate of r and a burst-size of b. Packets that exceed these rate and burst parameters would be considered out-of-profile, and packets that conform to them would be considered in-profile.

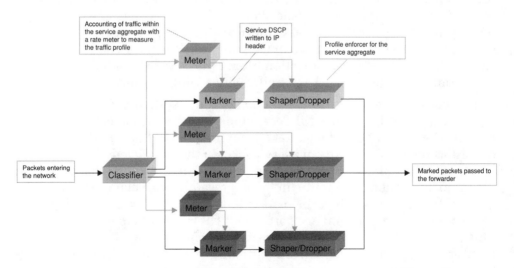

Figure 4.19 Logical view of packet classification and traffic conditioning.

Additional rules need to be added to the example to indicate the action to take regarding the disposition of out-of-profile packets. Out-of-profile packets can be re-marked to a lower relative priority. The profile meter does this by instructing the packet marker to mark each packet with an in- or out-of-profile DS codepoint, depending on the current match of the traffic stream to the profile state. Alternatively, the profile meter can pass directives to a traffic shaper to cause the packet in the stream to be delayed, then released when it conforms to the metered profile. As a final potential action, the profile meter can instruct the admission filter to discard the packet altogether.

The DiffServ Marker

The first question that should be asked is, "Who marks the packets?" Typically, three options are offered in answer to this question: First, that the end-system mark the packets with their desired DSCP, indicating the service quality desired. These packets can simply be policed at the network ingress to comply with the negotiated TCA. Packets in excess of the negotiated TCA could simply be re-marked with an appropriate lower-value DSCP, mapped to their corresponding PHBs, and subsequently forwarded as dictated by the PHB. This can be thought of as the "trust, but verify" method. The second option is to make no assumption about the capability of the end system to mark packets—the DiffServ network ingress node rewrites the DS packet mark unconditionally, to correspond to the traffic classification and the associated profile meter directive dictated by the network's admission policy. The third option is a hybrid approach; a pair of end systems exchange RSVP messages relating to an intended flow, and the RSVP message crosses the DiffServ boundary. A marking that approximates the RSVP service class is matched against the RSVP traffic classification filter. This uses RSVP admission policy to determine whether to mark the traffic with a DiffServ marking.

In either case, verify or rewrite, one additional system is needed to complete the description of the DiffServ admission system, namely the packet marker. The marker takes the packet stream from the traffic classifier, and sets or verifies the DS field value of every packet to the associated DSCP, or codepoint. The codepoint used may be a single constant, indicating that a uniform PHB is to be applied to all traffic in this aggregate service class. Alternatively, the profile meter may direct the marker to use one of a group of codepoints, corresponding to a selection of a particular PHB from a group of related PHBs. This profile meter-directed selection of a particular codepoint from a codepoint group could be a selection of a particular queuing precedence level, a jitter tolerance level, or selection of a discard precedence.

DiffServ-Capable Router

The general operation of a DiffServ-capable router performing traffic classification and conditioning is shown in Figure 4.20. As packets enter the network, the first-hop DiffServ node enforces the TCA and marks the packets in a fashion consistent with the desired PHB. There are three generic routers within the DiffServ architecture. The ingress node is responsible for the ingress conditioning and marking functions. The interior nodes are responsible for triggering per-hop behaviors based on the DSCP of the packets (see Figure 4.21), and the egress nodes may be responsible for further packet marking and conditioning, based on the network's egress policies and the identity of the next-hop network (see Figure 4.22).

Per-Hop Behaviors

A per-hop behavior is the means by which a DS node allocates resources to behavior aggregates. PHBs may be specified in terms of some priority to a particular network resource relative to other PHBs (simple queuing priority, for example), or in terms of link characteristics, such as delay, loss, and jitter.

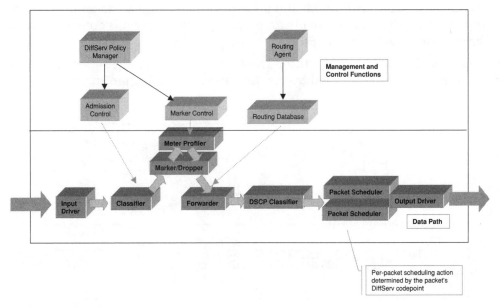

Figure 4.20 DiffServ first-hop router.

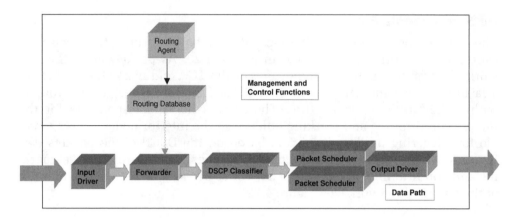

Figure 4.21 DiffServ interior router.

PHBs can either be a single behavior or they can be grouped, in which case each PHB within the group has a positioning relative to other PHBs.

PHBs are implemented on DS nodes according to settings of the basic performance tuning tools we examined in Chapter 3, "Performance Tuning Techniques." The PHB directs the node to respond to the packet in terms of appropriate buffer management and packet scheduling that is consistent with the service outcomes of the PHB.

This section examines the three groups of PHBs and their service implications currently being studied by the IEF DiffServ working group.

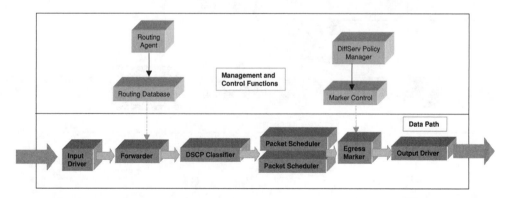

Figure 4.22 DiffServ egress router.

While there are 3 PHB sets being defined at present, it is not envisioned that there will only be just these three within the DiffServ architecture; there is no shortage of potential PHBs. The designated code space for listing PHBs is 16 bits in length. From this set, each network operator selects a smaller collection—no more than 64 PHBs, a limit resulting from the 6-bit length of the DSCP field in the IP packet header. The network operator then maps these PHB *encodings* to a collection of DSCPs for use within the network. Note there is no common core set of PHBs that must be supported by every DiffServ-capable network, and no PHB has an assured networkwide scope. This selection process ensures that each network operator is capable of selecting PHBs that are unique to that particular network, and not supported by any neighboring network.

Within such a scheme, is it conceivable that every service operator will select a unique set of PHBs to support? While this fragmentary service outcome is possible, it is considered a very remote possibility. It is more likely that a selection of common "supported" PHBs and the setting of an associated set of DSCPs will be undertaken by an industry forum. Such an action would allow end host software vendors to create service platforms that, by default, interoperate with vendors' network equipment and constrain the service provider to a more limited set of services consistent with such a forum's choice of core service points.

Class Selector PHB Group

The DiffServ architecture includes some level of backward compatibility to the IP TOS field. The DiffServ specification recommends the use of the 000000 DSCP as a mapping to the best-effort PHB service class. In this way, all packets not requiring any particular service attributes from the DiffServ network continue to be treated on a best-effort basis without the need to alter host software.

In addition, the DiffServ specification attempts to preserve the semantics to the TOS IP Precedence field (the first 3 bits in the old-style IP TOS octet). Therefore, a DSCP value consisting of values expressed *only* within the first 3 bits (yielding 8 possible values) of the DSCP are recommended to map to a set of PHBs that correspond to a set of relative priorities for the traffic. This is intended to allow older implementations of the IP Precedence functionality to coexist with DiffServ implementations. Therefore, a packet header DSCP that contains xxx000 (where x is either 0 or 1) is, by default, nominated to be a reserved DSCP value. These DSCPs are commonly referred to as the *Class Selector Codepoints*. Naturally, they have corresponding Class Selector PHBs, which interpret packets with higher values in the Class Selector Codepoint field as possessing a higher relative priority. Likewise, lower values in this field possess lower relative priority, expressed as a relative probability of lower packet delay and packet drop.

There are a few restrictions, however, on the corresponding PHBs for these Class Selector Codepoints. The PHBs mapped to be the eight Class Selector Codepoints must yield at least two independently forwarded classes of traffic (there may, of course, be more than two). And PHBs selected by a particular Class Selector Codepoint should give packets a probability of timely forwarding not lower than that given to packets marked with a Class Selector Codepoint of lower relative priority. For example, a packet with a Class Selector Codepoint/DSCP of 001000 should be afforded a higher relative priority than a packet marked with a default DSCP of 000000, and should be granted a priority in transmission. Additionally, RFC2474 suggests that PHBs selected by distinct Class Selector Codepoints should also be independently forwarded; this implies that packets marked with different Class Selector Codepoints may be reordered. This reordering can be realized by any number of queuing disciplines, such as strict priority queuing (PQ), weighted fair queuing (WFQ), class-based queuing (CBQ), weighted round robin (WRR), or other similar strategies.

Expedited Forwarding PHB

The intent of the Expedited Forwarding PHB is to support a behavior aggregate of low loss, low latency, low jitter, assured bandwidth end-to-end service through a DiffServ network [RFC2598]. This service has been referred to as resembling a virtual leased-line service across a heterogeneous IP network, or in other words, providing similar performance to that of a traditional point-to-point synchronously clocked leased line between the same end points. The service environment is intended to support a high-assurance voice or video channel, or to emulate the characteristics of a synchronously clocked leased line as a mechanism to support constant-quality virtual private networks.

The observation driving the EF PHB is that loss, latency, and jitter are all the result of queuing effects within the network. Constructing a premium service for this behavior aggregate implies that traffic within this service class must be exposed to minimal, if any, queuing delays in transit. As long as the aggregates' arrival rate at any node is less than the minimum sustainable departure rate, then an EF service is supportable. This minimum departure rate must be independent of any load imposed by all other traffic. The timebase over which this must remain true comprises a time interval of a link MTU at the given EF service class rate. To ensure that the arrival rate is always within such boundary conditions within the network, the collective actions of all admission policy meters at all ingress points must ensure that the amount of traffic admitted into the network is within the minimum departure rates for this service class at all interior network elements.

Within the network, the queuing function can be implemented by a strict priority queuing discipline, giving the aggregate service class the highest possible priority. With priority queuing, EF delay and jitter are minimized and the service outcome is as close to the service specification as possible. Of course, denial-of-service (DoS) risks inherent in priority queuing are present, and reliance is heavy upon the collective actions of the admission function to prevent such overload conditions. An interior node can adopt a more defensive position by using class-based queuing to enforce a bandwidth allocation to EF at every node. With this approach, a slightly higher level of jitter will be imposed on the EF service class.

Being able to take a small proportion of network traffic and emulate a dedicated circuit through priority queuing is not a surprising outcome. If a class of traffic is conditioned so that its rate within all points in the network is well within the network's capacity, and if this class of traffic is assigned absolute queuing priority, then it will experience minimal jitter and no loss. The real challenge here is not to achieve the service outcome, but to condition all the traffic inputs to conform to such a well-behaved model. The fundamental prerequisite of the EF PHB is based upon the capability to control the amount of available network capacity, as compared to the amount of EF traffic entering the network. In order for the EF PHB to be implemented in a functionally sound manner, the amount of aggregate traffic entering the network can never be greater than the amount of bandwidth capacity available to the EF service on any given link. Therefore, strict adherence to subscriber rate and burst contracts must be enforced at the network ingress point, and all network ingress points must adopt such a strict admission policy. The lack of coordination between admission systems and between the admission systems and the per-path load states within the interior of the network imply the need for a very conservative admission policy to ensure that EF PHB traffic remains within the intended service profile. This then implies that the admission policy must strictly enforce rate, with out-of-profile traffic being discarded by the traffic filter or marked to a non-EF PHB codepoint.

However, as an isolated mechanism, this is not an effective end-to-end premium service. All that is happening is that packet loss is being shifted from the interior of the network to the traffic-conditioning ingress elements. To be a true end-to-end premium service, the sending host, the path element from the host to the network ingress point, and the path element from network egress to the destination host must also be aware of the existence and characteristics of this premium service flow.

Assured Forwarding PHB Group

The Assured Forwarding (AF) PHB Group is a collection of PHBs that offer a high level of assurance that each packet will be delivered, as long as the traffic flow conforms to a given service profile, or TCA [RFC2597]. Though a traffic source may exceed its TCA, and the network will accept the excess traffic, the excess traffic may have a higher probability of being discarded within the network. However, if delivered, all packets, including the excess rate packets, remain ordered—at least to the level of ordering being preserved within each micro-flow. In EF, the excess (or out-of-profile) packets may be discarded or may be delivered out of sequence. In contrast, the AF service definition specifies that out-of-profile packets will be discarded, due to a higher drop precedence, or delivered with the same service precedence as all other packets within the same micro-flow. This behavior is very similar to Frame Relay service behavior, where the Frame Relay system preserves frame order and has a two-level discard preference, using the DE bit as the preference indicator.

There are four proposed AF classes within the AF PHB group, and three possible drop precedence levels in each class, making some 12 individual PHBs in total. Each AF class is allocated a certain level of bandwidth and buffer space in each DiffServ node. A node experiencing local congestion within an individual AF class resource allocation may choose to discard packets within that AF class, in which case it must preferentially discard packets with a higher drop precedence PHB. There may be some form of relative packet delivery priority between the AF PHB groups, creating a tiered set of services of the so-called Olympic variety, with gold, silver, and bronze service levels.

Every DS node must be configured to allow some minimum level of local buffer space and bandwidth to each AF class. This may be a fixed configuration, or the node may be configured to offer a greater level of resource to an AF class—if there is available capacity from other AF classes or from other PHB groups. The binding requirement is that if no buffer space is available to an AF class, then packets must be discarded in order of their drop precedence level, as marked by the PHB.

Admission traffic filtering may control the amount of traffic admitted at each AF level, and may include traffic shaping, packet discard, and alteration of any marked discard preferences. The major restriction on the ingress function is that this filter is also constrained from reordering packets from the same micro-flow. One proposed admission and marking scheme is a two-rate, three-color marker (trTCM). A packet is marked red if it exceeds some peak information rate, yellow if it exceeds some committed information rate, and green otherwise [ID-diffserv-trtcm]. These colors can correspond to various drop precedence levels, with red being the highest precedence and green the lowest. The trTCM can be used to mark an IP packet flow, where different, decreasing levels of assurances are given to packets that are green, yellow, or red. For example, a service may, under load, discard all red packets, because they exceeded the peak rate; forward yellow packets within the best-effort scheduling class; and forward green packets with a low drop probability, preserving the order of all packets that are not dropped.

The PHB behavior associated with the AF service class is minimization of long-term local congestion events, while allowing short-term burst traffic. The intended service clients for such a service are TCP-based applications, in which individual packets within a micro-flow may be dropped, but packet reordering within the network may present false congestion signals to the sending TCP stack.

To support the AF service class, each node needs to implement some form of bandwidth allocation to each AF service class, with an associated service buffer size. One approach is to use CBQ across the AF service classes, with the weightings adjusted to reflect the relative bandwidth allocations made to each service class. Another approach is to use the same fixed number of service classes and use WFQ as the queuing mechanism, as a means of ensuring a higher level of accuracy in the per-service class bandwidth allocations. To undertake the queue management within each service class, weighted RED can be used. The RED weighting associated with each packet would correspond to the packet's discard preference as marked in the DSCP (see Figure 4.23).

Differentiated Services and Network Resource Management

It is in the area of network resource management that the theory and practice of the DiffServ architecture diverge most notably. The actions of the admission traffic conditioners are intended to control the level of network resources consumed by these behavior aggregate traffic streams. If there is no feedback between the current network state and these admission conditioners, then the conditioners are in a difficult position of having to accept traffic with no assurance that the service response can be delivered to the traffic. In the absence of such feedback, the conditioners need to adopt a very conservative admission policy, ensuring that only a small proportion of the total network resources are committed to these service classes. This increases the probability that the TCAs will be achieved at all times.

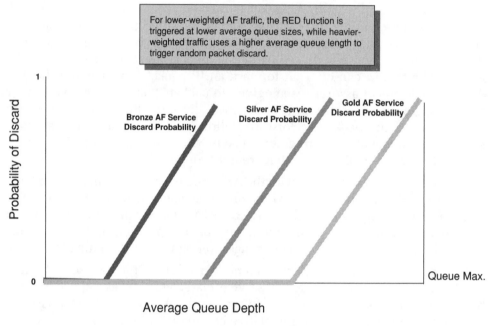

For lower-weighted AF traffic, the RED function is triggered at lower average queue sizes, while heavier-weighted traffic uses a higher average queue length to trigger random packet discard.

Probability of Discard

1

Bronze AF Service
Discard Probability

Silver AF Service
Discard Probability

Gold AF Service
Discard Probability

Queue Max.

0

Average Queue Depth

Figure 4.23 Weighted RED configuration to support the AF service class.

If there is some form of active feedback between the network operating state and the ingress traffic conditioners, the potential exists for more ambitious TCAs, which allow greater volumes of traffic to be admitted to the network within the associated service class marking. The feedback channel allows the admission conditioners to do some level of dynamic adjusting to changes in network state that may affect the capability to service the behavior aggregate at the required service level.

The DiffServ architecture specification notes that "there is a wide range of possible control needed" [RFC2475]. From a cynical perspective, this statement could be described as a case of artful evasion. Considerable complexity is involved in distinguishing between transient and persistent local resource contention events at any single DiffServ node. Additional complexity is introduced during the communication of this local condition to every admission control point in such a way that allows the admission profile meter to react. The admission profile meter has to determine whether such a condition has any impact on its capability to meet the associated TCAs; and if so, it then must determine how to adjust the profile meter to restore the integrity of the TCA.

RFC2475 notes that "the precise nature and implementation of the interaction between these components is beyond the scope of this architecture." This is perhaps unfortunate, given that feedback control mechanisms are at the heart of any QoS architecture. To omit these elements from the base architecture specification would appear to be a curious action on the part of the IETF working group studying Differentiated Services technologies. On the other hand, no currently available feedback technologies are candidates for standardization as yet, and the working group has decided, with some pragmatism, to study the most understood aspect of the architecture at this point in time, and leave the more elusive aspects for later.

There is always tension between the work of a group or committee and that of an individual. Individual efforts are often marked by a consistency of design style; that is, a small number of design elements are applied consistently to achieve a particular outcome. Often, however, such a design is not all-inclusive, and the areas of omission may be deliberate or inadvertent. Nevertheless, the strength of an individual effort is in its style consistency. It is rare for a large committee to create similar outcomes. Group efforts typically operate in an inclusive mode, accepting all contributions to the effort as equally valid and valuable. The risk of group efforts is a loss of design coherence, potentially resulting in something that simply will not work as intended at the outset. The IETF Differentiated Services effort falls into the latter category; it has received many contributions, causing concern that the outcome is not as coherent nor as robust as is necessary to support a diverse and scalable multiservice network service platform.

Interestingly, such a network resource management function could be undertaken through the adoption of QoS routing tools; admission meters could indicate the currently available level of resource within each service class allocation at each DS node, through the operation of the routing protocol. While this is not a fine-grained feedback control loop, the relatively coarse level of granularity adopted by the DiffServ architecture may be well adapted to a similarly coarse level of granularity of feedback provided within updates of a routing protocol. It is not even necessary for such a QoS routing system to use any other than normal unicast forwarding in the interior of the network. The primary intent of the use of such a routing protocol is not to create diverse forwarding paths within the network, but to enable the interior of the network to communicate its local DiffServ resource availability state back to the admission traffic-conditioning meters.

The outcome of such a model is similar to that of choosing MPLS as the DiffServ BA service platform. In the MPLS model, at every ingress, each BA is mapped to a label-switched path (LSP). The LSP is created with the service attributes of the BA, then it is passed to MPLS to maintain the path and its associated service parameters. If a network event compromises the BA, it is an MPLS task to rebuild the LSP over another path or to report LSP failure to DiffServ admission system.

The interesting aspect of both approaches is that any dynamic change that might affect the action of the admission policy control system will not compromise an entire BA to all potential destinations. But a network state change might compromise BAs operating over a particular set of network paths, and it is only these subcomponents of the ingress BA meter and profiler that should make dynamic adjustments to its admission policies. Including feedback of network state information to the ingress elements suggests that the network ingress element does not use a single profile and meter for each BA. Instead it appears that a more robust approach is to use a distinct BA profile and meter for each distinct network path, or MPLS LSP in the case of MPLS-supported DiffServ. In this way, the ingress element can react to dynamic network changes that affect one forwarding equivalence class without making any unnecessary changes to unaffected BA flows.

Differentiated Services and the End-to-End Service Model

The missing element within the Differentiated Services architecture as it is currently specified is the reproduction of the end-to-end service model. The end application needs some form of negotiation with the network, either by dynamic discovery or through some form of policy negotiation, that the network will be capable of sustaining the required service profile demanded by the application. Invariably this will not be as strict a negotiation as required by the Integrated Services architecture, but nevertheless it makes little sense for an application to proceed with a premium service profile if the network is incapable of meeting the associated service demands.

There are three parts to such a negotiation: one between the host application and the admission profile meter at network ingress, a second between the admission system and the network itself, and a third between the path-specific network egress point and the receiving host (or, in the multicast model, a number of such negotiations between all relevant egress points and the set of multicast receivers). (See Figure 4.24.) Unlike the IntServ model, this negotiation

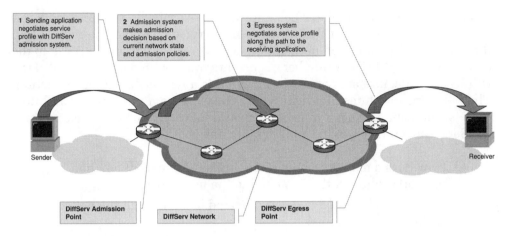

Figure 4.24 End-to-end service negotiation.

may not be a single transaction prior to commencing the application. The lack of a per-flow state within the network to support DiffServ implies that dynamic changes within the network will happen without the explicit knowledge of the ingress systems. To support the integrity of the service, it is necessary to use some form of continuous monitoring of the dynamic state of each network element and to feed back service quality information back to the ingress points and ultimately back to the application itself. Such negotiation and control mechanisms remain firmly within the realm of further research activity.

Accordingly, it is likely that the initial approach to DiffServ deployment will be very conservative in nature, with statically configured end policies and with admission meters that can assume that only a small proportion of network resources are available to service any premium service class. Consequently, the premium service traffic will be, to some extent, isolated from transitory local congestion events that occur within the network.

Observations on the Differentiated Services Architecture

When examining the DiffServ architecture, a critical difference between enterprise networks and service providers within the public Internet emerges. It is reasonable to state that there will be a somewhat fragmentary approach to DiffServ service support within the public Internet environment; any transition into DiffServ-aware networks will not take place as a universal synchronized event. Rather, DiffServ-aware networks will be constructed in a piecemeal fashion across the public Internet. It is therefore relevant to define clearly the anticipated mode of behavior when a service flow is initiated by an end system. In the case of DiffServ, when a service class cannot be honored, the anticipated action upon network ingress is to quietly reset the DSCP to the default best-effort PHB. A number of currently unspecified components of the DiffServ architecture are required to make this a reasonable course of action. The following questions highlight what is missing from the DiffServ architectural specification. These questions would have to be answered to turn DiffServ into a utility service on the public Internet.

- How does a network know of the ingress policies of neighboring networks when traffic is passed to the neighbor?

- How is this knowledge communicated back through the network to the ingress systems?

- Should an admission system mark a packet for some form of premium service if it is aware that that the selected path uses a neighbor that is incapable of carrying the service class?

- How can an admission system take into account not only the local neighboring DiffServ interconnection policies, but also any remote interconnection policies that may be on the current end-to-end path for the packet?

- How can a local system even identify which remote third-party interconnections may be used in delivering the packet, unless there is some form of signaling mechanism?

- If such a signaling system exists, how should it communicate routing state changes back to the original ingress system to alert it to a change in end-to-end service transparency?

- How can an aggregated flow provide any form of service level to individual end-to-end applications, given the interaction of elastic congestion-managed and inelastic real-time flows that may be placed into a single aggregate class?

Clearly, much remains to be done to lift DiffServ out from a carefully managed localized service platform onto a more robust, widely deployed technology platform for multiservice environments within large service networks. This does not imply that DiffServ is so immature that it is not ready for use in today's service networks. The elements of admission classification, profilers, packet marking, and various queue management systems that can support resource allocation are a reality in network equipment today. With the caveat that network admission policies adopt a very conservative approach to admission of premium service flows, a DiffServ-based network can be constructed today, in both small and large service environments. Further refinements in the DiffServ architecture will allow greater levels of resource commitment through more active forms of network resource management.

Combining Integrated and Differentiated Services

The discussions over the strengths and weakness of both the IntServ and DiffServ architectures lead to the conclusion that neither approach offers comprehensive and robust solutions for supporting a multiservice network platform. The IntServ model, while providing a very high level of assurance of per-flow resource management, has significant scaling issues, and the DiffServ model has a very weak approach to resource management, though possessing significant scaling properties.

The realization that neither model is sufficient has resulted in a proposal to combine the two models, using IntServ control mechanisms at the edge of the network and DiffServ within the network core [ID-intdiff]. The proposal has been further motivated by the observation that some form of end-to-end QoS negotiation is required to support the needs of certain applications, such as video-on-demand and various time- or quality-critical applications. The proposed model is conceived as a collection of IntServ-capable networks as the edge feeder networks and a DiffServ-capable network as the common service core system (see Figure 4.25).

It is proposed that applications use RSVP to request that their flows be admitted into the network. If a request is accepted, it would imply that there

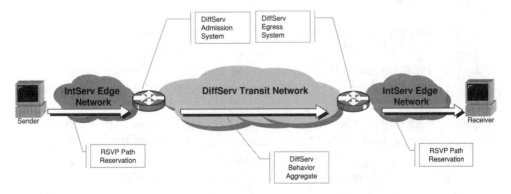

Figure 4.25 Combining IntServ and DiffServ networks.

is a resource reservation within the IntServ-capable components of the network, and that the service requirements have been mapped into a compatible aggregate service class within the DiffServ-capable network. The DiffServ core must be capable of carrying the RSVP messages across the DiffServ network, so that further resource reservation is possible within the IntServ network upon egress from the DiffServ environment.

The approach is that the DiffServ network will use MF admission classifiers, where the MF classification is based on the IntServ flow specification. The service specification of the IntServ resource reservation is mapped to an aggregate DiffServ behavior aggregate, and the MF admission filter will mark all such packets into the associated aggregate service class. It is also possible for the host to condition the traffic in advance of transmission into the network, and to mark the packets within the flow with the associated commonly defined DSCP values, so that the admission filter can operate via a behavior aggregate admission classifier.

The end-to-end QoS model requires that any admission failure within the IntServ network be communicated to the IntServ network, and from there to the end application via RSVP. This allows the application to take corrective action to avoid failure of the service itself. If the service agreement between the DiffServ network is statically provisioned, then this static information can be loaded into the IntServ boundary systems, and IntServ can manage the allocation of available DiffServ behavior aggregate resources. If the service agreement is dynamically variable, some form of signaling is required between the two networks to pass this resource availability information across to the IntServ environment.

The mechanics of an end-to-end QoS required in such a model are envisioned as follows:

1. The sending host generates an RSVP *Path* message, describing the intended traffic profile of the sending application.
2. The *Path* message is carried across the IntServ feeder network, invoking standard RSVP *Path* message processing and path state to be installed within the network elements.
3. The *Path* message is carried transparently through the DiffServ transit network, and passed to the terminating IntServ stub network.
4. The *Path* message is carried across the IntServ terminating network, again invoking standard RSVP *Path* message processing.
5. The receiving host generates an RSVP *Resv* response back to the sender.

6. This *Resv* message is passed back through the terminating IntServ network, invoking reservation processing.

7. At the interface to the DiffServ network, a DiffServ admission control system (DACS) is invoked, and requested to compare the current DiffServ resource availability against the requested service profile. If admitted, the DACS would record this reservation and the associated DSCP in the RESV message, then pass the *Resv* message back through the DiffServ transit network.

8. The *Resv* message is passed through the IntServ feeder network to the host, invoking reservation processing.

9. The host is then permitted to commence the traffic flow; it can set the DS field in the flow packets to the DSCP value, which maps to the relevant DiffServ service type specified in the *Resv* message.

The passing of the DiffServ DSCP value all the way back to the host, and the explicit marking of the flow packets with this DSCP value, are not strictly necessary in such a model; and if an agent at the boundary of the IntServ feeder network undertook this packet marking prior to passing the packets into the DiffServ network, the host now sees what is an end-to-end IntServ environment. From this perspective, the DiffServ network component takes the form of an instance of a specific type of media within an IntServ network [ID-diff-svc]. Of course, since the DiffServ environment carries a broader range of service models than the IntServ architecture, it is not entirely clear how to encode a behavior aggregate specification within an RSVP object, so some additional specific service descriptor objects would be required in the RSVP specification to support such a hybrid service model.

The QoS Management Architecture

A number of constraints are placed on a client's access to elevated or premium network services. Not all clients may be able to simultaneously select premium services for their traffic, as the network may not be large enough to cope with the increased level of demand on resources. Some clients may not have executed a service level agreement with their provider to permit them to request premium services in the first place. Other clients may have a service level agreement that permits the entrance of premium traffic, but with an associated traffic profile that limits the amount of such traffic. As there may be additional costs in the provision of specific service responses on the part of the network, the network provider may want to charge another fee for the use of premium services, and the admission system should initiate a reporting process to enable accounting records to be generated by the system.

Whichever service architecture is chosen by the network operator, the network operator and the clients of the network must have the ability to monitor and control the use of network resources. This control is normally based on a framework of access policies. The description of these policies may include the identity of the client; the identity of the application; the intended traffic profile; the bandwidth requirements; the delay, jitter, and loss tolerances; security considerations; time of day or day of week; ingress points; and others. A critical part of any policy is that of admission control, to implement the network's provisioning policy. From the perspective of the end user, an overloaded premium service is little different from an overloaded best-effort service. In either case, the application is not being given the amount of network resources it needs in order to execute correctly. For this reason, there are two quite distinct parts to any QoS architecture: managing a network's response to service requests and managing the demand placed on the network by controlling the admission of service requests into the network.

In the QoS architectures reviewed here, it is assumed that when a network element is requested to make a decision to admit a traffic flow and commit to its associated resource requirements, the network element can do so in a way that is consistent with the network's administrative policies and with the current operating state of the network. Part of the problem with this assumption is a matter of policy control. Without such policy control, an admission system may treat all requests equally, thereby indiscriminately admitting all resource reservations until all reservable resources have been exhausted. At that point, all subsequent requests are refused.

The role of policy control is to remove the significance of the time and order of the arrival of requests and use additional criteria to allow for the admission and maintenance of resource reservation requests. One way of representing this dimension of policy control is to associate a relative priority with every reservation request. This priority value represents the importance of the reservation, relative to other reservation requests that are to be serviced from the same resource pool. This priority may be fixed or may vary over time. The reservation may have a request priority, to be used at the time the reservation request is processed, and a maintenance priority, to be used once the request is granted. As long as there is adequate capacity in the network to accommodate all requests, there is no need to refer to the priority of the request. When a request would result in over-subscription of the resource, the policy system could compare the request with the maintenance priorities of existing reservations, rather than simply rejecting the request. If there is a sufficient number of lower-priority reservations that can be released to accommodate the higher-priority request, the policy system may elect to preempt these lower-priority reservations and reallocate the resource to the new request. Network state changes may induce a similar process of policy negotiation, where displaced reservations attempt to restore themselves across new network paths, potentially displacing existing lower-priority reservations. One way of representing this policy of reservation priority is through a set of classification rules maintained by the network manager. Alternatively, the request itself could be tagged with the associated reservation policy, presumably triggering some premium tariff from the accounting system.

To implement this functionality a copy of all of the network's administrative policies could be placed on every network element, to ensure that all updates to the policy are circulated in a consistent and timely fashion. But in a large network with a distributed resource state to maintain, this is probably not a viable approach. An alternative management approach is to place all the network's administrative policies in a single repository, or in a small set of such repositories, and configure the network elements to refer all admission decisions to a central decision engine that uses the policy repository. The IETF has been studying the elements of a Policy-based Admission Control architecture, whose design includes a policy decision engine that can exert policy control over the network's QoS admission decisions [ID-rap-frame]. Its immediate area of application is in processing RSVP admission requests within the IntServ QoS framework, although the intended generality of this approach would lend itself to any form of signaling control for admission systems.

The Policy Model

Before looking at the model of policy control for QoS networks, it's necessary to introduce the terminology used to describe this material. In particular, three terms are very commonly used in this area, and deserve mention here:

Policy. Refers to a combination of rules and services, where the rules define the criteria for access to various network services and their associated level of resource consumption.

Policy Decision Point (PDP). Describes a network element where policy decisions are made.

Policy Enforcement Point (PEP). Denotes a network element where policy decisions are enforced.

Policies are not, in themselves, mechanisms. The management architecture does not define specific network service responses, nor does it moderate the support for particular policies. One feature within the policy is important; that is, that policies allow preemption. A previously installed state can be removed in order to accept a new resource reservation. Also, policies may involve two parties (bilateral) or many parties (multilateral). Policies should also include explicit support for monitoring the policy state and for reporting on resource usage. Obviously, such information is essential for accounting and billing functions. A policy control architecture must be able to validate the originator of the request, and the transactions that occur between the PEPs and the PDPs, to ensure that the system is not vulnerable to abuse.

The envisaged mode of operation within an IntServ reservation framework is the following: The PEP receives a resource reservation message. It then generates a request for a decision, and passes this decision request to a PDP. The decision request may also include relevant information relating to the original request, as well as the current state of the PEP, to assist the PDP in making the decision. The PDP may, in turn, consult external information sources, such as an LDAP-based directory, a Radius database, or some SNMP-based query of the current network operational state in reaching its decision. Once the PDP has reached a decision, this is communicated to the PEP. Not all policy interactions are based on the PEP initiating the transaction. A policy configuration change may be undertaken by configuring the PDP with the change, which in turn will communicate the changed policy state to the PEPs.

This delineation of function into two discrete components is intended to assist with the definition of the policy architecture. It is possible to co-locate the PEP and the PDP in a single unit; or there may be multiple PDPs, each responsible for a certain class of decision. This is best illustrated with an example, which is given in Figure 4.26. An RSVP Resv request arriving at node A's input processor is passed through the local controller to the RSVP control module. To make a decision on this request, the message is passed to the node's PEP module. The PEP generates a request to the PDP, and awaits a response. Once the PDP responds to this request, the PEP then either directs the RSVP module to reject the request or instructs the RSVP module to accept the request. If the request is accepted, the RSVP module loads the associated configuration into the classification module, associates a scheduler action with the service class, and records the updated resource reservation level.

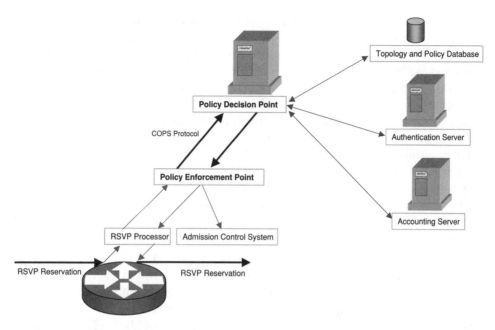

Figure 4.26 RSVP and policy control.

Types of Policies

Policies may take many forms. The identity of the requestor (*who*), the time of day or day of week (*when*), the location of the request (*where*), and the nature of the request itself (*what*) are all relevant factors in reaching a decision relating to each policy request. Any policy descriptor will include some, if not all, of these qualifiers within a policy ruleset. In addition, a policy may have an associated value, or priority (*how much*), where higher-valued policies take precedence over those of lower value. In such cases, decisions made on the basis of a high-priority policy will have a higher probability of acceptance.

Policies, and the decisions made in accordance with these policies, may be permanent, revocable, or negotiable. A permanent decision imposes a commitment that cannot be altered until the service has terminated. The implication of a permanent decision is that it is not possible to admit variable priorities of policies. If the network state at the time of the decision is such that the request can be admitted, then the decision cannot subsequently be revoked if a higher-priority request arrives, nor can low-priority decisions be revoked if the network state changes to limit the amount of available resources. A revocable decision is one that the network may subsequently reverse, at its discretion. In this way, a network can permit a service request associated with a high-priority policy to preempt previous decisions that were associated with a lower-priority policy. Preemption poses some synchronization questions, particularly if a sequence of PEPs is involved; a decision to preempt a previous decision at one PEP node must be correctly distributed to all related systems that are also supporting this decision. The intermediate position between these two extremes is a negotiable decision. A negotiable decision is one where the network may accept a request via a policy, then, under changed circumstances, may offer the requesting application an altered service level. Within the current QoS architectures there is no well-defined mechanism to support such dynamic negotiation relating to established services. However, it is a subject where we can expect some development with the successive refinements of QoS architectures.

In terms of the parties described by policies, there are two major forms: policies between an end client and the network provider, and policies between two network providers. This second category is no less important than the first, for, "[L]ack of sophisticated accounting schemes for inter-ISP traffic could lead to inefficient allocation of costs among different service providers" [ID-rap-frame]. Such a policy decision framework is a critical component of interdomain QoS agreements, both for policing interprovider QoS resource reservation and for accounting for such QoS requests and the associated traffic.

Common Open Policy Service

How does a PEP talk to a PDP? One protocol currently being examined by the IETF is a simple query-response protocol, called the Common Open Policy Service (COPS) protocol [ID-cops]. The need for reliable information exchange influences the transport choice, and COPS is a transaction protocol layered above a TCP transport session. The PEP opens up a session with its PDP, sends decision requests to the PDP, then awaits decision responses. The PDP can also use this session to send unsolicited decisions to the PEP, changing previously approved states as a part of a preemption capability.

One way of looking at COPS is as a configuration manager, where the configured state of the network elements (or parts of that state) are under the control of a remote configuration manager. In response to an RSVP request, the PEP passes the request to the PDP. If the PDP's policies permit this request, a configuration fragment is passed to the PEP for local installation. The configuration fragment may include a local classification filter description, coupled with a scheduling action and, potentially, a forwarding action.

Such distributed decision-making systems present synchronization issues, however, and the PEP/PDP model of policy enforcement is no exception to this general rule. The COPS protocol includes explicit commands to allow a PDP to reissue all active installed elements into a PEP, thereby setting the PEP to a synchronized state with the PDP. The COPS protocol is a single-operation per-protocol data unit. Each operation can be accompanied by an attached object.

Rather than looking at the details of the protocol's fields, we'll look at the basic message exchange between a PEP and a remote PDP. The PEP generates a *Client-Open* message to the PDP, informing the PDP of the PEPs identifier. The client may also use extension objects to inform the PDP of any client capabilities. In addition, the *Client-Open* message may reference a previous PDP for which the PEP is still holding active cached decisions, as an aid in synchronizing the PEP state with the PDP. The PDP responds with a *Client-Accept* message to accept the connection. The protocol uses a keepalive message exchange to ensure that both sides of the exchange are aware of the other's state; the keepalive interval is part of the accept response. The PDP can also specify a preferred accounting timer interval, indicating that the PEP should send accounting reports to the PDP no more frequently than the specified accounting interval.

Once the PEP and the PDP have set up this state, the PEP is in a position to pass requests to the PDP for resolution. The *Request* message is used for that purpose. The *Request* message includes a PEP identifier, which enables the PDP to refer to this request in the subsequent decision response and in any subsequent modifications to this decision. The *Request* message includes the context in which the request is being generated. This may be an admission control request, a resource allocation request, or a request for configuration data to be downloaded to the PEP. The request may include a reference to the PEP's input and output interfaces as they relate to this request; and if there is a local PDP module, the PEP may include the output from the local PDP in the remote PDP request. The request also holds the client-specific information for which a decision is being requested. In the case of COPS support for RSVP, the Client-Specific Information field of the COPS message includes all the objects in the received RSVP message, including all policy control extensions to RSVP [ID-cops-rsvp, ID-rap-rsvpext].

The PDP will respond to the PEPs request with a *Decision* message. The *Decision* response is one of *Install*, *Remove*, or *No Decision*. In addition, the *Decision* message may include stateless configuration data that can be applied by the PEP, replacement configuration data, or new configuration data to be loaded by the PEP to implement the decision.

The PEP will then respond to the PDP with a Report State message, to inform the PDP of the success or failure in implementing the decision, allowing the PDP to update its view of the PEP's state. The report object can also include an accounting field, and the PEP can generate periodic accounting reports per installed decision if so requested by the original decision.

In the normal course of events, the installed reservation state will be torn down, in which case the PEP must inform the PDP that this particular state has been completed. The *Delete Request State* message is used by the PEP to communicate this to the PDP. This message is also used to reset a request if the decision contained an unknown object.

There is one other message exchange in the COPS protocol, which is intended to allow the PDP to synchronize itself with the state of the PEP. The PDP may send a *Synchronize State Request* message to the PEP, referencing a single decision state or the entire PEP state. In response to this message, the PEP should reissue *Request* messages corresponding to all installed decisions, or the particular decision references in the *Synchronize State Request* message. Once the PEP has reissued all such requests, the PEP sends a *Synchronization State Complete* message to the PDP to inform it that the entire state has been sent.

The PDP can instruct the PEP to associate a priority with each installed policy decision. This priority value can be dynamically updated by the PDP in subsequent *Decision* messages. If this flow is then preempted by the installation of a higher-priority decision, the local decision point of the PEP may direct the PEP to remove the flow, in which case the PDP is informed of the removal by a *Delete State* request, with a reason code indicating preemption. The PDP may use an internal algorithm or a directory lookup to establish the priority of a reservation. Alternatively, the preemption priority and the defending priority (priority after installation) can be carried as a policy attribute of the RSVP message, allowing the RSVP source to determine the relative priority of the request [ID-rap-preempt]. In addition, the PEP can provide information to the PDP relating to the identity of the requestor, in such a way that the PDP can validate the identity claim [ID-rap-rsvpid]. Of course, additional mechanisms for securing the PEP/PDP connection are available, and IP Security (IPSEC) or use of a Transport Level Session (TLS) security protocol are also options here.

COPS and IntServ

One model of the Integrated Services architecture has the RSVP *Resv* message treated identically by all nodes along the path from the receiver to the sender.

Within this model there is no particular delineation between those RSVP nodes that lie in the interior of a network and those that lie on a boundary between one network and another. This is not a good working model, given that these nodes will use somewhat different criteria when assessing whether to accept the resource reservation.

It is more efficient to provision the policy decision for interior nodes to assume that if an RSVP request is passed to the node, then the node can assume that the policies of the network and the RSVP request are in alignment, and the network is willing to accept the request. In this case, the decision process used by the interior node is one of resource availability. If there are sufficient available reservable resources to accommodate the reservation request, then the request is honored and the *Resv* message is passed one hop upstream to the neighboring node. If the node's available resources are fully committed, the node will generate a *ResvErr* message and reflect it back in the direction of the reservation initiator.

The decision-making framework on the network boundary is more involved, as the node has to compare the details of the RSVP request with the prevailing network admission policy in order to make a decision as to whether to admit the RSVP request at all. This decision may be independent of the current level of available resources within the network; or this level of resource availability may form part of the policy criteria relating to the admission decision. The decision may depend on:

- Where the request was passed from to reach the network, in the case where two networks agree to honor each other's RSVP requests.
- The identity of the source of the request, in the case where the network has a service agreement with a client to honor RSVP requests generated by the client.
- The identity of the destination of the request, where the network has an agreement with a client to send traffic in accordance with a receiver's RSVP request.

COPS is seen as assisting the network admission points in making consistent admission decisions regarding RSVP requests. The emerging provisioning model for IntServ networks using COPS is for the RSVP request itself to provide all the information describing the request, both in terms of the details of the resource requirements and in terms of the identity of the requestor and the policy associations that may be attached to the request. The decision process then becomes one of matching the network's current state and policies to the information provided in the RSVP request.

COPS and DiffServ

The COPS operation model we've examined so far is a transaction model, where each RSVP request is treated as a separate transaction by the PEP, and is referred to the PDP for a decision. This management model is inappropriate for a DiffServ architecture, where there is no per-flow admission decision that has to be made by the PEP. Some study has been done regarding how to apply the COPS model of PEPs and PDPs to the DiffServ architecture [ID-rap-pr]. The result is an approach that uses the PDP to update the PEP in response to external events that trigger a change of service level agreements or a change in admission policy. Here, the COPS service plays the role of a distribution mechanism, allowing a single change in policy to be reflected in updated admission criteria across all network ingress points.

The mechanism proposed is to define a *provisioning* COPS client [ID-rap-pr]. The client uses a decision server, which, in turn, uses a general policy structure. This general policy structure is analogous to an SNMP Management Information Base, and is termed a *Policy Information Base*, or *PIB*. The PDP periodically updates the PEP with a new policy structure, and the PEP uses this downloaded decision framework as its admission policy for all traffic. Further updates to this local decision framework are not initiated by resource requests, nor by a change in traffic state or network state; they are initiated by a change in the externally defined network admission policies.

The periodic update of a formally defined information base proposed here is very similar to the SNMP management model, raising the obvious question: "What's wrong with using SNMP to support external policy inputs to control admission systems?" The reasons include the requirement of a validated state maintained between the PEP and the external decision processor [ID-rap-pr]. If the admission system is reset, the decision controller should be aware of the need to reload the operational state. In addition, SNMP has no concept of resource locking. The PEPs may be prone to operational inaccuracies as a result of unsynchronized changes caused by multiple PDPs attempting to manipulate a single PEP. SNMP is also criticized because of its use of UDP as a transport protocol. It is noted that complex policy updates may require a sequence of updates, and a reliable transport protocol allows the policy update to be conducted over a shared state between the PDP and the PEP. The COPS protocol addresses these concerns through the use of a single control channel with a shared state supported by a single TCP session.

The COPS protocol itself does not require any changes to support this mode of operation. A COPS provisioning service is established when an admission system is started. The initial exchange with a PDP is via a Client-Open message and a PDP's Client-Accept response. The PEP can then request the download of an admission configuration via a Configuration Request context object within a COPS Request message. This configuration request includes details of the PEP's hardware configuration and its location and identity within the network. The Decision message response of the PDP includes a download of all provisioned policies that are relevant to this managed network element. Policies are described in the manner of generic filter specifications and associated profile and metering specifications. The PEP admission system maps these generically stated policies into local admission parameters and local configuration options. Any changes in the policy state at the PDP are communicated to the network element via an updated Decision message. Similarly, any local changes to the network element are communicated to the PDP, which may, in turn, respond with an updated policy configuration.

The model is a little broader than that of a PEP admission filter and a PDP. In a hybrid form of per-flow provisioning and aggregate provisioning, it is noted that applications may communicate their service requirements to the PDP, which, in turn, would identify the relevant admission point for the data flow and make the necessary adjustment to the admission system at that point to admit the flow and to nominate the associated DSCP to apply to the flow to meet the requested service profile. The admission point's confirmation of this decision can serve as confirmation to the application that the service request has been accepted by the network. If the admission classification was installed specifically for this flow, the COPS Report State messages are able to report on the level of usage of this filter via an accounting object. Teardown of this per-flow imposed state could be accomplished by a subsequent Decision message to remove the relevant admission specification. No changes are required to the COPS protocol to support this mode of operation, although some additions are required to the protocol data objects to carry the configuration data.

The design issue with this proposal is the requirement for the PDP to reference a named structure to be able to retrieve policy elements and pass them to the PEP. This structure is the aforementioned Policy Information Base, or PIB [ID-cops-pib]. The reader is referred to this reference for further details of the syntax of the PIB and its relationship to an SNMP MIB. The key aspect of this work is an attempt to impose a systematic structure of describing the construction of traffic classifiers, router interfaces, DiffServ codepoints, queues and queuing disciplines and their associated service characteristics, and a structured approach to relating these components in a way that allows a policy control point to synchronize the behavior of a DiffServ admission system through a precise specification of the actions of the DiffServ entry point. The issue of whether to use SNMP or the framework of a PIB with COPS had not been resolved at the time of writing.

DIAMETER

An alternative approach to using policy constraints in making QoS admission decisions is described in the DIAMETER proposal. This work builds on the existing Radius toolset that is used in making network admission decisions for dial-up users. The general model of a Radius environment is one where the network collects information relating to the request, then consults a remote directory for a decision on the request. The decision response may be a simple yes or no, or the yes may contain additional parameters. These parameters allow the network entry point to include some dynamically configured attributes, customizing the entry point to suit the needs of this particular user. This client/server interaction is very similar to the requirements for QoS admission control. The functional difference between a Radius environment and the one proposed for QoS admission is the remote directory system's capability to communicate change information to the client, which would be a server-initiated transaction. It is not the intent of this book to describe DIAMETER in detail, as work is evolving at the IETF. The reader is referred to [ID-diameter], [ID-diameter-qos], and [ID-diameter-resmgmt] for a more detailed description of the current status of this work in progress.

It is useful to compare and contrast the approach of the DIAMETER client-initiated transaction model with the shared state incremental update model used by COPS applications. In DIAMETER, each RSVP message arriving at the ingress to the network generates a DIAMETER bandwidth request to the DIAMETER server. The server responds with a bandwidth response that includes a traffic profile specification and a QoS priority. Preemption through prioritization is not explicitly addressed within this model of interaction, although it could be added quite readily. Within the Differentiated Services model is perhaps a more fundamental mismatch between the externally triggered DIAMETER transaction model and the self-contained model of a DiffServ admission system. [ID-diameter-qos] proposes that an admission system triggers a DIAMETER request every time it recognizes a new data stream with a DiffServ codepoint that is known to the network. The DIAMETER bandwidth request to the server does not specify a particular traffic profile, but it may include details of the data stream, such as the source and destination addresses, the port addresses, and the DSCP of the stream. The DIAME-TER server is expected to respond to this request with a new DSCP to mark on this data stream, together with an associated traffic profile specification. This reintroduction of a per-flow admission system into the Differentiated Services architecture is outside the typical range of the aggregate approach to admission systems normally associated with DiffServ admission systems.

Management Observations

Management architectures could certainly be described as the weak point of current quality of service systems. No doubt there is much yet to be developed in terms of an architecture approach to QoS management; and while the approaches of COPS and DIAMETER show promise, much has to happen to turn these drafts into robust architectures.

The management architecture has to solve a number of issues simultaneously. It is expected to provide tools to assist in admission control to both the Integrated Services and Differentiated Services architectures. These admission control systems are expected to describe who has access to QoS services within the network, along with the basis of such an admission decision. More challenging than the issue of identification is the matter of managing network resources. As we have observed, in providing specific service responses to various service flows, the network's resources have to be allocated according to a set of network policies. This is a significant departure from the passive network that supports the best-effort service model, where various active end systems effectively equilibrate their resource demands among themselves. With the introduction of active resource management within the network, the network must first be able to calculate an inventory of available resources. Once the amount of available resources is known, the network must then react to the service requests for access to these resources. In addition, the network has to police the consequent traffic profile to ensure that the actual profile is within the parameters of the original request. Complicating this picture is the matter of attempting to synchronize admission behavior across a number of admission systems. It also must be noted that network resource measurements are qualified by a location within a network where the measurement is performed. A fully informed admission system is aware not only of the internal path of a service micro-flow, but also of the resource level along the path, to allow the admission system to make an accurate decision as to whether the service request can be accommodated.

The additional dimension of policy is composed of scale, cost, and responsiveness. In the same way that micro-management is often criticized as a poor management practice, operating a large-scale network by making thousands, or even hundreds of thousands, of micro-flow decisions per second is a poor management approach. A pragmatic balance must be struck between accuracy and scalability. Large systems necessarily rely on a principle of acceptable approximation, because accuracy of operation may not be a feasible or cost-effective approach to the problem.

It seems likely that any architecture will use fine-grained negotiable policies as the network nears the end application. Toward the aggregated core of many networks, and across aggregated interprovider network boundaries, the policies are necessarily far more approximate, and tend to be permanent and not based on any micro-flow state. At this point, there appear to be two distinct approaches, one for the network edge and one for the network center. It appears that the developmental objective is to use a management structure to pull these approaches close enough together so that a single service management regime can operate across the end-to-end network path. Of course, precisely what form a single-service management regime may take when the end-to-end path straddles multiple network providers and reaches from one end organization to a different organization at the other end is currently open to a wide range of speculation. Defining workable cross-domain management systems is a major challenge within the scope of this effort.

References

[AF-lane] *LAN Emulation over ATM Version 1.0 Specification,* The ATM Forum Technical Committee, af-lane-0021.000, January 1995.

The ATM forum specification for ATM emulation of Local Area Network characteristics.

[AF-mpoa] *Multi Protocol Over ATM (MPOA) Specification Version 1.0,* The ATM Forum Technical Committee, af-mpoa-0087.000, July 1997.

The ATM Forum specification for carrying IP datagrams across ATM networks.

[AF-tm] *Traffic Management Specification Version 4.0,* The ATM Forum Technical Committee, af-tm-0056.000, April 1996.

The specification of traffic management within the ATM architecture.

[Clark 1988] *The Design Philosophy of the DARPA Internet Protocols,* D. Clark, ACM SIGCOMM '88, August 1988.

The description of the end-to-end model of Internet service architecture.

[Clark 1997] *An Approach to Service Allocation in the Internet*, D. Clark and J. Wroclawski, diffserv.lcs.mit.edu/Drafts/draft-clark-diff-svc-alloc-00.txt, July 1997.

This document, originally published as an IETF Internet draft, described an approach to service allocation through the use of a resource manager that works in cooperation with network ingress elements to ensure that all premium service traffic admitted to the network is within the scope of available network resources to support the traffic's service demands.

[Durham 1999] *Inside the Internet's Resource reSerVation Protocol: Foundations for Quality of Service*, D. Durham and R. Yavatkar, John Wiley & Sons, May 1999.

A useful and clearly written description of the Resource Reservation Protocol and its role within the Integrated Service Architecture for delivering QoS services.

[Huston 1996] *Path Precedence Discovery*, G. Huston and M. Rose, www.telstra.net/gih/pprec.html, December 1996.

A proposal to use a mechanism similar to Path MTU discovery to undertake dynamic discovery of end-to-end IP Precedence support across a heterogeneous network layer.

[IEEE-802.1P] *Supplement to MAC Bridges: Traffic Class Expediting and Dynamic Multicast Filtering*, IEEE P802.1p/D6, May 1997.

A proposal to refine the MAC-layer bridge specification to include the capability to include packet-defined priorities and dynamic multicast filters.

[IEEE-802.1Q] *Draft Standard for Virtual Bridged Local Area Networks*, IEEE P802.1Q/D6, May 1997.

A draft standard specifying an encapsulation method for supporting virtual local area networks across a common local network platform.

[IEEE-802.1D] *MAC Bridges*, ISO/IEC 10038, ANSI/IEEE Std. 802.1D-1993.

The specification of the MAC-layer bridging mechanism used to connect to local area 802.1 networks.

[IEEE-802.3] *Carrier Sense Multiple Access with Collision Detection (CSMA/CD) Access Method and Physical Layer Specifications*, ANSI/IEEE Std. 802.3-1985.

The specification of the Ethernet 802.3 local area network standard.

[IEEE-802.5] *Token-Ring Access Method and Physical Layer Specifications*, ANSI/IEEE Std. 802.5-1995.

The specification of the Token Ring 802.5 local area network standard.

[Nichols 1997] *A Two-bit Differentiated Services Architecture for the Internet*, K. Nichols, V. Jacobson, and L. Zhang, November 1997. Available at ftp://ftp.ee.lbl.gov/papers/dsarch.pdf.

A simple approach to differentiated services support by packet marking, using 2 bits from the IP header to mark premium traffic and the outcome of an ingress profiler.

RFCs

Request for Comments documents (RFCs) are documents published by the RFC editor. They are available online at www.rfc.editor.org.

[RFC791] *Internet Protocol, DARPA Internet Program Protocol Specification*, J. Postel, ed., RFC791, Standard RFC, September 1981.

The specification of version 4 of the Internet Protocol.

[RFC793] *Transmission Control Protocol, DARPA Internet Program Protocol Specification*, J. Postel, ed. RFC793, Standard RFC, September 1981.

The specification of the Transmission Control Protocol component of version 4 of the Internet Protocol.

[RFC1112] *Host Extensions for IP Multicast*, S. Deering, RFC1112, Standard RFC, August 1989.

A description of the host functionality required to support IP multicasting.

[RFC1122] *Requirements for Internet Hosts—Communication Layers*, R. Braden, RFC1122, Standard RFC, October 1989.

This RFC is one of a pair that defines and discusses the requirements for Internet host software. It covers the communications protocol layers: link layer, IP layer, and transport layer. Its companion, RFC1123 covers the application and support protocols.

[RFC1349] *Type of Service in the Internet Protocol Suite*, P. Almquist, RFC1349, Proposed Standard RFC, July 1992.

This RFC defines some aspects of the semantics of the Type of Service octet in the IP header. The handling of IP Type of Service by both hosts and routers is specified in some detail. This document has been superseded by the IETF Differentiated Services DS field definition.

[RFC1633] *Integrated Services in the Internet Architecture: An Overview*, R. Braden, D. Clark, and S. Shenker, RFC1633, Informational RFC, June 1994.

This RFC describes the components of the Integrated Services architecture, a proposed extension to the Internet architecture, and protocols to support real- time traffic flows through service quality commitments.

[RFC1812] *Requirements for IP Version 4 Routers*, F. Baker, RFC1812, Proposed Standard RFC, June 1995.

This document defines and discusses requirements for devices that perform the network-layer forwarding function of the Internet protocol suite.

[RFC1889] *RTP: A Transport Protocol for Real-Time Applications*, H. Schulzrinne, S. Casner, R. Fredrick, and V. Jacobson, RFC1889, Proposed Standard, January 1996.

The specification of the Real-Time Protocol, intended to allow a UDP data stream to include timing information in the data stream, supporting the integrity of transferring time-related data across a network.

[RFC1932] *IP over ATM: A Framework Document*, R. Cole, D. Shur, and C. Villamizar, RFC1932, Informational RFC, April 1996.

Diverse proposals describe how to carry IP network traffic over ATM platform networks. This document categorizes such proposals to assist in further refining IP-over-ATM approaches. The intent of this document is to clarify the differences between proposals and to identify common features, to promote convergence to a smaller and more mutually compatible set of standards.

[RFC2022] *Support for Multicast over UNI 3.0/3.1-based ATM Networks*, G. Armitage, RFC2022, Proposed Standard, November 1996.

This describes an ATM mechanism to support the multicast needs of Layer 3 protocols in general. It also addresses its application to IP multicasting in particular.

[RFC2113] *IP Router Alert Option*, D. Katz, RFC2113, Proposed Standard, February 1997.

The specification of an IP Option type that alerts transit routers to more closely examine the contents of an IP packet. This is useful for, but not limited to, new protocols that are addressed to a destination but require relatively complex processing in routers along the path.

[RFC2139] *RADIUS Accounting*, C. Rigney, RFC2139, Informational RFC, April 1997.

The description of a protocol intended for carrying accounting information between a network access server and a shared accounting server. Current use has broadened this purpose to a more general role of carrying both accounting and user profile data.

[RFC2205] *Resource ReSerVation Protocol (RSVP) Version 1 Functional Specification*, R. Braden, L. Zhang, S. Berson, S. Herzog, and S. Jamin, RFC2205, Proposed Standard, September 1997.

This RFC describes version 1 of RSVP, a resource reservation setup protocol designed for an integrated services Internet. RSVP provides a mechanism to establish resource reservations for multicast or unicast data flows.

[RFC2208] *Resource ReSerVation Protocol (RSVP) Version 1 Applicability Statement: Some Guidelines on Deployment*, A. Mankin, F. Baker, B. Braden, S. Bradner, M. O'Dell, A. Romanow, A. Weinrib, and L. Zhang, RFC2208, Informational RFC, September 1997.

This informational document describes the applicability of RSVP, along with the Integrated Services protocols and other components of resource reservation, and offers guidelines for deployment of resource reservation.

[RFC2211] *Specification of the Controlled-Load Network Element Service*, J. Wroclawski, RFC2211, Proposed Standard, September 1997.

The specification of the network element behavior required to deliver controlled-load service in the Internet using the Integrated Services architecture. Controlled-load service provides the client data flow with a quality of service that closely approximates the service quality that same flow would receive from an unloaded network element, but uses capacity (admission) control to assure that this service is received even when the network element is overloaded.

[RFC2212] *Specification of Guaranteed Quality of Service*, S. Shenker, C. Partridge, and R. Guerin, RFC 2212, Proposed Standard, September 1997.

The specification of the network element behavior required to deliver a service in the Internet with the attributes of guaranteed delay and bandwidth, using the Integrated Services architecture. Guaranteed service provides firm bounds on end-to-end datagram queuing delays.

[RFC2215] *General Characterization Parameters for Integrated Service Network Elements*, S. Shenker and J. Wroclawski, RFC2215, Proposed Standard, September 1997.

This document defines a set of general control and characterization parameters for network elements that support the IETF integrated services QoS control framework capabilities. General parameters are those with common, shared definitions across all QoS control services.

[RFC2225] *Classical IP and ARP over ATM*, M. Laubach and J. Halpern, RFC2225, Proposed Standard, January 1998.

This proposed standard defines an application of classical IP and ARP in an Asynchronous Transfer Mode (ATM) network environment configured as a logical IP subnetwork (LIS). It considers only the application of ATM as a direct replacement for the "wires" and local LAN segments connecting IP end stations and routers operating in the "classical" LAN-based paradigm.

[RFC2332] *NBMA Next-Hop Resolution Protocol (NHRP)*, J. Luciani, D. Katz, D. Piscitello, B. Cole, and N. Doraswamy, RFC2332, Proposed Standard, April 1998.

This document describes the NBMA Next-Hop Resolution Protocol (NHRP), which can be used by a source station (host or router) connected to a nonbroadcast, multiaccess (NBMA) subnetwork to determine the internetworking layer address and NBMA subnetwork addresses of the NBMA next hop toward a destination station. NHRP is intended for use in a multiprotocol internetworking layer environment over NBMA subnetworks.

[RFC2381] *Interoperation of Controlled-Load Service and Guaranteed Service with ATM*, M. Garrett and M. Borden, RFC2381, Proposed Standard, August 1998.

This document provides guidelines for mapping service classes and traffic management features and parameters between Internet and ATM technologies. The service mappings are used for providing effective interoperation and end-to-end quality of service for IP Integrated Services networks containing ATM subnetworks. The discussion and specifications support the IP Integrated Services protocols for guaranteed service (GS), controlled-load service (CLS) and the ATM Forum UNI specification, versions 3.0, 3.1, and 4.0. Some discussion of IP best-effort service-over-ATM is also included in this document.

[RFC2382] *A Framework for Integrated Services and RSVP over ATM*, E. Crawley, L. Berger, S. Berson, F. Baker, M. Borden, and J. Krawczyk, RFC2382, Informational RFC, August 1998.

This document outlines the issues and framework related to providing IP Integrated Services with RSVP-over-ATM. It provides an overall approach to the identified problems and related issues.

[RFC2474] *Definition of the Differentiated Services Field (DS Field) in the IPv4 and IPv6 Headers*, K. Nichols, S. Blake, F. Baker, and D. Black, RFC2474 Proposed Standard, December 1998.

This document describes an approach to differentiated services enhancements to the Internet Protocol, intended to enable scalable service discrimination in the Internet without the need for per-flow state and signaling at every hop. The document proposes a small, well-defined set of building blocks to deploy in network nodes, using packet marking as the trigger to active per-hop service responses.

[RFC2475] *An Architecture for Differentiated Services*, S. Blake, D. Black, M. Carlson, E. Davies, Z. Wang, and W. Weiss, RFC2475, Proposed Standard, December 1998.

The architecture description for the differentiated services enhancements to the Internet protocol. This architecture achieves scalability by aggregating traffic classification state, which is conveyed by means of IP-layer packet marking using the DS field. Packets are classified and marked to receive a particular per-hop forwarding behavior on nodes along their path. Sophisticated classification, marking, policing, and shaping operations need only be implemented at network boundaries or hosts. Network resources are allocated to traffic streams by service provisioning policies that govern how traffic is marked and conditioned upon entry to a differentiated services-capable network, and how that traffic is forwarded within that network.

[RFC2597] *Assured Forwarding PHB Group*, J. Heinanen, F. Baker, W. Weiss, and J. Wroclawski, RFC2597, Proposed Standard, June 1999.

This document defines a general-use differentiated services per-hop-behavior group called Assured Forwarding (AF). The AF group provides delivery of IP packets in four independently forwarded AF classes. Within each AF class, an IP packet can be assigned one of three different levels of drop precedence. A DS node does not reorder IP packets of the same micro-flow if they belong to the same AF class.

[RFC2598] *An Expedited Forwarding PHB*, V. Jacobson, K. Nichols, and K. Poduri, RFC2598, Proposed Standard, June 1999.

This document describes a differentiated service per-hop behavior (PHB) called Expedited Forwarding. The document shows the generality of this PHB by noting that it can be produced by more than one mechanism; it includes an example of its use: to produce at least one service, a virtual leased line. A recommended codepoint for this PHB is given.

IETF Internet Drafts

IETF Internet drafts are work-in-progress documents. They are valid within the IETF process for a maximum period of six months from the date of submission.

Internet drafts are not normally referenced, but those cited here are the best pointers to current research and developmental efforts, and so have relevance to this topic of quality of service and network performance. Typically, these documents follow the Internet Standards process and are published as RFCs sometime in the future. The current collection of Internet drafts, along with pointers to the RFC documents, can be found at www.ietf.org references.

[ID-cops] *The COPS (Common Open Policy Service) Protocol*, J. Boyle. R. Cohen, D. Durham, S. Herzog, R. Rajan, and A. Sastry, IETF Internet Draft, draft-ietf-rap-cops-06.txt, February 1999.

This document describes a simple client/server protocol for supporting policy control over QoS signaling protocols and provisioned QoS resource management. The model underlying the protocol specification does not make any assumptions about the methods of the policy server; it is based on a policy server returning decisions to requests from policy clients.

[ID-cops-pib] *Quality of Service Policy Information Base*, M. Fine, K. McCloghrie, J. Seligson, K. Chan, S. Hahn, and A. Smith, Internet Draft, ietf-mfine-cops-pib-01.txt, June 1999.

This draft defines a set of policy rule classes for describing quality of service policies. This document structures QoS policy information as instances of policy rule classes. A policy rule class (PRC) is an ordered set of scalar attributes. Policy rule classes are arranged in a hierarchical structure similar to tables in SNMP's SMIv2. As with SNMP tables, they are identified by a sequence of integer identifiers (an object identifier). For each policy rule class, a device may have zero or more policy rule instances. Each policy rule instance is also identified by a sequence of integers where the first part of the sequence is the identifying label of the PRC. Collections of policy rule classes are defined in PIB modules.

[ID-cops-rsvp] *COPS Usage for RSVP*, J. Boyle, R. Cohen, D. Durham, S. Herzog, R. Rajan, and A. Sastry, IETF Internet Draft, draft-ietf-rap-cops-rsvp-05.txt, June 1999.

This document describes the format of client-specific information fields for COPS when supporting COPS policy services in RSVP environments.

[ID-diameter] *DIAMETER Base Protocol*, P. Calhoun, IETF Internet Draft, draft-calhoun-diameter-07.txt, November 1998.

The DIAMETER base protocol is intended to provide a framework for any services that require AAA/policy support. The protocol is intended to be flexible enough to allow services to add building blocks (or extensions) to meet their requirements. This draft specifies the message format and transport to be used by all DIAMETER extensions.

[ID-diameter-qos] *DIAMETER QoS Extension*, P. Calhoun and K. Pierce, IETF Internet Draft, draft-calhoun-diameter-qos-00.txt, May 1998.

This document describes a simple client/server model for supporting QoS policies. A router that supports RSVP or one of the proposed differentiated service schemes will require a policy database and a means to access it. This document describes the extensions to a protocol that are based on Radius, called DIAMETER. The document describes the elements of the protocol and some examples of interactions based on it. The document does not describe the policy database or policy enforcement.

[ID-diameter-resmgmt] *DIAMETER Resource Management Extensions*, P. Calhoun and N. Greene, IETF Internet Draft, draft-calhoun-diameter-resmgmt-03.txt, February 1999.

DIAMETER is a policy protocol used between a client and a server for authentication, authorization, and accounting of various services. Examples of such services are for dial-up users (ROAMOPS), RSVP admission policies (RAP), Fax-over-IP (FAXIP), voice-over-IP (IP Tel) and integrated services. This document defines a set of commands that allow DIAMETER servers to maintain session state information.

[ID-diffserv-frame] *A Framework for Differentiated Services*, Y. Bernet, J. Binder, S. Blake, M. Carlson, B. Carpenter, S. Keshav, E. Davies, B. Ohlman, D. Verma, Z. Wang, and W. Weiss, IETF Internet Draft, draft-ietf-diffserv-framework-02.txt, February 1999.

This document provides a general description of issues related to the definition, configuration, and management of services enabled by the Differentiated Services architecture.

[ID-diffserv-trtcm] *A Two-Rate Three-Color Marker*, J. Heinanen and R. Guerin, IETF Internet Draft, draft-heinanen-diffserv-trtcm-01.txt, May 1999.

This document defines a two-rate three-color marker (trTCM), which can be used as a component in a DiffServ traffic conditioner. The trTCM meters an IP packet stream and marks its packets based on two rates—peak information rate (PIR) and committed information rate (CIR)—and their associated burst sizes to be either green, yellow, or red. A packet is marked red if it exceeds the PIR; otherwise, it is marked either yellow or green depending on whether it exceeds or doesn't exceed the CIR.

[ID-diff-svc] *Integrated Services-over-Differentiated Services*, P. Ford and Y. Bernet, IETF Internet Draft, draft-ford-issll-diff-svc-00.txt, March 1998.

A proposal to map Integrated Services functions to a Differentiated Service network, adopting the approach that a Differentiated Services network appears as an instance of a specific media type to the Integrated Services network.

[ID-intdiff] *A Framework for End-to-End QoS Combining RSVP/IntServ and Differentiated Services*, Y. Bernet, R. Yavatkar, P. Ford, F. Baker, and L. Zhang, IETF Internet Draft, draft-bernet-intdiff-00.txt, March 1998.

A proposal for end-to-end QoS services that assumes the use of Integrated Services mechanisms to negotiate QoS at the edge of the network and the use of Differentiated Services mechanisms in the core of the network.

[ID-isl802-frame] *A Framework for Providing Integrated Services over Shared and Switched IEEE 802 LAN Technologies*, A. Ghanwani, J. Pace, V. Srinivasan, A. Smith, and M. Seaman, IETF Internet Draft, draft-ietf-issll-is802-framework-07.txt, June 1999.

This document describes a framework for supporting Integrated Services on shared and switched LAN infrastructures. It includes background material on the capabilities of IEEE 802-like networks with regard to parameters that affect Integrated Services, such as access latency, delay variation, and queuing support in LAN switches. It outlines a functional model for supporting RSVP in such LAN environments.

[ID-isl802-map] *Integrated Services Mappings on IEEE 802 Networks*, M. Seaman, A. Smith, E. Crawley, and J. Wroclawski, IETF Internet Draft, draft-ietf-issll-is802-svc-mapping-04.txt, June 1999.

This document describes mappings of IETF Integrated Services over LANs built from IEEE 802 network segments that may be interconnected by IEEE 802.1D MAC bridges.

[ID-isl802-sbm] *SBM (Subnet Bandwidth Manager): A Protocol for RSVP-based Admission Control over IEEE 802-style Networks*, R. Yavatkar, D. Hoffman, Y. Bernet, and F. Baker, IETF Internet Draft, draft-ietf-issll-is802-sbm-08.txt, May 1999.

This document defines SBM, a signaling protocol for RSVP-based admission control over IEEE 802-style networks. SBM provides a method for mapping an Internet-level setup protocol such as RSVP onto IEEE 802-style networks. In particular, it describes the operation of RSVP-enabled hosts/routers and link-layer devices (switches, bridges) to support reservation of LAN resources for RSVP-enabled data flows.

[ID-isl-slow] *Providing Integrated Services over Low-Bitrate Links*, C. Bormann, IETF Internet Draft, draft-ietf-issll-islow-06.txt, June 1999.

An architecture for providing integrated services over low bit-rate links, such as modem lines, ISDN B-channels, and sub-T1 links. The main components of the architecture are: a real-time encapsulation format for asynchronous and synchronous low bit-rate links, a header compression architecture optimized for real- time flows, elements of negotiation protocols used between routers (or between hosts and routers), and announcement protocols used by applications to allow this negotiation to take place.

[ID-isl-svcmap] *Network Element Service Specification for Low-Speed Networks*, S. Jackowski, D. Putzolu, E. Crawley, and B. Davie, IETF Internet Draft, draft-ietf-issll-isslow-svcmap-08.txt, May 1999.

This document defines the service mappings of the IETF Integrated Services for low bit-rate links, specifically, the controlled-load and guaranteed services. The approach takes the form of a set of guidelines and considerations for implementing these services, along with evaluation criteria for elements providing these services.

[ID-rap-frame] *A Framework for Policy-based Admission Control*, R. Yavatkar, D. Pendrakis, and R. Guerin, IETF Internet Draft, draft-ietf-rap-framework-03.txt, April 1999.

This document is concerned with specifying a framework for providing policy-based control over admission control decisions in a QoS-enabled environment. In particular, it focuses on policy-based control over admission control using RSVP as an example of the QoS signaling mechanism.

[ID-rap-pr] *COPS Usage for Policy Provisioning*, F. Reichmeyer, S. Herzog, K. Chan, D. Durham, R. Yavatar, S. Gai, K. McCloghrie, and A. Smith, IETF Internet Draft, draft-ietf-rap-pr-00.txt, June 1999.

This draft introduces a new client type to the COPS framework to support provisioning with policy input. This client model is based on the concept of a Policy Information Base. The area of applicability for this model includes policy input into DiffServ admission control systems.

[ID-rap-preempt] *Signaled Preemption Priority Policy Element*, S. Herzog, IETF Internet Draft, draft-ietf-rap-signaled-priority-03.txt, February 1999.

This document describes a preemption policy element for use by signaled policy-based admission protocols. The preemption priority defines the relative weight of a flow within a set of flows that are competing for admission to the network.

Rather than admitting flows by order of arrival, preemption priorities allow the network to support the highest-priority subset of these flows.

[ID-rap-rsvpext] *RSVP Extensions for Policy Control*, S. Herzog, IETF Internet Draft, draft-ietf-rsvp-ext-06.txt, April 1999.

This document presents a set of extensions to RSVP to support generic policy-based admission control. The extensions include the standard format of policy data objects and a description of RSVP's handling of policy events.

[ID-rap-rsvpid] *Identity Representation for RSVP*, S. Yadav, R. Yavatar, R. Pabbati, P. Ford, T. Moore, and S. Herzog, Internet Draft, draft-ietf-rap-rsvp-identity-04.txt, July 1999.

This document describes the representation of identity information in the policy data object extension for RSVP for supporting policy-based admission control in RSVP. The goal of identity representation is to allow a process on a system to securely identify the owner and the application of the communicating process, and convey this information securely in RSVP messages. The document also describes the use of this identity information in an operational setting.

chapter five

Performance Engineering

Making it work is important. Making it look pretty is an optional extra.

We have examined the basic tools of performance engineering and the various architectures that have been proposed to support well-defined service responses from an Internet network. But a toolkit without a problem to use it for is often a dangerous situation. As the saying goes, "When all you have is a hammer, everything looks like a nail." So in this chapter, we will look at various applications for which QoS mechanisms may be of some benefit to the network engineer.

The first question to ask is: "What outcomes can be achieved by using these performance architectures within the network?" As we will see, no single performance architecture will address all possible service requirements, as each has its strengths and weaknesses. But even when one architecture cannot achieve everything, it's useful to know what the set of potential requirements are. The intended outcome of performance engineering is to modify the service response of the network, altering the default uniform best-effort service response to add capabilities, which include:

◆ Customizing performance for a network application.
◆ Emulating a dedicated bandwidth circuit.
◆ Mitigating the effects of network congestion.
◆ Providing a premium quality service.
◆ Providing service quality to multicast services.

First we will examine a number of practical requirements for quality of service, and for each requirement, the engineering approach that can produce the appropriate outcomes.

QoS Components

All engineering activities undertaken to construct QoS networks require the configuration of all active network elements to respond to QoS signals, if end-to-end services are to be supported. Routers have to be configured with packet classification functions and selective queue management capabilities, together with selective packet discard responses. This enables the router to enforce various service disciplines, as well as to perform traffic policing and traffic shaping roles. Applications, hosts, and routers need to support a common signaling environment, so resource allocation requests can be passed through the network. Most important, there is a requirement to support resource management across the entire network, so that dedicated resource requests are not compromised through over-subscription and so that end-to-end service profiles are maintained under conditions of varying load.

Different network elements play different roles in the support of QoS. For example, many Internet service networks use a core and edge design whereby traffic is collected from client networks and client hosts at edge routers. The aggregated traffic is passed to core routers, and transmitted across the backbone of the network through large high-capacity circuits. The traffic is then passed to the destination client network or host or to the next-hop transit network through an egress edge router. In such a design, the ingress edge routers undertake the functions of traffic classification, admission control, and signaling management. The ingress edge routers must enforce a resource management regime such that the service levels can be maintained across the network. The interior core routers undertake queuing control to manage prioritization of traffic and congestion through selective packet discard mechanisms. The egress routers enforce egress policy with respect to transit networks or customer service level agreements. This service architecture of a boundary admission control process and a simple resource management response in the interior of the network is shared with many ATM and Frame Relay service architectures where the service profile is determined through admission control. Traffic entering the network is compared against this service profile, and packets passing through this profile filter are either marked as *in-profile*, or *out-of-profile*, depending on the relative states of the packet flow and the profiler. The interior elements of the network can then use this *per-packet marking* to determine a selective discard process as a part of congestion management. The engineering rule of thumb is that uniform deployment of all component elements of QoS functionality is undesirable. The most appropriate engineering approach should ensure that each network router is configured with just those elements that enable it to undertake necessary local QoS tasks.

This QoS engineering effort is of benefit not only to a narrow sector of Internet service networks. No network is so extravagantly over-provisioned that every application and every business requirement can be met without some level of contention within the network. Enterprise networks need to allocate network resources to align with business imperatives; the so-called mission-critical applications for an enterprise may require some form of preferential response from the network; the various multiprotocol and multiservice loads of the network may be subject to some level of relative prioritization to ensure that each application receives the amount of network resource it requires to function correctly. ISP service networks are looking for means to broaden their market opportunities through QoS. They want to be able to offer various levels of service response from the network with an associated range of price points, along with higher-value service responses to higher-valued applications, such as voice and video support. At the same time, ISPs must be able to offer such QoS responses across very large and diverse network platforms, so scalability is as critical an issue as performance tuning.

Service Models

There are three basic service models that can be applied to a traffic flow:

◆ Best-effort service (or *variable* service)
◆ Differentiated service (or *better* service)
◆ Integrated service (or *constant load* service)

The suitability of the use of these models depends, to an extent, on the application that generated the flow in the first place. Nonelastic real-time applications, such as interactive voice or streaming video, typically assume a constant service response from the network. Bandwidth emulation services, such as those commonly associated with virtual leased lines, can be supported effectively using a premium differentiated service model, while adaptive applications, such as the file transfer protocol, FTP, assume a best-effort service response from the network, and optimize their behavior accordingly. It is also useful to remember that the deployment of premium and constant services is not without cost, not only for development, but for the additional detail within the operational environment; consequently, the broader service environment will entail still additional cost from operating the network.

Best-Effort Service Profile

This is the base level Internet service profile. The application is able to pass data into the network using any load profile, without any form of notification to the network. The response from the network is to deliver the load to its destination, but no service undertaking is made with respect to delay, variability of delay, or levels of packet loss.

Best-effort service is the outcome of the normal action of routers, without any form of admission control policy. By default, routers implement a simple FIFO queuing discipline (as described in Chapter 3, "Performance Tuning Techniques"); and when the queue is saturated, packets are discarded (implementing a *tail drop* packet discard method).

Differentiated Service

The differentiated service profile adds some QoS functionality to the network, but it does not attempt to maintain end-to-end QoS paths. Network admission policy is introduced, encompassing classification, packet marking, and, potentially, some form of traffic-conditioning or -shaping function. The classification and marking are intended to be consistent with the chosen QoS admission policy, and the marks are configured to induce various per-hop behaviors in the interior of the network. There is no application signaling within the differentiated services environment, as the admission controls are configured by the network operator.

The per-packet markings trigger differentiated service responses from the network's interior routers. The service response may be queuing precedence or queuing service class selection, in which cases the packet mark would be used in conjunction with priority queuing or weighted fair queuing mechanisms respectively. Alternatively, the packet mark may be some drop preference level, potentially used in conjunction with weighted random early detection on the router, or as a simple drop precedence as an alternative to tail drop.

The advantage of the differentiated service approach is that it decouples the admission mechanisms from the interior per-hop behavior, allowing a single service response to scale to a large number of aggregated traffic flows.

Integrated Service

Integrated services adds the application to the service delivery architecture. The intent is that the application request a specific service profile from the network before sending any data. The application and the network share some common service signaling mechanisms. The application is expected to generate the service request, then await a positive confirmation from the network that a path whose characteristics match the service request has been created within the network. Only then should the application commence sending data along this network path.

The signaling used to support this service model is based on the Resource Reservation Protocol (RSVP). This protocol sets up a unidirectional end-to-end path by forwarding a service profile as a call request, and having the call reservation response follow backward along this path, locking down the path and the associated reservation state. The network has to dynamically maintain an end-to-end state within the network. The state is necessary to ensure that the reserved network capacity is not over-subscribed to subsequent reservations. It is also necessary to ensure that if the network state changes in such a way that the forwarding path is broken, the path must be reestablished.

The call setup actions include establishing an admission control filter, in order to police the requested service profile. The service path also has an associated distributed classification state, so that all packets generated by the application that pass through the admission filter as part of the QoS flow are queued and forwarded according to the path specifications.

In the interior of the network, two queuing responses can be used with this model. Guaranteed service profiles require some form of preemptive queuing, either by conventional priority queuing mechanisms, or by adjusting the weights of a weighted fair queuing system. A controlled-load service does not require the same levels of preemptive response from the interior routers of the network, and some level of network-imposed jitter and packet loss can be tolerated on a controlled-load service stream. The queuing response from the interior routers to a controlled-load service request would normally be via a weighted fair queuing system, or some variant of class-based queuing. In the case of TCP streams, the same outcomes can be produced by varying the weights of a weighted random early detection queue management environment.

There are three generic approaches to the application of performance engineering within an Internet network, corresponding to these three service models.

The first approach is to undertake reactive *oversupply service engineering*, whereby the network is periodically redimensioned to carry the observed load levels with some margin of over-supply. Given the highly self-similar nature of Internet traffic patterns, this approach exposes the network to transient congestion peaks, so some form of admission traffic-shaping is very useful with this approach. The underlying assumption here is that it is cheaper overall to dimension all parts of the network to meet current load profiles than it is to deploy a set of policy-driven admission systems and a related set of service management features in the interior of the network. This approach to service quality engineering has been a relatively constant feature of the LAN environment, where tracking of demand has evolved from a common bus cable to a hub, and then to a switching hub, with the increase in available capacity from a shared 10Mbps, to a switched 10Mbps, to 100Mbps, to full duplex variants of 100Mbps, to the current LAN technology of 1Gbps-switched LANs. To date, in the absence of deployment of service performance management systems in Internet networks, this same approach is a feature of Internet service networks. The service network is dimensioned to meet the current load profile, without the explicit statement of any service level agreements or quality and performance guarantees. In a wide area network (WAN), this approach often leads to a cyclical service behavior: An increment in network capacity allows the network to operate in a mode of over-supply for a period, and then as the load levels expand to fill this space, the network performance starts to deteriorate, calling for another increment in capacity (see Figure 5.1).

The second approach involves *state-based systems*, or flow-based systems. These systems use an initial service request via a signaling negotiation across the network, followed by a traffic flow, only if the network accepts the service request. The request is for a specified service level from the network; associated with the request is an indication of the intended volume and load profile of the subsequent traffic flow. This request is normally mapped to an RSVP interaction. The outcome of the RSVP signaling is the addition of a new state to the network. This state is a distributed state, linking each active network element on the selected end-to-end path. On each network element, the state includes a filter, used to recognize packets that are members of the service flow; a service directive of some form of queuing precedence and drop preference; and a forwarding directive to pass the packet along the predetermined path. The ingress component of the path may include an admission filter, to police the requested traffic profile; the ingress element may also undertake path integrity control, to ensure that any changes within the network that compromise the integrity of the path cause a new path to be established.

The overhead of such state-based systems include an initial time penalty to the application, because it must await a minimum of one round-trip interval before it will receive a path confirmation and proceed with passing traffic into the network. Such a start-up latency is a prohibitive performance penalty for very short transaction-based applications. From the network perspective, the

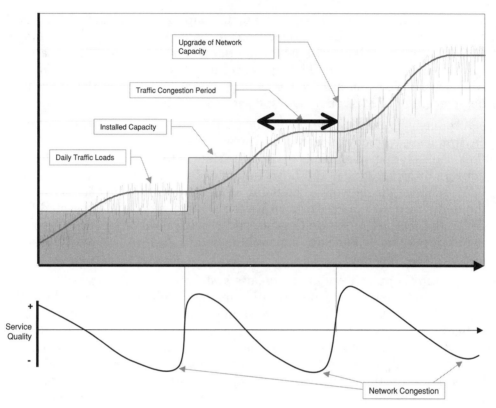

Figure 5.1 Cyclical network service levels.

implications of supporting state-based systems are more significant. Each active network element must allocate some amount of local resource to store the classification condition, the corresponding traffic service directives, and the forwarding instruction, together with the resource reservation quantity. In addition, the network has another traffic component in managing the integrity of paths with associated resource reservations. Such overheads are not without incremental cost, and this *scaling cost* is at the heart of the consideration to determine whether to deploy state-based or stateless QoS architectures. Alternatively, if the network uses an ATM service platform, one approach is to set up an ATM SVC for each QoS request, and use ATM to match the requested service parameters. But such an approach raises the issue of mapping the burst characteristics of an Internet traffic profile to the ATM service environment, plus the associated requirement to map QoS requirements to ATM queuing behavior for correct operation of the service. The proliferation of SVCs to support such micro-flows is another relevant consideration, as are the associated demands on the PNNI protocol to correctly manage the ATM network's available resources. Moreover, this approach assumes the ubiquitous deployment of ATM as an end-to-end service environment.

A refinement of this approach is that of traffic engineering within an MPLS environment, whereby the granularity of a flow is extended to that of a *macro-flow*. This alternative attempts to limit the state of dynamic change within the network by aggregating micro-flows into one of a number of discrete channels on entrance to the network, then using a soft state within the network to pass the traffic through the network within the parameters of the aggregated flow. Given the lack of an explicit end-to-end service reservation within such a model, it is arguable that such a model fits within the third approach from the end host's perspective.

The third approach involves *stateless systems*, or nonflow-based systems. These do not use the explicit reservation model to deliver service. Overall, the approach can be regarded as an exercise in preemption. An application is able to mark packets with a service indicator. This service request is confirmed or denied at the point of admission to the network, through the actions of some form of admission filter. Alternatively, these service indicators can be added to the packet at the time of ingress to the network, using a group of rulesets as the policy mechanism. Each active network element uses these coded service indicators to deliver some level of preemptive service to the packet as it transits the network. Many overheads of the state-based system are eliminated by using this service architecture. There is no setup latency for the service request, nor a per-element requirement for memory or processing overhead, nor any additional resource management traffic level supported within the network. Of course, not all is gain; the loss is that the service client now has no means of guaranteeing a particular service response, nor can the client request some determined level of network resource to support its application. The best the client can depend on is some level of service that approximates a guaranteed service profile, but that is prone to some level of congestion impact.

The granularity of the service client has some impact on the choice of a service architecture. Services can be applied at the level of micro-flows, corresponding to individual TDP or UDP traffic sessions. This may be a Web page request and response or a Domain Name System zone transfer. Within TCP, micro-flows are readily associated with a sequence of packets from the initial SYN handshake to the closing FIN sequence. Within UDP, such micro-flows are more difficult to identify from a network perspective, given the absence of any protocol start and close indicators; but they can be seen as a sequence of packets generated by a single instance of an application, directed to a constant destination IP and port address. In network terms, a micro-flow can be classified by a sequence of packets with a constant address vector of:

```
(source IP address, source port, destination IP address, destination
port)
```

Within a large provider network, there may be hundreds of thousands or even millions of micro-flows being created per second. When using such a fine level of granularity within a service delivery architecture, the scaling properties of the architecture become a major concern.

Note that in tunneled environments, and where IPSEC is used to encrypt the IP packet payload, the TCP and UDP port addresses are not visible to the network, and the source and destination addresses may not necessarily represent the actual end addresses, but may instead represent the tunnel end points, or the IPSEC end points. As a general observation, micro-flows are not always visible to the network, and in such cases the granularity of the service response can, at best, be directed toward servicing aggregates of micro-flows, or macro-flows.

With macro-flow responses the granularity of the service model becomes coarser. A macro-flow service response granularity can be a path between two networks, where the common classification is a pairing of a set of address prefixes. This would correspond to a set of packets with source IP addresses from one set and destination IP addresses from the other set. This level of granularity matches the traffic flows normally associated with support of the virtual private network (VPN) service model. If the model is relaxed slightly to match source or destination addresses from within a single address set, this would correspond to a per-client service level guarantee (SLG) application. As the service granularity becomes coarser, there are fewer concerns about the scaling properties of the service architecture. However, the accuracy of the service response also becomes coarser, thus the intrinsic value of the service declines.

The stateless service models have a relatively broad range of applicability. In contrast, considerations of the scale and value of QoS at various levels of service granularity imply that state-based service models have a more focused range of applicability. In corporate and campus networks, the scale is of a size that makes micro-flow support feasible. Larger ISP service networks would see the optimal service application point for state-based services models at the level of VPN support, and similar aggregated traffic service applications, and will avoid the imposition of micro-flow state onto the network.

Service Engineering

Meeting the need for some form of premium or consistent service response to a selected subset of network traffic requires the imposition of a number of traffic control mechanisms into the network. These mechanisms would be used to:

- Manipulate queuing systems to reserve some amount of bandwidth to certain classes of traffic, or to provide some form of prioritized response to classes of traffic.
- Control the packet discard selection function to provide limited loss probability to classes of traffic.
- Provide a differentiated response to various traffic classes in the face of local congestion events.
- Shape the load profile of various traffic classes.

These mechanisms in the network establish control over:

- Router functions of classification, queuing, forwarding and discard, to generate outcomes of traffic differentiation, shaping, and policy-based resource sharing.
- Traffic entering the network to impose a network-defined classification marking, admission policing and shaping, and to implement resource allocation policies.
- Signaling mechanisms to coordinate a service response across all active network elements.

This is a relatively complex agenda for a large and diverse service network, and is complicated by the additional requirement to simultaneously support best-effort traffic, using a model of distributed resolution of resource contention that drives this class of traffic into a condition of critical resource oversubscription.

Generally, the engineering response to such requirements is to place all the service-intensive functions as the edge of the network. Traffic classification, shaping, packet marking, configuration management, and signaling control functions are placed at the ingress points to the network, where the individual ingress traffic flows are relatively low in proportion to the amount of processing capability that the network can deploy to manage the function. IP backbone engineering normally uses an aggregated backbone of a small number of major trunk circuits, rather than a finely distributed mesh of smaller circuits. The implication of this engineering model is that the interior of the network uses high-speed dedicated switching control functions, so that the service management functions that can be undertaken cost-effectively in the interior of the network are limited to congestion management and avoidance functions, implemented through simple controls imposed on the queuing function. This specialization of function is shown in a typical Internet service provider access center design, shown in Figure 5.2. The systems that terminate customer circuits are configured with admission control systems, and use a policy server to synchronize the operation of the unit with the network's admission policies. Admitted traffic is marked with a service

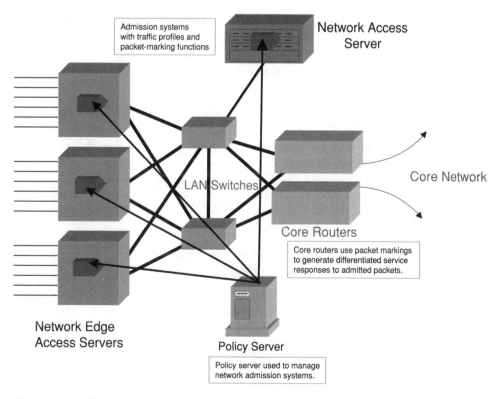

Figure 5.2 ISP access Point of Presence design.

selector, as defined in the DiffServ architecture, allowing the core routers to use this marking when forwarding the packet. This engineering approach to supporting QoS-enabled Internet networks can be summarized by the phrase *smart edges, dumb core.*

QoS Service Applications

The question to ask at this juncture is: "What outcomes can be achieved by using these tools in the network?" No single performance architecture can address all possible requirements; as pointed out in Chapter 4, each has its strengths and weaknesses, and therefore its limitations for use. We will now examine a number of practical requirements for quality of service, and for each requirement, explore the engineering approach that can produce the appropriate outcomes.

There are a wide range of possible QoS requirements in Internet service networks, but they tend to be variants of a smaller collection of generic cases:

- Customizing performance for an instance of a network application.
- Emulating a dedicated bandwidth circuit as a response to a macro-flow.
- Providing a consistent quality service by mitigating the effects of network congestion.
- Shaping network traffic to a service profile.
- Providing a premium quality service through traffic prioritization.

In this section we will look at each of these generic cases in conjunction with the engineering responses that can produce these outcomes.

Per-Application Services

In order to support per-application services, the granularity of the service delivery mechanism is at the level of individual applications (or micro-flows). The ideal interaction between an application and a network willing to provide per-application services follows this sequence:

1. The application passes its service requirements to the network, indicating the level of service requested, and the destination point (or points, in the closed group multicast case) for the service.

2. The receiving applications and the network respond to the initiating application, indicating whether the service request is granted.

3. The application generates traffic to the nominated destination points; the application enforces the originally specified service profile, while the network may reshape or police the traffic to the service profile.

4. Immediately prior to completion, the application may inform the network that any resources associated with the service request can be released.

This interaction is shown in Figure 5.3.

Not all applications can state their service requirements so clearly, and there is some allowable latitude for a slightly less rigid interaction between the application and the network. There is a second way of classifying appli-

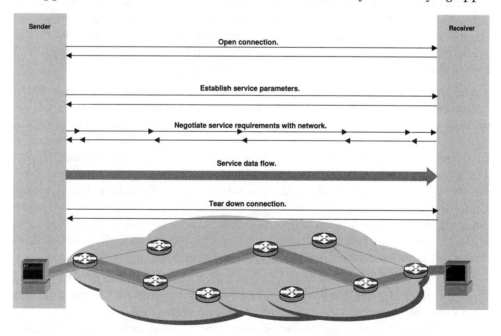

Figure 5.3 Per-application service delivery.

cations: by the manner of their request to the network for a service response. There are two forms of service requests:

- Premium best-effort service
- Consistent service

The per-application requests being generated by premium best-effort applications generally take the form: Please provide disproportionately greater levels of resource to this identified traffic flow. Put more simply, this is the go-faster request. This is a per-application premium service request, where the request is to provide some form of elevated service level. It may be for greater bandwidth, lower loss rates, lower levels of induced jitter, or lower delay. The essential attribute is that an absolute level of service is not specified; rather, the request is for a better response than would otherwise be the case if the request were not made.

As an example, a streaming voice application may be tolerant of a variable level of jitter, packet reordering, and delay, up to certain thresholds. One way of ensuring that the network does not distort the data to exceed these threshold values is for the streaming audio sender to request a premium best-effort service. In this case, there may still be jitter, reordering, delay, and some level of loss at the receiver, but the probability that the level of signal distortion will exceed the receiver's tolerance is reduced considerably.

An alternative request is the consistent service response, where the application requests a specified amount of network resource, regardless of the amount of concurrent network activity. In simple terms, this is the constant speed request. Here the application request is for the network to allocate a fixed level of dedicated bandwidth, a maximal jitter rate, or a maximal packet loss rate, regardless of the amount of traffic being carried by the network. Such service responses can be phrased as a minimum performance level, where higher levels of performance can be provided to the application basis if additional capacity is available within the network; it is more common, however, to see a constant performance request, where the requested profile is a precise allocation request.

As an example, if a user wanted to conduct a video call, the desktop video application might generate a consistent service request to the network for bandwidth to match the encoding of the video signal, thereby triggering the remote video application to undertake the same request in the opposite direction. The application might also indicate its level of tolerance to packet loss and induced jitter as part of the service profile. If the network accepted these service requests, both video applications would receive a favorable response, allowing the call to proceed at a determined quality level.

Which family of applications are we talking about in this category of per-application service responses? It makes little sense to group an elastic non-real-time application within such a strict service model. The family of applications that can make effective use of such a per-application service response are nonelastic real-time applications using predictive constant rate traffic models. Such applications are normally associated with voice and video streams, and similar multimedia applications with a predictive load profile. The application characteristics we are looking for are those with a reasonable duration and that carry a high-value payload, where the lack of network distortion of the payload is part of the value of the application's network transaction. Examples of such applications include:

- Interactive voice, or voice-over-IP
- Video streaming, or IP video
- Multimedia real-time conferencing
- Real-time telemetry

A number of aspects of these application families should be highlighted as general to per-application performance systems.

Predictive Applications

The application must be "enabled," so that it is aware of its service requirements in advance of its activity. Such predictive capability is not a property of all application types. Whereas voice and video applications generate a data stream based on the encoding algorithm used, many other real-time applications generate a data stream that varies in response to external events. Various forms of remote sensing systems, for example, generate data loads in response to the amount of remote activity, while other forms of remote visualization applications generate data loads in response to user activity.

Applications that use TCP do not normally have a predictive load profile. TCP is an adaptive protocol, where the traffic rate (or network load) generated by a TCP application varies according to the prevailing network conditions, rather than the application itself. Given this adaptive capability, there is no predictive load associated with a TCP application, and a TCP application may be restricting its potential performance if it nominates a service profile. Any service profile associated with a TCP-based application can only be effectively interpreted by the network as an enforceable maximum load profile, where the network is allowed to enforce the specified rate in the event of network congestion. This is in contrast to a UDP-based application, where the network is requested to regard the application's load profile as the minimum sustainable load level required to support the correct operation of the application.

QoS-Enabled Applications and Networks

The application must use a service negotiation protocol that is common to both the application and the network. The application must be *quality-enabled*, to negotiate its service requirements with the network in advance of commencing the data flow. This assumes that there is a signaling capability enabling the application to signal its service requirements to the host platform. In turn, this platform has to signal these requirements to the next network element on the path, and so on through to the application invoked at the destination. The signaling coordinates the network's response to the application's request and requires that every active element in the path from the source to the destination be configured to deliver a QoS response in response to this signal.

In a controlled corporate network environment, it is possible to undertake uniform deployment of QoS mechanisms. In very large corporate environments, in virtual private networks, or in the Internet itself, this is definitely not the case. Any application, and the signaling mechanism that it uses, must adopt behaviors that detect those instances where a network element is not QoS-enabled, and take sensible default actions in such a case.

Network State

This service model requires the application and the network to maintain a shared state. To effectively guarantee the throughput of a particular load profile, the network must hold a state for the duration of the application. The path through the network must be held steady (or *pinned*); if the path breaks, the entire end-to-end path must be reestablished or the application must be informed of the path failure. The ingress to the network must operate an active admission function, admitting those packets from the application that fall within the service load profile and marking them. Each router on the end-to-end path must recognize packets from this application that have the admission marking and handle them in accordance with the negotiated service profile, using an appropriate queuing precedence function and a forwarding decision per the QoS path specification. When the application finishes, this state has to be cleared from the network.

Requiring a distributed state to be maintained within the hosts and the network is perhaps a critical point of weakness in this model. One of the major strengths in the Internet service model is that the end-to-end Internet application makes no assumptions about the state of the network. Each packet is transmitted as a separate network transaction, and the application decides if and when a packet should be retransmitted, just as the application takes responsibility for determining the profile of load generated into the network. This model allows for very simple, and cheap, Internet networks. The introduction of a per-application state into the active elements of the network does have considerable overhead, however. The application will take longer to start, as it has to undertake a service profile negotiation with the network, a process that will take a minimum of one full round-trip time to pass the service request into the network and receive an indication that a service path has been created for the application. Within the network there is the additional overhead of resource reservation; this is also a resource efficiency constraint, in that a network is constrained from over-subscribing on resource requests. Furthermore, the routers may incur additional packet identification and processing overhead, depending on the method of QoS chosen to support this function.

Unicast and Multicast

This model of request and response has some problems in adapting to the multicast application, where a single sender is attempting to set up a service state to multiple receivers simultaneously, using a branching path. Such potential users of multicast QoS are video conferencing applications, where each sender has multiple, real-time receivers. An application may need to generate a single initial service request and have the service request branch at the same points where the multicast data stream would need to branch. The response to the initial service request would be a set of responses, each indicating acceptance or refusal; concomitantly, there may be a number of nonresponses, indicating a possible non-QoS hop to a particular destination. Obviously, a complete set of acceptances allows the application to proceed, whereas a mixed set of responses poses a dilemma for the application, in that some receivers may not obtain the appropriate service quality. The additional factor to bear in mind is whether the multicast group is statically or dynamically defined. In a dynamically defined multicast group, the set of receivers may not be known by the sender at the time the sender commences setting up the service application. This would tend to place the onus on the receiver to set up the service path across the network to join the existing multicast group.

From these considerations, it would appear that the best way to address this model is to pass the service request function to the potential receiver. If a receiver wants to join a multicast group, the receiver should pass a request toward the sender, specifying a service profile that should be associated with the traffic that will be received. If all the points between the new receiver and the join point of the multicast group are capable of supporting the receiver's service request, then the request can be admitted; otherwise, the request will be denied. This reversal of the service request roles, where the receiver initiates the request, is an important feature of the multicast environment.

We now have two models. For the unicast environment, the sender initiates the service request, then awaits some form of confirmation before proceeding to send traffic. For the multicast model, the receiver generates a service request, and a confirmation of the service request can be immediately followed by the multicast traffic flow. This distinction appears artificial, however, raising the question of whether these two model are different for any good reason. The reason is in the difference in the routing model. In the unicast model, traffic is not necessarily symmetric, and the receiver may take a different network path to reach the sender than the sender takes to reach the receiver. If the sender generates the request, the request will be passed into the network along the same path as that to be used by the subsequent traffic flow. If the receiver generates the request, the request may not pass along the path that the sender will use to reach the receiver. Multicast uses *reverse path forwarding* (RPF) to pass multicast traffic from the multicast group to the receiver, taking the reverse direction to unicast traffic. In this case, initiating the request from the receiver will ensure that the request takes the same path as the subsequent traffic, albeit in the opposite direction.

Engineering Considerations

The sum of these service considerations is that a per-application service model may be useful in a managed network environment that is not overly large. In such environments, where applications and network features can be deployed in a coordinated fashion, this per-application resource reservation model allows the network's resources to be rationally allocated on a demand basis. This ensures that, once a resource commitment is made, other traffic will not affect the operation of the application. Currently, applications that could benefit from such a negotiated service model include video streaming, voice and video calls, virtual reality applications, and similar nonelastic real-time applications where the load profile is predictable to some extent.

An Integrated Services Architecture Approach

For a consistent service requirement, the most suitable approach would be to use the guaranteed service model of the Integrated Services architecture. The network, and the application, would need to use the RSVP to dynamically reserve network bandwidth to meet the application's service requirements. The application, the hosts, and the network routers would all need to be configured to support RSVP requests and to maintain the path state. The application could then generate a data stream to flow data along this *reserved* path, with the understanding that the network would not distort the quality or timing of the flow beyond the limits specified in the original service request.

While the state-based Integrated Services response is technically feasible in a singly managed network, such a per-application guaranteed service response on the part of the network has significant issues associated with scalability in larger multiprovider network domains. The concerns relate to the potential number of paths that have to be maintained across a very large network, along with the ensuing cumulative performance overheads in the per-hop packet classification function, together with the demands on router memory and network traffic in path maintenance. One potential engineering response is to use the Multi-Protocol Label Switching (MPLS) network architecture. Each label-switched virtual circuit can be associated with a per-application data flow, where the flow itself is established within the network using RSVP to reserve network resources to match the application's load profile. MPLS can address some aspects of the interior switching load by pushing the traffic classification function to the network ingress. But this still leaves scaling issues unresolved in such a model, given that a very large network would be prone to consume all the per-flow label address space at the major backbone switching points.

The uncertainties associated with per-flow scalability have led to some reluctance to deploy IntServ per-application QoS measures in many networks. This becomes a circular condition: Applications will not include IntServ per-application service requests until there is widespread deployment of IntServ within the network, and network operators will not deploy IntServ configurations until there is a widely used family of applications that can make use of the facility.

At issue here is the widespread deployment of per-application service response systems, and its associated costs and benefits. Probably, at this stage, the sum of the costs far outweighs the assessment of benefit—which is another way of saying that the price of the service will be beyond the expectation of the potential users of the service, so that usage level will be low. One answer is to chip away at the costs, to make this more attractive to users. An alternative approach is to use a cheaper QoS mechanism that offers lower operational costs with comparable benefits. This is the direction taken by the Differentiated Services architecture.

A Differentiated Services Architecture Approach

In the context of per-application service responses, the Differentiated Services architecture assumes that the application has the capability to mark the packets to indicate the desired service response and that the network elements will attempt to meet that implied service request. As we have already seen with the DiffServ architecture, the application's service requirements can best be phrased in quite broad terms, relative to the best-effort base service level; they are moderated by other traffic that may also be present in the same aggregate service class. This model's weakness is the lack of firm commitment by the network to this implied service level; therefore, the individual application has no indication that the service handling is being met. The application can set the appropriate DS field service bits in the IP packet header, but there is no feedback that all the routers in the end-to-end path will be enabled for differentiated service. Consequently, there is no feedback that the requested service is being provided. A slightly firmer commitment to servicing this traffic can be made as the DiffServ network introduces an admission policy, limiting the amount of traffic load admitted into the network within any particular service profile, to a level commensurate with the amount of resources available to service such a load in the interior of the network.

In a network with multiple internal routes, such an admission policy generates some complications in implementation. For a service resource to be managed with a very high level of assurance, the admission system needs to know how much resource has not yet been committed on the path to the destination; it then undertakes a resource reservation to match the new demand. Such a state-based approach is not available in a Differentiated Services architecture, so the admission control systems must instead make relatively conservative decisions regarding resource availability. From an application's perspective, the introduction of admission control into the service model leaves the situation unchanged. Given that there is no service quality feedback to the application in any case, the application is not aware whether the request service is being denied by the admission control mechanism or by congestion within the network. In either case, the outcome, in terms of compromised service quality, is similar. Damage to the application's traffic flow from an active admission system is identical to damage to the application's traffic flow from a congested network. The application is unaware of the location of the damage. From the network's perspective, the introduction of admission control mechanisms makes the outcome somewhat better. Overload of one service class at one network access point will be trapped by the local admission control, and the remainder of the network will function without any cross-impact. Without admission control, the overload condition would occur in the interior of the network, affecting other flows using the same service class on the path and, potentially, other service classes on the same overloaded router.

The conclusion is that it is possible to support per-application service responses using a differentiated service model, but that the individual application must not rely on being able to use a constant service profile. The application can make use of an aggregated premium service, where, subject to competing demand levels, it can request a better than base best-effort service response. Strengthening the application service model is a subject of further research. An attempt is being made to combine the attributes and behaviors of both Integrated Services, as a edge architecture and interprovider signaling mechanism, and Differentiated Services, as a means of resource management within the interior of the network.

Dedicated Bandwidth Services

The previous section addressed the service needs for individual instances of applications. Another class of service provides some form of service response to a collection of services; it is referred to as an *aggregate*. This service environment is typically used to support a virtual private network (VPN); here, some defined service level is provided to the VPN traffic. The most common objective of these services is to emulate the characteristics of a leased line across an Internet service network. This is commonly referred to as a virtual leased line (VLL). The characteristics of a leased line include constant delay and constant bandwidth, so the emulation of a leased line includes the objective to minimize induced jitter and to ensure that the network provides a constant bandwidth response (see Figure 5.4).

In the case of a VLL, this is the same basic structure as a Frame Relay virtual circuit, or an ATM virtual circuit, where a path with fixed ingress and egress points is maintained across the network. The VLL circuit also has a number of service characteristics, including the minimum committed bandwidth of the circuit, the delay of the circuit, and the imposed jitter bounds for the circuit. Individual applications that pass traffic with the VLL may experience varying levels of service, because the service arrangement does not extend to the interaction of traffic flows within the VLL. The service arrangement is constrained to refer only to the aggregate of all such traffic.

This is a particular instance of a broader class of QoS applications that include *multientity link sharing*, where a link is divided between a number of clients, and each client is given a fixed base level of the link's resources. This is very similar to Frame Relay networks, where each Frame Relay SVC has a Committed Information Base rate that is intended to represent the minimum throughput provided to all traffic within the SVC. Such systems may also be used for *multiservice sharing*, where each class of applications is provided with a fixed base level of resources. This may be used to limit the proportion of network resources consumed by a Web server or to provide a consistent service level to a trunk voice-over-IP (VoIP) switch, where the switch is provided with an aggregate service equal to the total number of simultaneous calls to be supported by the switch.

A number of features of aggregated service requests are worth noting, before we examine the engineering operations that support such a service.

Aggregate Service Quality

The network will not provide any service undertakings for individual applications within an aggregated service class. For example, if the data load within a VLL exceeds its service profile, the admission control may attempt to enforce the requested service profile. Packets within the aggregate data flow may be marked with a high discard preference; or, if the admission system is used to enforce a service profile, packets may be dropped or delayed at the admission control point. Those applications that have experienced this

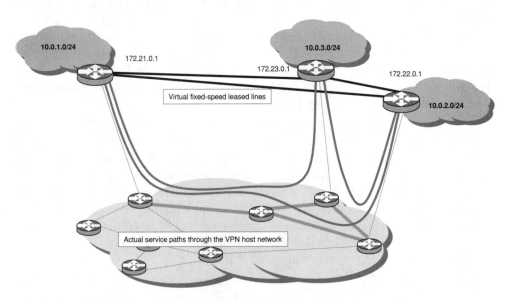

Figure 5.4 The virtual leased line model of VPNs.

packet drop or delay will suffer some form of impaired service quality even though the aggregate service class is being properly supported by the network. This leads to a somewhat anomalous situation: While a network is providing the correct level of service to a aggregated data flow, the applications within that flow may still receive very poor service due to a high level of contention within the aggregate.

For this reason, the dedicated bandwidth aggregated flow service requests still require careful attention to capacity engineering. While it is useful to create a service request for a collection of data flows, it is still necessary to perform some form of resource management within the data flow to ensure that the appropriate service policies and preferences are undertaken.

Static and Dynamic Service Structures

Per-application service requests must be managed dynamically; in contrast, it is possible to use statically configured service structures in many instances of an aggregated flow service request. Often, where the links form part of a VPN, component VLLs are used to support a *tunnel* that links two interior locations across the service provider's network. In such cases, the tunnel can be configured statically, as can the associated service profile for the tunnel's VLL.

Static tunnel service profile configurations of VLL tunnels incur significant administrative overhead, caused be two common objectives. First, not only must the configuration allow the VLL to operate with some form of assured resources from the VPN host network, the interior of the VLL tunnel may be using private address space. As shown in Figure 5.5, the static configuration requires access lists on each router in the end-to-end path. In each access list the traffic corresponding to the VLL is identified, and the next hop on the path is identified. This structure does not use the dynamically configured forwarding table due to the static service reservation associated with each hop of the tunnel's path. This statically configured next hop operation is an instance of policy-based routing. Some form of preferential service is invoked to ensure that the packet is not delayed in flight nor discarded unnecessarily. In addition, the ingress element should include some form of admission control to ensure that the load profile of the VLL meets the assigned service parameters. This requires considerable operational coordination; it is also prone to failure following dynamic changes in the network state, as the next-hop calculations are statically configured, rather than based on the current state of the routing system.

In related applications that use bandwidth allocation without the additional functions of tunneling, there is still the option of using some form of manual configuration to support the application. The actions are similar to the VLL

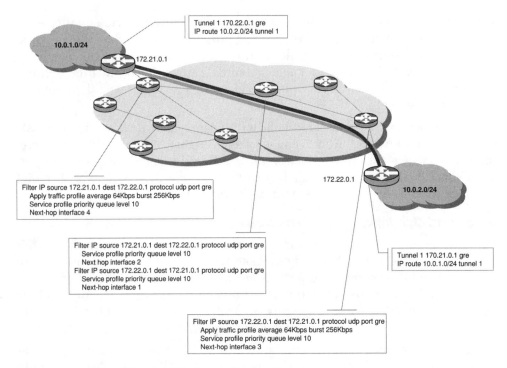

Figure 5.5 Virtual logical links and tunnels.

case, where the traffic is classified and passed through an admission profile system to ensure that the traffic conforms to the bandwidth allocation. The packets can then be marked with the appropriate service classification and passed into the network.

Aggregate service systems do not have a well-defined dynamic service lifetime. Generally, no single application flow defines the start and end points for the service, so that the network admission systems are invoked for as long as the service contract is in force. Normally, however, such aggregate service contracts have well-defined entry points, corresponding to the locations of the client systems that form part of the aggregate service class. These systems often have well-defined network entry and exit points, so that the service contract is specified as a set of transit paths between entry and exit points, with an aggregate bandwidth requirement for each path. Static configuration of the admission systems is often a viable option for the network operator. Dynamic systems should be used in conjunction with a more involved approach to service delivery, one that encompasses these longer-lived service requirements as well as short-term dynamic service requests.

That is not meant to imply that dynamic systems cannot be constructed for aggregated service flows. A dynamic tunnel configuration, for example, requires the use of a tunnel "agent," an instance of an application at either end of the tunnel. The tunnel agent takes the configuration parameters and then issues a resource reservation request in conjunction with its remote peer, using, for example, RSVP as the reservation protocol. The state of the tunnel at both ends is linked to the state of the RSVP reservation. If there is a network change or a policy preemption, the RSVP path will be lost, in which case the tunnel application can use local policy to determine whether to attempt to refresh the reservation or drop to a best-effort service level.

Engineering Considerations

The approaches taken to support dedicated bandwidth within the network interior can be very similar to those required for support of dynamic per-application service requirements. If the tunnel model of bandwidth service is assumed, the tunnel itself assumes the role of the trigger application; the service support structure is then the same as the application-initiated service request model. The model that is intrinsically different is best illustrated by the VPN *cloud*. Here the service is defined by the aggregate ingress capacity admitted into the network at each point. No particular bandwidth is associated with any individual path between an entry and an exit point for the VPN.

AN INTEGRATED SERVICES ARCHITECTURE APPROACH

One approach is to use the Integrated Services architecture and a guaranteed load service. In this case, the tunnel initiator becomes the RSVP service requestor, and the end point of the service is the tunnel's end point. The service profile requested for the tunnel would correspond to the requirements of a constant bandwidth circuit. In a provider-operated VLL, the admission filter for the service would be the tunnel ingress point; in a client-side tunnel, the admission filter would be the point where the tunnel enters the service provider network.

Aggregate ingress-defined services (the cloud VPN) present a greater challenge to the Integrated Services architecture. The most conservative approach is to assume a worst-case traffic spread and perform a series of RSVP reservations for the total ingress capacity to each of the possible egress points. Such an approach can offer the most robust service guarantees when used in con-

*A*n IP tunnel is an increasingly common tool used within the Internet environment to support service layering. The general principle of a tunnel is quite simple: The IP packet is carried as a payload of an enclosing IP packet. At the ingress of the tunnel, each packet is encapsulated within a new IP packet by adding an additional header to the packet. The new packet specifies the IP address of the tunnel end point, and the IP payload is the complete original packet, including the original IP header and IP payload. At the destination point, the outer IP header is stripped, and the inner IP packet is then routed normally again.

Behind this very simple process are a number of critical design questions:

> **How much of the original IP header should be mapped to the encapsulating header?** *The protocol number is usually replaced by a protocol number indicating that this is a tunneled packet. If a network uses the protocol number to distinguish between TCP congestion-controlled packets and UDP packets, it may treat encapsulated TCP packets incorrectly.*
>
> **Should the source address of the encapsulating packet remain the original source address or that of the tunnel ingress point?** *If the tunnel packet encounters an error condition that generates a return IP address, the choice of packet source address will determine the place where the ICMP message is to be sent. If it is the tunnel ingress point, then the ingress point may not be able to remedy the problem. If it is the original source point, the different*

junction with a guaranteed service reservation, but it is also the most wasteful in use of reserved network capacity. In such service environments, it is more efficient to use a controlled-load service request, reducing the level of commitment on reserved network resources.

Given the long-held nature of such aggregate services, and the relatively small number of such services in relation to the number of application invocations, the scaling properties of an Integrated Services engineering solution are less of a concern. Nevertheless, long-held RSVP services do impose some constraints on the network's engineering. To accommodate a topology change within the network, not only does the network have to manage the transient load due to a shift in routing, but there would be an additional period of disrupted service while a collection of affected RSVP paths reestablished a stable end-to-end path capable of supporting their service load. As the number of *destination IP address may confuse the source's ICMP processing, as the end application will not be aware that the end-to-end path includes one or more tunnel components. This is a critical issue when an application chooses to use MTU discovery, because the ICMP error messages are critical in determining the available MTU on an end-to-end path.*

Should the original packet's TTL field be copied to the encapsulation level? *If so, from the source's perspective, the packet simply disappears from view without the sink point being visible to the source. The encapsulation adds a further 20 bytes to the total packet size.*

Should the encapsulation packet allow fragmentation? *If so, the tunnel egress must perform packet reassembly, because the interior IP headers are not replicated within the fragments of the tunnel packet. It is advisable that tunnel packets not permit fragmentation, because the end application may be subject to silent packet discard if the packet size is too large for a network segment within the tunnel.*

If the tunnel is operating in a secure mode by encrypting the original IP packet upon entrance to the tunnel, should exposing any part of the interior IP packet's header compromise the intended security? *In such cases, there is a strong argument for mapping no interior IP header values up to the level of the encapsulating IP packet.*

such RSVP paths increases, the time it takes to reach a stable point of equilibrium of load balancing will inevitably become longer, exacerbating the impact of any instability in the underlying network.

A DIFFERENTIATED SERVICES ARCHITECTURE APPROACH

The alternative is to use the Differentiated Services architecture, in a mode of Assured or Expedited Forwarding. The Expedited Forwarding rate is defined so that the departure traffic rate from any node is no less than a configured rate. An Expedited Forwarding system attempts to operate within a fixed resource level defined across the entire Differentiated Services network. When used with explicit path bandwidth allocation services, such as VLLs, the problem with this approach is that all the various paths that start at this ingress point are forced to share this Expedited Forwarding admission profile. The individual paths are not assured of a particular resource allocation from the Expedited Forwarding admission function. To make such an approach work, the ingress function would need to initially qualify each aggregate stream by an individual admission filter, using an MF classifier and associated filter, then pass this stream to the marking function. This collection of per-stream admission filters structure is slightly different from the conventional Differentiated Services admission system model.

Assured Forwarding does not appear to be as relevant to this service requirement, given that it does not provide firm bandwidth allocation commitments to the aggregate service. For point-to-point path bandwidth allocation functions, Assured Forwarding is not an appropriate engineering approach, but for an aggregate ingress-defined service, the VPN cloud, the Assured Forwarding approach is quite suitable. Given that there is no firm bandwidth commitment for an individual path, a high-quality assured service can be used to support such a service model.

An MPLS-Based Approach

The final alternative approach examined here is based on Multi-Protocol Label Switching (MPLS), where a dedicated bandwidth circuit can be emulated using an MPLS path through the network. In this case, the quality of service parameters associated with the path match the profile of the requested dedicated circuit. The long-held nature of the path and the constant service requirements are a relatively clean fit to the set of MPLS capabilities. The approach is similar to the Integrated Services architecture with guaranteed service profiles, where the tunnel initiator can use an RSVP request to set up an MPLS label-switched path (LSP) with resource characteristics that match the bandwidth allocation request.

Similar to Integrated Services, MPLS does not lend itself readily to the VPN cloud. MPLS must create explicit LSPs that correspond to a complete mesh of ingress-to-egress paths. As with Integrated Services, the conservative approach is to configure each LSP to be capable of sustaining the entire load from each ingress path, which may correspond to an extravagant use of reservable network resources.

Congestion Management

The application environment associated with congestion management is a *better-effort* service, meaning that the service offered is no worse than the normal best-effort default Internet service model; and when the network resources are under stress caused by traffic overload, the better-effort traffic is handled to avoid the worst effects of congestion collapse. This performance management regime is shown in Figure 5.6.

This performance application is typically associated with a TCP application with some time-critical transfer function. The goal is to ensure that the TCP transport does not enter into a mode of repeated slow start and retransmission time-out, and that there is a sufficient level of packet delivery to maintain the TCP end-to-end timers even in the face of very high congestion load along the data path.

Not all TCP applications will be susceptible to a better-effort service. Short-duration TCP sessions, such as individual Web page fetches or T/TCP trans-actions, are limited in their performance more by the latency between the sender and the receiver than by the factors of transient congestion load. Long-held TCP transactions, including file transfers, detailed image transfers, remote backup operations, and remote data sensor downloads, are the ideal candidates for this type of performance application, as these sessions are the most significant contributor to congestion load in the first place. Managing their behavior offers the best approach to overall congestion management. TCP sessions, particularly those passing along a long-delay, high-bandwidth network circuit, tend to operate in a repeated burst cycle, where the steady state of the transfer is to burst a sequence of packets into the network at twice the bottleneck speed, then switch to an idle state to await a matching

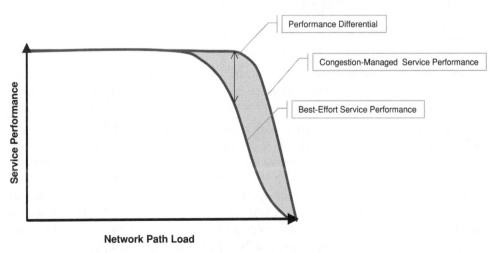

Figure 5.6 Congestion-managed service performance.

The reference to TCP as the only rate-adaptive transport protocol in the IP suite may be correct from the strict perspective of the Internet standards, but there is strong interest in further development of other IP tools that allow the reliable and timely transfer of data, in particular in the areas of authentication, authorization, and accounting (AAA) [ID-diameter-reliable] and in signaling systems to support the interworking of PSTN and VoIP services to create a robust, lightweight, reliable transport protocol [ID-sigtran-rudp, ID-rsgp]. This has been extended into support for the signaling requirement for multimedia gateway controllers [ID-megaco-tudp]. In addition, there is increasing interest in creating tools to develop reliable UDP streaming applications [ID-purdet]. It is envisaged that such UDP applications would be able to reduce their sending rate on the basis of some form of local congestion information collected regarding the state of the network path [ID-cm]. The number of Internet drafts that describe current study in this area is a good indication of the interest in this topic.

Any efficient, reliable transport protocol needs to adjust its sending rate in the face of path congestion, reducing the sending rate when packet loss is experienced and probing a higher rate when reliable transmission is experienced. One school of thought is that the protocol overhead of TCP is the minimal overhead required in order to support a secure initial handshake and maximize throughput efficiency for reliable data transfer. There is considerable interest in exploring the validity of this claim, in particular whether measures can be taken in the client or the server to create lightweight transport protocols with similar properties.

sequence of ACK packets from the burst. The amount of data transmitted at each burst is a function of the sender's perspective of the current congestion state of the network, as well as the receiver's capacity to accept further data.

This better-effort service makes less sense in an inelastic UDP-based application, where the UDP application will continue to pass packets into the network at a rate determined by some external system. In these cases, attempts to manipulate the queuing systems at the point of congestion-based queue overload will have only minor impact on the received quality of the UDP-based application, given that it will continue to pass packets into the network at a rate unaltered by the network path's congestion state. Congestion from inelastic UDP-based applications is generally not mitigated through dynamically applied discard measures, because the application is, in general, insensitive to such actions. If some control over inelastic UDP-based applications is desired, then widespread deployment of some form of weighted fair queuing or class-based queuing may be the most effective response. This would require some form of Differentiated Services response that classifies inelastic data flows into a discrete behavior aggregate. Such a measure could limit the amount of UDP traffic that would pass across any congestion point.

Recognizing Congestion-Managed Flows

This service profile is intended to operate on congestion-managed traffic flows, so necessarily, the network ingress must be able to identify those flows that are subject to end-to-end congestion management. One approach is to use the protocol identifier in the IP header to classify all those packets with a protocol identifier value of 6 (TCP) as being part of a congestion-managed flow.

There are two problems with assuming that all congestion-managed traffic flows use TCP and that all real-time inelastic data flows use UDP. Not every TCP-managed flow has a protocol identifier of 6 in the outer IP header. IP packet encapsulation, as found with various forms of tunnels and payload encryption, use a different protocol identifier in the outer IP header, regardless whether the payload IP packet is part of a TCP flow or not. Not only might the use of TCP be hidden from the network through packet encapsulation, but, as we have already noted, not all UDP-based packets carry inelastic real-time data streams. Ideally, there should be some form of uniform packet marking to indicate that a packet is part of a congestion-managed transport protocol or is a component of an inelastic real-time UDP data flow. For congestion-managed packets, it would also be helpful to the network if the sending system were able to indicate a relative discard preference of the packet within the context of the flow, given that packet discard is commonly used by such protocols as a signal of path congestion.

The original TOS field of the IP header did not carry such semantics, but the more flexible packet-marking facilities that are part of the Differentiated Services architecture can provide this capability. Once the network can recognize which packets form part of end-to-end-managed flows, it can group these flows to maximize network efficiency and to achieve some relative level of service differentiation.

Here, for the sake of readability, we will continue to refer to the generic class of congestion-managed flows as TCP flows, and inelastic real-time data flows as UDP flows.

Interaction of UDP and TCP

Part of the challenge to the network operator is to attempt to generate some form of predictability in service outcomes from a network that operates with a self-similar traffic pattern. An element of the uncertainty factor in such networks comes from the interaction between TCP and UDP. More precisely, this is the outcome of the interaction between end-to-end congestion-managed traffic flows and inelastic real-time data flows.

This interaction, when left unmanaged, can lead to a wide variance of service outcomes. A large-volume TCP flow with a long RTT, a large transmission window, and high-speed connectivity into the network can burst into all available buffer space at a bottleneck point. This burst exhausts the bottleneck buffer for all concurrent flows. Continued operation in this mode can lead to all concurrent flows synchronizing themselves against the primary flow's RTT, because the cycle of packet delay and loss will coincide with the RTT of the dominant TCP flow (see Figure 5.7). Alternately, a collection of inelastic real-time flows can saturate the egress capacity of the bottleneck point, in which case the real-time flows generally overwhelm the bottleneck buffer. This ensures that all concurrent congestion-managed flows operate in a mode of a single packet exchange followed by a retransmission time-out. Even when the real-time flow rate is less than the egress capacity, but is still the dominant traffic component, the TCP session is heavily damped. Whenever the session starts to open its window, the queue overflows, and the TCP session is forced into a time-out and a slow-start restart (see Figure 5.8).

If predictable service levels are desirable, there is strong motivation to separate the handling of these types of transport protocols in each router. Inelastic real-time flows do not normally exhibit high levels of burst, and imposed jitter is an important service consideration. The router's response is optimized through the use a relatively short buffer for such traffic. Congestion-managed flows probe the current bottleneck capacity through a cyclical rate variation, which includes exerting pressure against the bottleneck point by injecting traffic at a rate of up to twice the egress rate. If the objective is to ensure that the network operates efficiently—so that if there is no long-term idle capacity along a path, and if there is a sender that can increase its sending rate—then the network should allow the sender to increase its sending rate. This is achieved by provisioning buffers for congestion-managed transport protocols so that the buffer size is of the same order as the bandwidth-delay product of the link it feeds.

This leads to a number of considerations when determining how to efficiently mix end-to-end congestion-managed elastic flows and inelastic real-time flows:

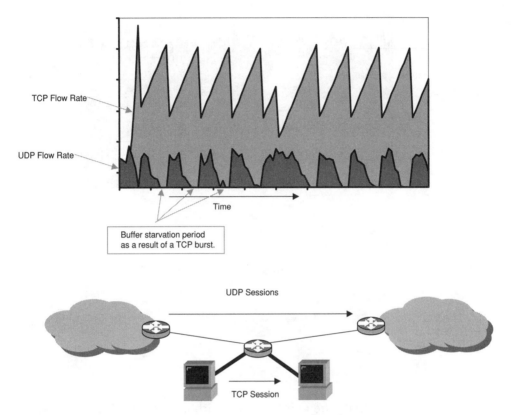

Figure 5.7 TCP and UDP queue interaction with a dominant TCP session.

Congestion-managed flows should be handled using a distinct queuing resource from the unmanaged flows. These distinct queues should be configured to a size about the same as the delay-bandwidth product of the output link. Service differentiation between classes of congestion-managed flows must be balanced against an overall desire to achieve maximum network efficiency. The greatest network efficiencies appear to be attainable by using a single queue resource and a differentially weighted service management algorithm for each service class, such as weighted RED and, where possible, weighted ECN.

Inelastic flows should be admitted through an admission profile system to ensure that the aggregate of all such flows does not cause sustained overload at any point within the network. Inelastic flow queues should be as long as necessary to perform basic contention resolution between multiple inputs and variable packet sizes, but not so long as to impose significant jitter on the flows.

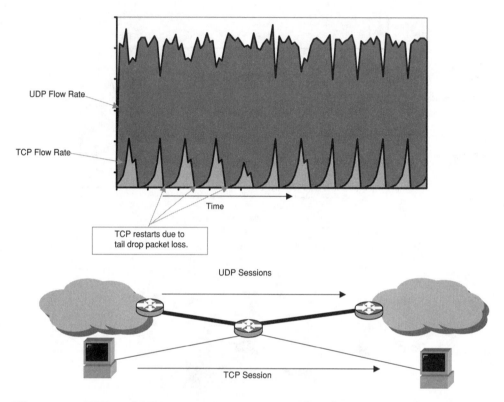

Figure 5.8 TCP and UDP queue interaction with a dominant UDP session.

Mixing the output of the two queuing structures can be managed by a WFQ or CBQ queuing discipline. Mixing these two sets of flows needs to be performed in a fair manner, corresponding to some determined network policy, in order to ensure that neither flow set starves the other of resources. WFQ or CBQ can undertake this operation of fair sharing, ensuring that both flow sets receive some set minimum service level.

Service Levels for Congestion-Managed Services

Given the adaptive nature of congestion-managed transport protocols, it makes little sense to impose an absolute bandwidth allocation against an individual flow or against an aggregate of flows. Congestion-managed flows do not operate at a constant sending rate. The congestion control process is one of integrating network congestion state information back to the sending algorithm. The result is a constant fluctuation of the flow rate; the rate is increased when no congestion is being experienced, and reduced when congestion is being signaled.

This implies that it is not practical to offer a congestion-managed service a zero packet loss or a zero jitter service level. Congestion-managed services will experience a constant level of loss and jitter due to the nature of their operation. As the sending rate increases, the network will fill with data. Due to the nature of the transport protocol, it uses a packet burst into the network. This burst is absorbed by a network queue, imposing a jitter element into the data stream as the queued data feeds into the output circuit. The sender will experience packet loss once the queue is filled. If the queue size is equal to the delay bandwidth product of the output circuit, this packet loss will only occur once the network itself is operating at maximal load.

So, if a service agreement cannot stipulate maximum levels of packet loss and jitter as a service offering to clients, what service options are available to the network operator? The best that can be reasonably achieved is *better* performance. If the two hosts are capable of sustaining the traffic flow, a premium congestion-managed service will sustain a better throughput than an equivalent best-effort service (a simulation of this situation is shown Figure 5.9). Such a relative claim has a number of substantial caveats. If the sending or receiving hosts have constrained local buffers, or the network path has a large RTT, the network may not be the constraining factor in throughput in the first place, in which case the differentiated services may make no visible change. Moreover, a well-tuned and recent TCP stack that is operating on a best-effort basis may outperform an untuned stack operating as a differentiated premium service in the network. Finally, the differentiated service levels may only be apparent over a significant duration, and there may be no visible difference in performance with short-term, low-volume traffic flows.

Engineering Considerations

The conclusion of this examination of the interaction of congestion-managed flows and inelastic flows is that any form of tangible service outcome needs to satisfy one major engineering constraint: TCP and UDP cannot be handled within the same queue by the router. Congestion-managed flows and inelastic real-time flows should be managed by distinct router queues. If maximal link efficiency is desired, queues for congestion-managed traffic flows should be configured with a queue length similar in size to the delay-bandwidth product of the link.

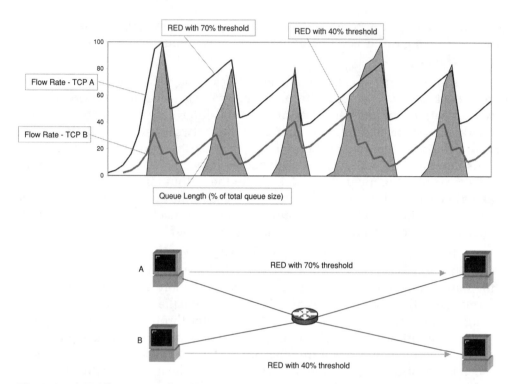

Figure 5.9 Differentiated congestion-managed flows.

RED AND ECN

An effective approach to management of congestion-managed transport applications is the deployment of random early detection (RED) queue control on TCP packets. The intent of RED is to provide an early signal to the TCP session to reduce its packet burst rate, without causing the TCP session to lose all of its session-timing information as a result of tail drop in an exhausted queue. Obviously, if the flow is ECN-capable, as indicated by an ECN bit in the packet header, the ECN-experienced bit should be set as a preferred alternative to early packet drop.

Both RED and ECN can perform acceptably close to the optimal form of congestion management for rate-adaptive flows. This optimal model uses both router queue length information and TCP flow state information. When the average queue length increases beyond some threshold level, the desired router response is to selectively mark a single packet from the burst with an explicit congestion notification indicator. If such explicit marking is not possible, the alternative is to selectively discard a single packet from the burst head of the major bursting TCP flow. The intent of the discard is to cause the trailing packets of the burst to generate a sequence of duplicate ACKs to be passed back to the sender. These duplicate ACKs cause the TCP sender to undertake a fast retransmit of the missing packet, then halve its burst size. This ideal behavior model demands significant levels of router processing. The router has to cache the TCP flow state and identify the contribution each flow is making to the queue pressure. When a flow is identified as a candidate for selective packet discard, the most effective discard function is to select a packet from the head of a packet burst rather than at the tail.

This optimal approach is not ideal in a large-scale operational environment, because the requirement to keep cached information on the state of every active TCP flow in every router incurs considerable overhead. A comparable performance outcome can be achieved with a much lower overhead simply by selecting packets at random to discard once the average queue length exceeds some threshold. For such large-scale environments, RED is a low-overhead, highly effective response to TCP-based congestion management.

A level of differentiation of service response in this area of congestion management can be achieved by weighting the packet drop probability differently within each distinct service class. Lower-weighted traffic classes will see packet drop at shorter queue lengths, and will effectively stabilize their transfer rates around a queue-limited throughput limit. Traffic in the heavier-weighted service classes will see packet drop only when the congestion window is at a size comparable to the full size. The outcome is an efficient TCP flow-mixing system that avoids the inefficiencies of congestion collapse, but produces differential service outcomes.

ECN offers the same functionality of signaling TCP sessions of impending queue saturation, without resorting to packet drop as the signaling mechanism. The advantage of the ECN approach is that it avoids the retransmission requirement and the burst of up to one window of data—which is part of the fast-retransmit recovery algorithm. ECN, when used in a fully ECN-capable network, allows the end system to distinguish between congestion and transmission errors. Wireless environments can be particularly prone to bit error bursts; and in the face of such errors, the correct response is not necessarily one of reducing the sending rate, but simply of retransmitting the errored packets in the existing data window.

TCP ADMISSION CONTROLS

Working on the common theme of signaling to the sender to reduce its burst rate, other TCP congestion avoidance approaches are also possible. One approach is to decouple the identification of discard-eligible packets and the RED discard function. For example, a marking-enabled TCP host stack can mark the leading packets of a large packet burst as discard eligible. A router under some threshold level of congestion load can, as a first-level response, discard those packets marked as discard eligible.

No host TCP stacks do such marking today. The alternative approach is to have some other network element mark the packets as a proxy agent for the stack. An admission filter at the edge of the network could use flow state caching to perform a similar function of packet marking. The uncoupling of packet marking and subsequent congestion-triggered packet discard is very similar in function to Frame Relay or ATM congestion control, but the packet-marking model required for efficient TCP operation is quite different. Admission filters based on a token-bucket model tend to mark the *trailing* packets of a packet burst as being out of profile. This practice increases the probability of TCP window collapse due to timing signal loss. The TCP admission filters would need to use a *mark at the head* approach to mark packet bursts, to ensure that subsequent packet loss will cause a fast-retransmit response from the sender.

UNRESPONSIVE FLOWS

Of course, some flows do not use end-to-end congestion control, or implement a congestion control algorithm in a way that attempts to behave unfairly under congestion conditions like a standard TCP congestion control algorithm. The problem posed by such flows is that as the contribution to total load made by such flows increases, the behavior of the network queues alters. Instead of the queues acting as a buffer to absorb transient burst conditions, the queues tend to operate in permanently-full mode. Queues that are permanently full act simply as a source of delay and loss, unable to absorb any form of transient burst. This is a form of congestion collapse, where large levels of network resource are expended on passing packets through a network that will be discarded before delivery.

This leads to the observation that the objective of active queue management could be extended not only to regulate those flows which are responsive to RED or ECN, but also to identify and impose some resource control on those flows which are unresponsive to such selective control mechanisms. The first question to be posed by such an objective is whether it is even possible for a router to identify such unresponsive flows. There are a number of approaches that have been proposed to undertake such a task, including rate matching, drop response, and disproportionate bandwidth usage [Floyd 1999]. Such tests are intended to allow the router to compare the behavior of each flow against a model of the bounds of a congestion-managed flow under similar circumstances. The first test advocates tracking the per-flow packet arrival rate against an upper bound, and identifying all flows whose rate exceeds such a bound. The second test is one to identify if a flow responds to a change in the packet loss rate by a proportionate change of its packet arrival rate. This test is based on the observation that in a long-help TCP flow, if the long-term packet loss rate increased by a factor of x, then the packet arrival rate of the flow should decrease by a factor close to the square root of x [Floyd 1999]. The final test is to identify those flows that are using a disproportionate share of the bandwidth under conditions of congestion.

The intention of these measures is to allow the router to enforce a resource constraint on unresponsive flows, in order to allow the congestion-managed flows to operate fairly and efficiently under conditions of load. The cost of such measures is the need to undertake some form of per-flow measurement within each router. At the edge of a network, where the number of discrete flows is relatively low and access bandwidth is often constrained in some fashion, such measures may prove to be a valuable adjunct to RED and ECN as a control tool. Within the interior of a high-speed network such per-flow measures may pose an unacceptable level of overhead to the router.

ACK MANIPULATION

Such approaches all have some latency associated with the congestion response, as the sender must, first, await the reception of trailing packets by the receiver, and second, await the reception of the matching ACK packets from the data receiver to the sender. This may take up to one RTT interval to complete. An alternative approach to congestion management responses is to manipulate the ACK packets to modify the sender's behavior.

The prerequisite to perform this manipulation is that the traffic path be symmetric, so that the congestion point can identify ACK packets traveling in the opposite direction. If this is the case, a number of control alternatives can mitigate the onset of congestion:

ACK pacing. Each burst of data packets will generate a corresponding burst of ACK packets. The spacing of these ACK packets determines the burst rate of the next sending packet burst. One approach to slow down the burst rate is to impose a delay on successive ACKs. This measure will reduce the burst rate, but not impact the overall TCP throughput. ACK pacing is most effective on long-delay paths, where the TCP burst behavior establishes a bottleneck point by exhausting buffer capacity, rather than by establishing a maximum path capacity use. ACK pacing spreads out the burst load, reducing the pressure on the bottleneck queue and increasing the actual data throughput. A good example application of this technique is on satellite circuits, where the RTT of some 600ms significantly stresses the transport protocol and the network equipment.

Window manipulation. Each ACK packet carries a receiver window size. This advertised window determines the maximum burst size available to the sender. Manipulating this window size downward allows a control point to control the maximal sending rate. This manipulation can be done as part of a traffic-shaping control point, enforcing bandwidth limitations on a flow or set of flows.

Both mechanisms assume symmetry of data flows at the control point, where the data and the associated ACKs flow through the same control point (but in opposite directions, of course). Both mechanisms also assume that the control point can cache per-flow state information, so that the current flow RTT and the current transfer rate and receiver window size are available to the service controller. ACK pacing also implicitly assumes that a single ACK timing response is active at any time along a network path. A sequence of ACK delay actions may cause the sender's timers to trigger, and the sender to close down the transfer and reenter slow-start mode. These environmental conditions are more common at the edge of the network, and such mechanisms are often part of a traffic control system for Web-hosting platforms or similar network service delivery platforms. As a network control tool, ACK manipulation makes too many assumptions, and the per-flow congestion state information represents a significant overhead for large network systems. In general, such manipulations are more appropriate as an edge traffic filter, rather than as an effective congestion management response. For this reason, the more indirect approach of selective data packet discard is more effective as a congestion management measure.

A DIFFERENTIATED SERVICES ARCHITECTURE APPROACH

The offerings of the Integrated Services architecture are not all that appropriate to the dynamic behavior of congestion-managed data flows, whereas there is significant potential in the aggregate management of the Differentiated Services architecture.

The first step is to deploy some form of DiffServ that explicitly places congestion-managed traffic into a PHB group that excludes all other forms of traffic flow. Within this PHB group, a number of service levels can be defined to allow for some gradation of relative service responses. As a further level of granularity within each service class, a drop precedence indicator can be used to allow a TCP to indicate to the network which packets within the flow can be discarded. Such marking would presumably mark packets at the head or in the middle of a packet burst, avoiding marking packets at the trailing edge of a packet burst.

It is no coincidence that this description matches almost exactly the structure of the DiffServ AF service class. The two-dimensional structure of the service class, with a tiered collection of service classes and a tiered discard precedence per service class, is well aligned to these requirements. The EF service description refers to a three-level discard precedence. It is unclear whether three levels are required for this application; it may be that only two levels are used in this instance.

One very important factor is that to support the end-to-end congestion management model it is necessary to extend the DS marking semantics to operate on an end-to-end basis. This does not imply that the DSCP values are constant end to end. The requirement is for the transport protocol in the host to mark a packet as being part of a congestion-managed data flow, and for the transport protocol to mark the appropriate packets within a packet burst with the ordering of discard precedence. This means that the DiffServ AF (TCP) admission systems should:

- Perform BA admission classification.
- Use a distinct discard precedence marking so that traffic conforms to a certain admission profile—normally to an intermediate level of precedence between unmarked in profile and the user-marked precedence.
- Attempt to shape traffic in preference to packet discard.
- Use a weighted RED service implementation for the PHB.
- Avoid pacing any non-CM traffic flows into the same service queues as the CM services.

The outcomes of this approach are not those of a constant rate bitstream. The attempt here is to bias the average rate of a number of flows, so that flows within an elevated service class receive a proportionally higher throughput rate than traffic within a best-effort service class.

Premium Services

The requirement of premium services is to offer different levels of service to different customers. The underlying motivation is not primarily technical, in that there is no requirement to elicit a particular service response from the network to match a particular application. Rather, the major motivation is a business factor: A service network wants to extend its market scope through provision of a broad range of price and service quality points from a single network platform.

Many networks provide a single constant grade of service to all clients; the telephone network is a good example. Every call receives the same service response from the network. Other networks also deliver a single grade of service, but it may vary over time, for example, a road network. Two cars traveling the same path at the same time of day will have similar trip times; but if they take the same trip at different times, the trip conditions will vary. The best-effort Internet is another example of this uniform service response.

Introducing various levels of premium service response is intended to provide a range of service responses simultaneously. The basic characterization of a premium service is that, at all times, it offers a superior service response over the base best-effort service. Continuing the road analogy, this corresponds to an express lane, on which one class of cars will chart faster trip times for the same trip than another class of cars traveling in slower lanes. In the context of network services, a premium service gives a class of traffic a level of priority over all other traffic. No particular application or session is typically associated with such a service, nor is there a particular network path to be managed as a premium service path. Rather, a premium service is normally associated with a network client, and the service offering is that the client's traffic is given some priority over all other traffic within the network (see Figure 5.10). This type of service is often well matched to a Differentiated Services architecture, where packets are admitted through some form of admission filter, with an associated service request written into the packet header.

Though simple to describe, the underlying engineering task involved with this service can be quite challenging. Posing a number of questions will help to illustrate these issues:

♦ Does the definition of the *client's traffic* refer to traffic sent by the client, or traffic received by the client, or both?

◆ In TCP traffic flow, which use the return ACK packets to time the sender's transmissions, do the ACK packets require the same level of service as the data packets?

◆ Is the premium service applied by the network, or is the client allowed to vary the requested service level?

There is no single correct answer to these questions; there are degrees of difficulty in attempting to formulate certain answers. Let's look at this service in greater detail, and offer potential answers to some of these questions along the way.

Marking Models

If the service refers to packets *sent* by the client of the premium service, the logical implementation of the service is to place a filter on the client's interface to the network, marking all packets that pass through the filter with a premium service mark. Such a system also allows the network service provider to control how much traffic is admitted within the premium service profile, because all the client's traffic passed into the network goes through this admission filter.

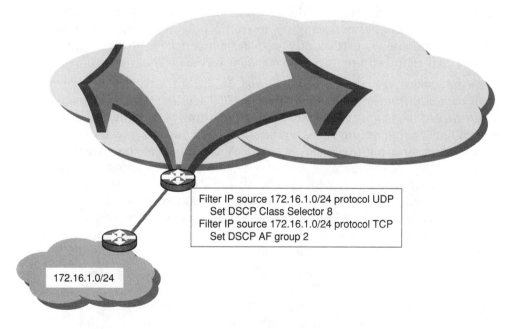

Filter IP source 172.16.1.0/24 protocol UDP
Set DSCP Class Selector 8
Filter IP source 172.16.1.0/24 protocol TCP
Set DSCP AF group 2

172.16.1.0/24

Figure 5.10 Premium services for a network client.

If the service refers to traffic *received* by the client, then the implementation is more challenging. The network service provider has to place admission filters at every network ingress point, and pass all incoming traffic through classification filters. Where packets match the address prefixes advertised by the client, the admission filter must be invoked to mark the packet with a premium service mark. In fixed-size networks, the overhead of such classification functions can be assessed. In a rapidly expanding ISP network, for example, where there may be many hundreds of ingress points, a comprehensive filter function may represent considerable overhead. Oddly enough, such a model still will not guarantee a premium service on all packets destined to the client. With so many admission points, there is no synchronized control over the volume of traffic admitted into the network destined to the client. In reality, it is possible for a collection of remote sources to overwhelm the client's connection to the network, causing packet discard and congestion at the client interface.

The pragmatic conclusion is that a premium service is most efficiently implemented as an admission policy imposed on traffic being passed from the client to the network.

Traffic Flow Models

Remember that this service model, if using an ingress marking of traffic from the client into the network, applies to traffic passing in one direction only. This has limitations, particularly with reference to TCP-based applications whose total round trip requires a consistent service level.

In the case of TCP-based applications, this unidirectional service model does pose some performance questions. A TCP sender paces its transmissions according to the timing of the ACK packets coming back from the receiver. In a unidirectional premium service model, the data packets may be sent using a premium service, but the ACK packets may be subject to the general level of delay and packet loss being experienced within the best-effort service class. The sender will then lock into this signal, and its transmissions will be based on the distortion experienced by the ACK packet train. It is preferable to return the ACK packets with the same service level as the received data packets. However, the receiver may not wish to request a premium service for its traffic at the instigation of a remote party.

The limitations of this service model are more evident when looking at the client receiving data via a TCP session. While the client's ACK packets are treated as premium packets, the packets containing the data payload are not marked as a premium service packet. Thus, performance of the client in receiving data will be little changed from a default best-effort service.

Alternatively, the network could elevate the service level of these reverse packets as part of the sender's premium service. But as just noted, such a reverse path premium service faces significant technical challenges to implement as the network becomes larger and the number of ingress points grows.

Profiling Premium Services

The next consideration is control over service selection. One approach uses a network-based definition of the traffic selected for premium service, passing all the client's traffic through an admission filter and selecting premium traffic based on a service level agreement. This network-based selection of premium service traffic poses some usage concerns. If a premium service is free on any incremental charges, in that there is no additional cost over the base best-effort service tariff, then there is a strong incentive for the client to place as much traffic as possible into a profile that is selected by the network as a premium traffic profile. If a premium service has an additional tariff element, then it is highly unlikely that a client would allow the network to unilaterally determine which parts of its traffic are subject to the higher tariff. If the client is to be subjected to an additional tariff element, it stands to reason that the client would want a say in the decision.

Alternatively, the client can undertake the marking of premium service requests on a per-packet basis. This allows the client to select a premium service response for only those applications for which there is some perceived value in making such a selection. In the light of the observations about usage models with network-based marking, it makes some sense to allow the client to vary the requested service level in conjunction with some form of additional tariff. The consequence is that the client will only see value in requesting a premium service level when the base best-effort service is under stress due to congestion load. The pricing of such a premium service is critical to managing this peak load profile. Under such load conditions, there is not an unlimited capacity to absorb a high volume of premium traffic.

Engineering Considerations

The task of applying a premium service level to classes of traffic from a client is best done using an approach based on the DiffServ architecture. Unless the client's traffic is constrained to a set of known transit paths, and the client's traffic fits a predictable load profile, the service profiles of the IntServ architecture are not particularly relevant to this situation. A client's traffic is passed through a network admission filter where the admission policy is applied. This admission policy may include a traffic profile meter, to limit the amount of premium traffic admitted into the network, or a traffic-shaping filter, to control the peak load of premium traffic at any time. The admission policy may mark all the client's traffic according to an agreed admission policy. Alternatively, the admission policy may be a limiting filter that allows the client to mark the packets up to a maximal level of priority as determined by the service agreement with the client.

Within the network, the necessary support for a premium service model can be accomplished using queuing management responses. For TCP traffic, a weighted RED response can provide effective differentiation of traffic to create a premium-level TCP service. For UDP traffic, a simple priority queuing scheme can be used, with the condition that the admission filters ensure that the amount of premium traffic is limited to a very small proportion of total traffic. A more robust service response can be obtained by using class-based queuing; or if flow-fairness of the premium-level service is an objective, weighted fair queuing can be used for UDP traffic.

This service response does not precisely correspond to any of the currently defined DiffServ offerings. The optimal way to bias the network to allow congestion-managed TCP flows to receive premium service levels—or, in other words, a greater relative allocation of network resource compared to other congestion-managed flows—is to use weighted RED or weighted ECN and a lower drop probability function against the premium service packets. This is similar to the desired service characteristics of the Assured Forwarding service class. For real-time UDP traffic flows, there are three basic priorities: reduce delay, reduce jitter, and reduce drop probability. All three can be achieved using a simple priority queuing model corresponding to the Class Selector DiffServ service model. Here, premium real-time traffic is given an elevated Class Section codepoint, triggering a priority-queuing PHB within the network.

Picking either the AF service class or the CS service class to support premium traffic models will result in a compromise in service outcomes for one class of traffic or the other. Congestion-managed traffic in a CS service priority-queuing environment will require strict admission control to ensure that there is no service denial on the part of a runaway premium TCP flow. But if there is strict admission control, how can the flow expand into idle network capacity when there is no other low-priority traffic that would be affected? If the admission control system uses a two-level marking system, under congestion load there is a strong probability that the low-priority traffic will be delayed and delivered out of order. There is also a big risk of the TCP sender sending duplicates of packets already in flight within the network, thus producing consequent session inefficiency simply because there are now two round-trip times interacting with the TCP control system. A two-level service response effectively creates two different services; and for TCP, the consequent requirement is to time two different services simultaneously: the low-jitter premium channel set of timers and the out-of-profile best-effort channel timers. Given that the TCP session is unaware of which data packets and which corresponding ACK packets correspond to which service channel, TCP will be forced in areas of inefficient operation due to the conflicting control information flowing back to the sender. AF for real-time UDP flows will not ensure that the flow receives priority within the packet scheduler so as to minimize jitter and packet drop.

The real issue here is that premium service means different network responses in each case. A premium response to an inelastic real-time data flow implies an objective of elimination of network queuing delays and packet drop. The objective of a premium service to such data flows is a precise reproduction of the relative timing of input packets at the output point. Though an implementation may not achieve this objective at all times, the measure of success in providing premium services in this case is of imposed delay, jitter, and loss. A premium response to an adaptive congestion-managed data flow implies that the data flow will equilibrate with other concurrent congestion-managed flows so that the premium flow will receive a greater relative proportion of network resources. The premium flow will go faster than other flows. The measure of success in providing this service is a statistical comparison of the premium flow behavior with concurrent best-effort flows, noting the relative throughput of each class of flow. Given these different service objectives, it should not be surprising that these objectives are constructed using different network responses. An ideal situation is one where the host is capable of marking packets in terms of their flow-control characteristics. This would allow the network to structure a premium response in terms of a CS priority-queuing response for premium real-time traffic and an AF relative weighting of packet drop for congestion-managed traffic flows. In the absence of such marking, the DiffServ ingress can structure an approximation of such a response by marking all TCP traffic from the premium client into an AF service class and marking all UDP protocol traffic into a CS service class.

Multicast Applications

Multicast is a one-to-many transmission model, where a single packet is simultaneously transmitted to a group of receivers through the use of a *multicast group address* as the packet destination address (see Figure 5.11). The applications of multicast transmission services include:

> **Network resource discovery.** Using multicast as a more efficient and better directed form of discovery than conventional broadcast.

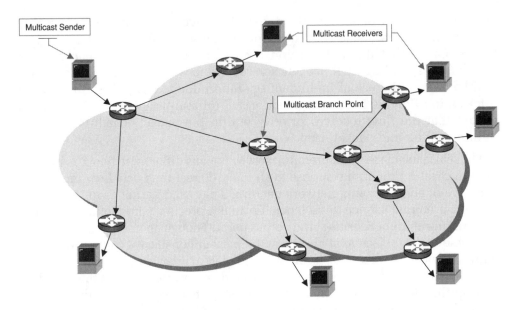

Figure 5.11 The multicast service model.

> **Content flooding.** Using multicast to implement real-time flooding within the network, rather than simulating flooding via a set of point-to-point relay paths.

> **Support of real-time collaborative tools.** Using multicast to support conference voice and video, as well as groupware applications, such as shared electronic whiteboards.

With these applications, in particular, with the latter class of real-time collaborative support tools, there is a definite requirement for a form of per-application multicast quality definition and service management.

Multicast services have been poised to move into the mainstream Internet service environment for some years now, but the necessary steps to complete the picture for widespread deployment have proved elusive. One explanation is that stabilizing the unicast IP traffic in the face of continued scaling of the network is proving to be an all-consuming engineering task. A more likely explanation is the relatively poor control over the service aspects of the multicast environment and the associated complexity of the multicast service.

The range of multicast routing protocols available today illustrates this issue. The design engineer has to pick between the Distance Vector Multicast

Routing Protocol (DVMRP), Multicast Open Shortest Path First (MOSPF), Protocol Independent Multicasting-Dense Mode (PIM-DM), Protocol Independent-Sparse Mode (PIM-SM) and Core-Based Trees (CBT) for an interior multicast routing protocol. The available interdomain multicast routing protocols include Multicast Border Gateway Protocol (MBGP), Multicast Source Discovery Protocol (MSDP), and Border Gateway Multicast Protocol (BGMP). The shared purpose of these routing protocols is to support a per-group state within the network that supports efficient packet transmission to all group members. The different approaches to this objective, as evidenced in the wide range of multicast routing protocols, is indicative of the level of inherent complexity in achieving such a goal. The challenge for multicast services is the need to complement this per-group routing state with a per-group service quality, giving all receivers access to a uniform quality signal.

Within the packet protocol itself, the only difference between a point-to-point unicast IP datagram and a multicast IP datagram is the use of a multicast Class D IP destination address in the packet header. Other fields of the IP header, including the DS field, are unchanged. While there are no changes to the IP packet header itself, there are some changes to the network and to hosts to support multicast applications. The multicast group is normally maintained as a distributed soft state within the network, where each multicast router is aware of the currently active groups within the network. Individual hosts announce their intention to join and leave a group via explicit calls allowing the network to adjust its group routing tables to maintain a minimal per-group spanning tree that covers all the members of the group. Most notably, the service architecture is altered quite radically by the use of a multicast service model. Any multicast application may have a number of multicast receivers with a significant breadth of available capacity and a difference in delays and jitter between a multicast sender within the group and the set of receivers. Attempts to impose a uniform service delivery model upon such an environment will always be very challenging.

Multicast Application Resource Requirements

As we have noted, service quality is an outcome of both engineering a differentiated response for selected traffic classes on the part of the network and ensuring that traffic load limiting is performed by the end host applications or by the network admission systems. An overloaded premium service is often indistinguishable from an overloaded best-effort service. The issue with support of multicast applications is of establishing bounds on the resource requirements of the multicast application.

In a multicast environment, the amount of network load is not solely under the control of the traffic generator; it is also a function of the number and relative location of the receivers. Each new receiver joining the multicast group may cause the multicast traffic flow to be duplicated at a branch point inside the network; the multicast flow is then directed over a new network path to reach the new receiver. The number of network paths carrying the multicast flow will grow as the number of multicast group members grows. The total network load then grows as a function of the number of discretely located receivers.

The model comprising a single sender and multiple receivers is a broadcast model. Multicast adds another dimension to the service model by allowing any group member to transmit; this transmission is permitted without necessarily coordinating with any other multicast group member. For example, a video conferencing service model may have every conference member transmitting a video and audio stream into the multicast group so that all can see each other simultaneously. In such a service model, the resource requirements quickly expand with the number of group members. In the worst case, the traffic load imposed on the network grows in proportion to the square of the number of active group members.

Congestion Control

A conventional model of congestion control has the sender reacting to path quality signals coming from the path to the receiver. There are a number of ways to deal with congestion control in a multicast group. One method is to use a lowest common denominator approach, whereby each sender reacts to all congestion signals coming from all receivers. The anticipated outcome of this approach is that the sender will limit its transmission so the traffic flow is moderated to characteristics of the most congested receiver path. For large multicast groups, the inflow of per-path or per-receiver congestion information to the sender may overwhelm the sender (an instance of the so-called multicast ACK-implosion problem). The approach of adapting the application quality to match the requirements of the worst case receiver may unfairly compromise the multicast application because all receivers will get a restricted signal in order to meet the path characteristics of a single receiver.

It may seem logical for each sender to negotiate service quality with the network, but the point is, the network load is influenced strongly by the number of receivers, not just by the characteristics of the sender. An alternative approach to congestion control is to dispense with dynamic control of the signal completely, and move to a fixed-resource allocation model whereby the multicast group itself is associated with a maximal resource load. Then congestion control is passed back to the senders within the group, as the senders should now signal each other as to how to allocate the available group resource to each concurrent sender. If the network is going to police this resource allocation, there is also an implicit signaling requirement between policing admission systems, where the sum of the resource requirements of all flows into the multicast group at all network ingress points should be no greater than the allocated group resource. Active research is ongoing to figure out precisely how to create this coordination of action between multicast hosts and network admission systems.

In the absence of such measures, currently there is little actual congestion control within multicast systems. Multicast applications typically generate inelastic traffic flows; the flow dynamics remain under the control of the sender, with no congestion management interaction with the network or with the set of receivers. This places the onus of multicast resource management on the network and its admission systems.

Engineering Considerations

Two potential interactions can be used to support service quality-based multicast applications in this environment: the interaction between multicast packets and the per-hop queuing systems and the interaction between the multicast routing system and the application environment to set up and maintain quality-based multicast paths through the network.

DIFFERENTIATED SERVICES

In the first case, the interaction with the per-hop queuing system, a multicast QoS environment can use mechanisms drawn from the DiffServ architecture. Multicast packets can be passed through an admission filter and marked with a service request corresponding to the network's multicast QoS policies. As the packet is passed through the network, the multicast branch points create packet headers with identical service request codes.

Within the router's queues, multicast packets can be handled similarly to unicast packets, using the service request to determine queuing priority and drop preference. This creates a level of better-effort service; the multicast service is positioned as being at some premium to the uniform best-effort service. As already noted, the unicast flow control mechanisms of TCP have no real meaning within a multicast environment. While priority queues, class-based queues, and weighted fair queues are all relevant to multicast QoS, weighted random early detection and related queue congestion notification schemes have no benefit to a multicast application. It would appear that the EF DiffServ behavior aggregate is too great a level of resource commitment, given the very strict requirements for jitter and loss associated with this service class. If the AF service classes are implemented using RED or ECN, then such congestion management responses will have little impact on the outcome of a multicast traffic flow. The most effective network mechanism is a combination of a strict admission control and some level of elevated scheduling priority. This would normally imply the use of a Class Selector PHB as a simple priority selector to ensure that loss and jitter of the multicast packet flow is minimized. To minimize the risks of denial of service to lower-priority traffic, the strict admission policy is intended to ensure that the amount of elevated priority traffic is limited to some extent. When applied to multicast traffic, the outcome of this Class Selector (CS) behavior aggregate service model is not a uniform service level from which all receivers obtain an identical service from the multicast sender. The differences in bandwidth, delay, jitter, and loss between receivers still remain under such a system; and while the CS multicast service may ameliorate the worst effects of local congestion on service quality, the service is still quite variable between receivers.

The associated admission control requirements of the DiffServ architecture is a more challenging problem. Given the dynamic nature of a multicast group, the admission gateway may not be able to establish the extent to which the sender's packets are replicated within the network and what the subsequent network load may actually be. A single multicast packet stream may generate a substantially greater load upon egress from the network. This load may vary from time to time depending on the dynamic state of the set of receivers. This is the most significant weakness of the multicast model when using a DiffServ QoS service model.

INTEGRATED SERVICES

The use of a QoS-based multicast path in a multicast version of RSVP is a viable technology approach within the Integrated Services architecture. Here, the architectural decision to use the receiver to generate the RSVP Resv message is extremely useful. If the model of joining a multicast group entails the new group member making a resource reservation in the existing group, each path within the multicast group is managed with an associated resource reservation. As the message is passed back along the path to the sender, the path routers make a resource reservation. When the path meets the branch point of the multicast group, the Resv message can stop, as there is no requirement to make any further upstream resource reservations.

This approach raises some serious issues. In the unicast RSVP model, the RSVP-maintained path could deviate from the currently selected lowest-metric path between the sender and receiver. As long as the resource reservation is maintained, there is no significant benefit in continually recomputing a resource reservation to ensure that the chosen network path matches the best-effort forwarding path. In the context of multicast, there is no way to be able to pin a multicast path along a route that does not match the current unicast routing model. The multicast forwarding algorithm forwards a multicast packet on every interface that is associated with a path to a multicast group member, with the exception of the interface that is associated with the unicast forwarding path to the packet's source address. If an RSVP-maintained multicast topology loses synchronization with the unicast topology, the result is an increased risk of multicast loops and, consequently, a traffic multiplier effect. Either the RSVP-maintained multicast paths must be recomputed in response to any change in the unicast network or the multicast routing system must be locked against the RSVP path maintenance operation.

Variations in approach are possible with this model. One method would enable a service profile to be stated when a traffic sender joined a multicast group. But this would mean that receivers could join and leave the group after the sender's join operation. This raises a number of interesting questions, including whether the sender would need to restate the service requirement upon each change in group membership. If the sender were unable to set up a service path to a single receiver, should the sender be allowed to proceed in any case, or should the group join operation fail? Should the group itself maintain an associated service state? Should any new members to the group be required to set up an RSVP-installed connection? In this case, the join operation would fail if it could not be completed within the stipulated service profile. The multicast group itself would maintain a uniform service quality—though individual senders would need to adapt their sending application to meet the common service requirements. Today, uniform service levels within a multicast group do not exist; therefore, a number of critical multicast applications relating to real-time conferences remain elusive.

MPLS and QoS Engineering

Multi-Protocol Label Switching plays an important part in many networks to support QoS services. We described MPLS in Chapter 2, "The Performance Toolkit," when we explored various forms of switching technologies, noting that the essential attribute of MPLS was support of label-switched paths (LSPs) across an Internet switching environment. Here we'll example the MPLS potential to support both IntServ and DiffServ network environments.

MPLS and the Integrated Services Architecture

There is a relatively obvious relationship between the functionality of MPLS and the requirements of the IntServ architecture. IntServ requires the establishment of a path between the ingress and egress points, where all the switching elements along the path can classify the traffic into a reserved flow and then undertake the appropriate scheduling and forwarding decision based on the flow's resource reservation and reserved path. Within the MPLS environment, this path can correspond to an LSP (see Figure 5.12). At each network element, the incoming label value or a combination of label and interface, not only can determine the egress interface and label, as in a conventional MPLS model, the label can also determine the scheduling actions appropriate to this flow [ID-mpls-rsvp]. The advantage for IntServ routers is that the reservation state and forwarding directive can be inferred from the incoming MPLS label, thereby reducing the overhead required to pass the IP header through a potentially very large classification filter. Because MPLS labels and RSVP paths both support unidirectional data flows and can support receiver-initiated path setup actions, there is great similarity between the two concepts. Within such a model, RSVP plays two roles: of the Label Distribution Protocol, to set up and maintain the LSP, and of the path resource reservation protocol, to maintain an associated resource allocation and scheduling discipline associated with the virtual circuit.

The modifications required for RSVP to support this association with MPLS are relatively modest; they consist of the addition of two RSVP objects, an *RSVP_Label* object to carry the MPLS label value and a *HOP_COUNT* object to enable TTL processing within an ATM label-switching environment. This association is sufficiently general to support both a unicast path model and the multicast case. In the multicast case, there is a requirement to maintain a label per sender. The operation of a unicast path reservation is quite straightforward. When the receiver receives an RSVP Path message, an RSVP Resv message is passed back toward the sender. At the boundary point of the MPLS network, the boundary router allocates a label and writes this into the RSVP message. At the upstream node, the label is stored as part of the reservation state for this flow. A new label is generated and written into the RSVP Resv message, and the message is passed upstream. In this way, RSVP assumes the role of a label distribution protocol.

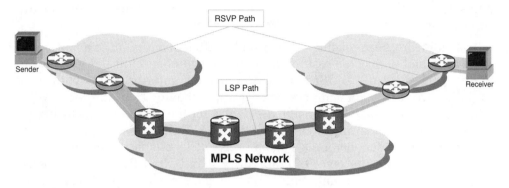

Figure 5.12 MPLS and RSVP.

The issue with the use of MPLS LSPs as a mechanism of supporting IntServ reserved paths is relatively unchanged in terms of the scaling properties of this approach. The scaling issue is in the proliferation of labels in the interior of large networks, where the use of IntServ on a per-application instance has the potential to create concentration points at which the number of labels required becomes a scaling problem.

MPLS and the Differentiated Services Architecture

This scaling concern is part of the rationale for DiffServ. DiffServ tries to create a core service network that deals only in aggregates of flows, rather than in individual micro-flows. Can MPLS, with its path-oriented approach to network flow management, offer any mechanisms to DiffServ, an architecture that specifically avoids a flow- or path-oriented approach to service quality? MPLS supports LSPs through the label-switched network, where each LSP corresponds to a Forwarding Equivalence Class (FEC) path. The number of LSPs supported within the network is of the same order as the complexity of the network's internal topology.

Why introduce DiffServ capability into MPLS in the first place? This is a reasonable question in that DiffServ actions are based solely on the DiffServ codepoint in the IP packet header, and directed to the router's scheduler, rather than the forwarding module. The rationale lies in the desire to combine DiffServ mechanisms and traffic engineering approaches so that DiffServ behavior aggregates (BAs) can be managed through a traffic engineering framework. In this manner, different BAs could be assigned to different physical paths through the network, optimizing the network load and ensuring that the service parameters of the BA are met by the network.

The most direct approach is to map each DiffServ codepoint to a new MPLS label-switched path. This results in one label-switched path per Forwarding Equivalence class per DiffServ per-hop behavior. Interior nodes within the MPLS network can infer the correct per-hop behavior from the local MPLS label without having to look inside the packet for the DS field. This technique is referred to as a label-inferred-PHS LSP, or L-LSP [ID-mpls-diffppp]. In large networks, this may consume a large amount of label space; and in terms of traffic engineering, the level of granularity may be too fine. An alternative approach is to group a number of PHBs that share some scheduling behavior into a single LSP that supports a scheduling aggregate (SA) per LSP. In this case, the additional information required to select the actual PHB can be encoded in the EXP field of the MPLS shim header, allowing a maximum of eight behavior aggregates to be carried in a single LSP. This is analogous to DiffServ itself, where the DiffServ code-point is carried in the packet header. Here the DiffServ behavior aggregate is also carried in the MPLS header, using both the label value and the EXP field to carry the inferred DiffServ service request. This technique is referred to as an EXP-inferred-PHS LSP, or E-LSP. The resulting capability to manage traffic flows on a DiffServ behavior aggregate basis is a very powerful tool, enabling the network to manage traffic flows that share a common service requirement in a single managed unit.

MPLS and Traffic Engineering

What does traffic engineering have to do with quality of service? Quality of service is concerned with the engineering of differential responses from the network to meet the service demands of clients of the network; traffic engineering is concerned with the optimization of traffic load on networks. The two areas of activity are much closer than such characterizations suggest, however. The need for QoS differentiation arises from the onset of localized congestion conditions that impact application performance. In this light, QoS differentiation can be seen as managed damage control in response to congestion. An alternative response is to attempt to avoid the congestion condition in the first place. We have already noted that one potential implementation of this objective is to over-provision the interior of the network so that the network is capable of passing all traffic loads presented to it. Another potential method is to use traffic engineering to mitigate the effects of congestion, and in so doing prevent the need for engineering differentiated responses to congestion.

The conventional Internet architecture uses a simple address-based network paradigm. Within this structure, each packet is forwarded within the network to the address specified as its destination address, using a routing-determined best path from the packet's current location to the destination location. In a richly meshed network with a relatively even network traffic balance between all network hosts, there is a reasonably good fit between the underlying mesh of connectivity and the network traffic patterns. But where there are traffic concentration points, this model breaks down. For example, large exchanges deployed at a small number of points on the network represent a significant traffic clustering point; such traffic clustering tends to disturb the model of an even mesh of traffic. Within a network, this destination address-based forwarding paradigm tends to aggregate traffic flows within trunk systems. The ingress to aggregate trunk systems represents a significant traffic cluster point. The clustering represents a threat to the service performance model, because sustained concentration of traffic can exhaust local network resources and cause local congestion. Traffic engineering is a mechanism for dispersing traffic flows so that the traffic concentration condition that signals the onset of congestion is avoided.

Traffic engineering is directed toward managing traffic flows so the network load is more evenly distributed; it addresses a number of traffic performance attributes. In a best-effort Internet, the major objective of traffic engineering is to reduce the probability of packet loss while minimizing jitter and latency and maximizing throughput and overall network availability and reliability. In a differentiated services network, the overall objective of traffic engineering is to meet the service parameters associated with each service class, while allowing for a more diverse spread of traffic flows across a broader span of the network. The limitation of routing control systems is that they will converge on a single shortest path. Where the path metric includes some element of feedback relating to path load, such as in QoS-aware routing systems, the outcome is often one of oscillation, where the selection of a single viable path and the consequent shift of traffic onto this path can cycle across multiple available paths. One approach is to allow the use of load sharing across unequal cost paths where the path metric is interpreted as a relative path bandwidth, and traffic is distributed on a collection of paths according to their ratio of path metric values [ID-ospf-omp, ID-isis-omp]. The system can be further refined by using a path metric that includes a factor of load so that the distribution function would tend to weight the more lightly loaded path with more traffic. Refinements to the operation of routing protocols pose significant challenges to the integrity and stability of the routing system.

An alternative approach to spreading the load is an overlay switching model, as is used in IP-over-ATM, IP-over-Frame Relay, or MPLS. Here a mesh of virtual circuits can be constructed across an underlying physical topology. The load carried by each virtual circuit and the actual routing of the circuit itself can be varied to suit the dynamic conditions of the network. In examining an approach to traffic engineering and MPLS, the basic element of traffic control is that of the *traffic trunk,* an aggregation of traffic flows that share a common classification [RFC2340]. The classification could be as simple as a common forwarding equivalence class, so that the traffic trunk shared an MPLS network egress point, or it could be a common forward equivalence class combined with a common service classification, such as those defined within the DiffServ model. Additional attributes may be associated with a traffic trunk, such as policy-oriented attributes of relative priority and fault restoration requirements.

In the same way that traffic trunks have service qualities, LSPs can be created with specific service attributes using the approach outlined with MPLS support of IntServ. In addition to service-specific LSPs, it is also possible to use explicit path setup to create parallel LSPs that share network ingress and egress points, but use diverse paths within the MPLS network. The traffic engineering task is to map traffic trunks to LSPs in accordance with a match of service attributes and available capacity, with the objective of meeting all the traffic service commitments and making efficient use of the underlying network resource. If both the LSP and the traffic trunks use explicit traffic profile attributes, then the match of traffic trunks to LSPs can be based on a computed best fit. The more likely scenario is where no explicit traffic profile is associated with traffic trunks, in which case the mapping of traffic trunks to LSPs can be dynamically controlled. In this situation, traffic trunks may be remapped to different LSPs in response to shifting traffic load patterns. The key requirement here is a feedback control system that allows the traffic engineering controller to receive the current level of performance on each LSP. In addition, the controller must be able to use this information to adjust the traffic trunk mapping to make incremental changes that are intended to reduce the load level on the affected LSP. This area of technology development is currently one of active study and research.

Low-Bandwidth Networks

Internet networks cover a very broad diversity of applications, ranging from IP-over-SONET systems operating at 2.4Gbps to very low-speed systems such as GSM data operating at 9.6Kbps. While the high-technology frontier of the Internet is pushing the speed limit of the Internet with Gigabit Ethernet LANs and 10Gbps systems using OC-192c SONET bearers, keep in mind that the majority of Internet users see it through the other end of a voice-grade analog modem operation. True, in theory, high-speed modems across high-quality analog circuits can operate at speeds of up to 56Kbps, but in reality, access speeds of between 21Kbps and 41Kbps are more typical in most access networks. As Internet products start to permeate the mobility area—including consumer electronics devices that combine telephone functions with pagers and personal computing functions—it is anticipated that there will be tens of millions of additional Internet access devices, each using relatively low-bandwidth wireless access systems. Many of these low-bandwidth systems operate with high levels of jitter and delay and higher levels of bit error rates, which are typically associated with high-speed fixed networks; for example, a third-generation wireless service offering 2Mbps access operates with a 200-millisecond delay. The delay is not primarily due to end-to-end signal propagation delay, but is a side effect of a signal encoding that uses both data compression and forward error correction.

If service performance is important for high-speed networks, it is of critical importance to low-bandwidth systems. Here the effort is not so much one of attempting to selectively mitigate the effects of congestion, but of tuning both the applications and the network to minimize the impact of the low-bandwidth network [ID-pilc-slow].

Managing Performance in Low-Bandwidth Environments

One of the primary concerns in a low-bandwidth environment is the relative size of the packet header to the data payload. This is most effectively illustrated in an interactive environment where a single-byte payload carries a 40-byte TCP/IP header.

Using the assumption that the number of simultaneous flows across the low-bandwidth link is relatively small, it is possible for the routers at either end of the link to hold a state description of the flow. This state description can maintain the source and destination addresses, the IP protocol number, and so on. If the sender has an index number for the flow, the sender can use it in place of many of the header fields. The receiver can then reconstruct the original packet and pass it on to the next hop. In this way, instead of sending the entire TCP/IP header, or the RTP/UDP/IP header in the case of real-time data flows, the sender can send an index pointer, together with a minimal set of header information. All fixed fields in the header can be omitted from the transmitted header, as can all fields that change in predictable ways or that can be computed from other field values. The remaining fields can be compressed so that the difference in value, rather than the absolute value, can be transmitted. This technique can compress a TCP/IP header to a minimum of 3 bytes [RFC1144]. Similar compression techniques can be used on UDP/IP data flows; and, where the Real-Time Protocol is used to time the data flow, the compression can extend to include the RTP header. The best-case RTP/UDP/IP header compression is 2 bytes [RFC2508].

It is also possible to realize some additional bandwidth efficiency through the use of data compression. The actual benefits of data compression are, of course, highly dependent on the nature of the application's data in the first place. There is a defined IP Payload Compression Protocol (IPComp) that can be used to compress an IP packet payload [RFC2393]. Of course many forms of payload are already carried in a compressed format (such as many compressed image formats) or are encoded in such a way that compression is impossible (such as in the case of many data encryption methods).

Further improvements can be made by some modifications to TCP. For TCP, the intention is to increase the speed of the slow-start algorithm that is attempting to reduce the number of round-trip time intervals to get the application to the point of efficient use of the network resource. The initial congestion window can be increased from the standard value of a single segment to higher values. Experimental evidence suggests that an initial value as high as three segments offers a level of performance improvement to the application and increases the use efficiency of the low-bandwidth link. Increasing the segment size is another way to improve the speed of this algorithm, although certain interactions between the segment size and network jitter and the link's bit error rate need to be factored in as well.

Many other potential adjustments to TCP have been investigated, and some show promise [ID-pilc-ltn]. A TCP sender uses a flow-control loop governed by the reception of ACK packets. If the receiver uses a delayed ACK mechanism, the sender will not necessarily see one ACK for every sent packet. As the ACK stream is used by the sender to estimate the packet stream's arrival rate at the receiver, delayed ACKs will give the sender a lower reading of this value. One approach to improve the sender's view of the receiver's arrival rate is to use the byte sequence counters within the ACK packets, so that the arrival rate in bytes, rather than packets, can be used to control the sender's behavior. Though effective in some environments, tests have also shown that this approach can lead to amplified burst characteristics. Nevertheless, refinement of this approach with an additional burst damping filter shows some promise. A simpler, yet surprisingly effective approach to the slow-start speed-up rate is to ensure that the receiver promptly ACKs the initial data segment of a new TCP connection, then activates a delayed ACK mechanism for subsequent packets.

An alternative to protocol modification is to ensure that the active network elements are correctly configured to efficiently drive low-bandwidth network systems. One of the major objectives is to ensure that the router's buffers are configured with output interface buffers that are the same order of size as the delay bandwidth product of the output circuit. TCP is attempting to establish the point of maximal network efficiency. It achieves this by driving into the network's buffers at increasing rates until the buffer signals that it is full. At this point, the network link controlled by the buffer is also operating at maximum efficiency (assuming a work-conserving scheduling discipline is being used on the output queue). As TCP cycles its sending rate at an interval of an end-to-end RTT, the bottleneck buffer will need to contain up to one RTT of data to ensure that the link remains constantly busy.

One of the compromises in modem access design is between cost reduction of the network access servers and the need to equip the network access server with adequate buffer space to drive the modem link at maximal throughput. There is also a service-level compromise to be made in the choice of buffer size. Large buffers introduce a significant jitter component into real-time data flows. While such buffers can improve link efficiency through assisting TCP sessions to operate efficiently, the quality of real-time services is compromised. One approach is to reduce the buffer size, but its use requires another, regarding managing TCP traffic flows. If the traffic through the bottleneck is symmetric, and the number of active flows is small, the router can potentially use a technique known as *ACK spacing,* shown in Figure 5.13 [ID-ackspace]. Here the sequence of ACK packets from each data flow is passed through an admission shaper so that the inter-ACK spacing is extended to a configured ACK rate. This ACK spacing causes the sender to space out the data transmissions so that the sender does not burst at a rate greater than the router's buffers can manage. Another solution to the jitter problem of large buffers is to pass UDP packets through a traffic profile filter that enforces a maximum average rate and a maximum burst rate, and uses a high-priority short queue for UDP. For TCP, the complementary approach is to use a RED/ECN-managed large queue. More flexible approaches to scheduling are possible by using DiffServ techniques.

All these approaches rely on the queuing discipline at the boundary of the high-speed network with the low-speed network. The queuing discipline performs the speed adaptation role and allows the transparent operation of single end-to-end transport protocols across the two networks. An alternative

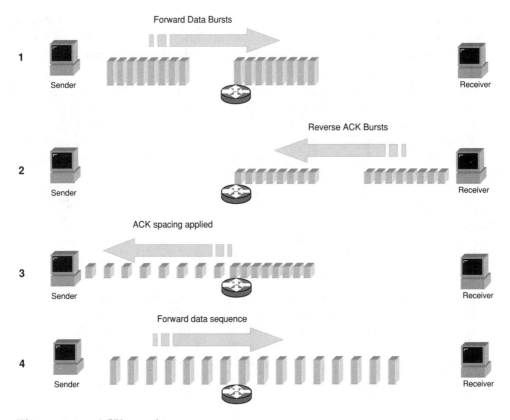

Figure 5.13 ACK spacing.

approach to low-bandwidth networks is to use split transport connections, whereby the interface to the low-speed network terminates a TCP connection; a second connection is set up across the low-bandwidth connection. This allows the speed adaptation to be explicitly configured to the desired functionality [Bakre 1995]. In addition, the second TCP connection can be explicitly configured to operate the low-speed connection at a level of maximal efficiency, using the lower end-to-end delay to adapt more quickly to the prevailing load conditions on the low-bandwidth path. This relies on symmetry of routing through the low-bandwidth gateway and only comes into play when the TCP payload is visible to the gateway router. When IPSEC is used, the IP payload is not visible to the gateway, and no "slave" TCP session can be set up.

Another approach to splitting the control loops into two discrete components is to use proxies at the application level. The most common of these is the use of Web proxy caches at the gateway [Huston 1999]. Here the proxy cache can split the transport connection into two components, serving the low-bandwidth client with an optimally tuned TCP stack. In addition, the proxy agent can cache content to allow subsequent requests for the same content to be served locally from the cache, thereby eliminating the delay component of the primary content fetch.

Managing Performance in Unreliable Environments

The assumptions inherent to the IP suite regarding the reliability of the network need to be reviewed in this environment [ID-pilc-error]. There are two levels of checksum within the end-to-end IP transmission model. Intermediate nodes perform forwarding based only on the integrity of the IP header, and bit errors within the IP payload are undetected until the packet reaches its destination. This architecture implies that packets with damaged payloads are passed through the low-bandwidth network unnecessarily. There is a role for the link layer to encapsulate the packet within a framing protocol that includes a checksum covering the entire packet to ensure that packets with payload bit errors are detected and discarded as quickly as possible.

In addition, there is some benefit in including forward error correction (FEC) encoding into the link-level protocol in environments with significant bit error levels. One of the first considerations is tuning congestion-managed TCP data flows in environments where the probability of bit errors is high. The TCP control system assumes that packet loss as a result of bit errors in the network is rare, as compared to the incidence of packet loss as a result of bottleneck congestion. Accordingly, the TCP control system reduces the sending rate as a response to packet loss. In an error-prone low-bandwidth system, such as wireless, this response is unhelpful to maintaining data throughput. The first essential step is to minimize the incidence of bit errors by using forward error correction within the link layer. While this does take up some additional bandwidth, there are substantial benefits in improved TCP throughput performance.

Also of benefit in such environments is the use of Explicit Congestion Notification (ECN) within TCP [RFC2481]. A pair of systems using ECN across an ECN-aware switching path can react to packet loss as a bit error condition while congestion is explicitly signaled through the ECN 2-bit field.

Choosing an appropriate MTU is critical for low-bandwidth unreliable systems. A large MTU allows the packet's headers to take up a relatively small proportion of bandwidth as compared to the payload. In addition, TCP uses an initial slow-start control mechanism that increases the sending rate in units of segments per round-trip time interval. A larger MTU implies a more rapid startup phase for TCP; however, a larger MTU increases the probability of a packet encountering a bit error. In addition, a larger MTU increases the jitter level within the system because the larger maximum packet sizes in a mixed traffic environment cause a higher level of packet quantization delay on low-bandwidth systems. A compromise would suggest the use of a smaller MTU, with header compression and TCP Path MTU Discovery to avoid excessive packet header overheads, as well as the overhead of packet fragmentation while using an initial TCP congestion window of two segments to speed up the TCP slow-start phase.

Managing and Measuring QoS

There are two basic motivations for QoS management and measurement, and before we examine measurement techniques in greater detail, we should explore these motivations.

The first motivation is of the network operator, whose primary management question is: "Is the network delivering the levels of service as intended in the network configuration?" A complementary measurement question is: "Is the network delivering the levels of service as specified in the various service level agreements with the network's clients?"

The second motivation is of the client. The service question being asked is: "Is the network delivering the level of service as specified in the service level agreement?" Equally important is the question: "Does activating the service level agreement represent value for money?" Or, to put it another way: "Is it worth generating a request for premium-level service for this transaction at this point in time, or will a best-effort service meet the transaction's requirements?"

Within the realm of network management is an almost boundless amount of information that can be generated from an operating network. The various management information bases (MIBs) defined within the Structure of Management Information (SMI) of the Simple Network Management Protocol (SNMP) architecture enable a network management station to gather element-based status information at arbitrary frequencies. The challenge is that a sequence of snapshots of element status values cannot readily be reconstructed into a comprehensive view of the operational status of the network as an entire system. The underlying approach of element polling systems is that the polling agent, the network management station, is configured with an internal model of the network; status information, gathered through element polling, is integrated to the network model. The correlation of the status of the model to the status of the network itself is intended to accurate enough to allow operational anomalies in the network to be recognized and flagged.

How can this approach be used to monitor and manage the service levels being delivered to clients of the network? The network element polling approach can indicate whether or not each network element is operating within the configured operational parameters, and alert the network operator when there are local anomalies to this condition. But such a view is best described as network-centric, rather than service-centric. An implicit assumption is that if the network is operating within the configured parameters, then all service level commitments are being met. This assumption may not be well-founded. Thus, there is a need to look at a service-centric view of network management and measurement. To do this, we need to consider the possible service measurements within a network and how such measurements can relate to the service levels being delivered to the network's clients.

Service Measurement Approaches

A network can alter the characteristics of a data flow in a number of ways. Packets within the data flow can be altered by delay, discard, or fragmentation. The delay can be experienced uniformly by all packets within a flow, or the delay may be more variable, leading to instances of packet reordering. In some anomalous situations, packets may even be duplicated. How can the network operator and the network client measure the levels of delay, discard, and fragmentation experienced within a service differentiated network?

The Network Operator's Perspective

A management system can poll each active network element to retrieve the number of packets dropped and the number of packets successfully forwarded. From these two data items, the relative drop proportion of packets can be calculated on an element-by-element basis, and a series of element measures can provide a per-path drop proportion by multiplying the individual packet-forwarding measurements. Packet fragmentation levels are a function of packet size and the MTU of each forwarding component of the path. Again, this can be calculated from polling the network.

Delay is somewhat more challenging to measure on an element-by-element basis using element polling. In theory, the polling system could periodically poll a router's output queue length and estimate the queuing delay based on

*The Simple Network Management Protocol (SNMP) is an informa-
tion structure designed to support the detail of network element
management. The overall intent of SNMP is to provide a standard way
of querying the status of network elements and for network elements to
report on their operational status. This allows a vendor-independent
network management environment to be operated by the network manager. The architecture of SNMP is based on a query-response model, in
which a management system generates a query within an SNMP-
defined query structure and the query is passed to the managed network element via a UDP network transaction. The network element
responds, also via UDP, with an SNMP-formatted response that corresponds precisely to the query. SNMP also can generate traps, which are
managed-element-initiated SNMP messages directed to a management
host to inform the host of an exception or alarm condition that may
require attention.*

*The security model provided with SNMP, both in versions 1 and 2,
are relatively basic, relying on simple password-based access mechanisms. SNMP version 2 was intended to have a more robust security
architecture, but the IETF Working Group fragmented; subsequently,
the IETF Network Management Area was disbanded. The outcome, as
far as SNMP was concerned, was far from ideal.*

*The Structure of Management Information (SMI) defines the struc-
ture of the data objects manipulated by SNMP. This is termed a
Management Information Base (MIB). MIB is defined using the ISO-
defined Abstract Syntax Notation One (ASN.1) notation. In addition to
the data objects, ASN.1 is used to define the specification of the set of
Basic Encoding Rules (BER) for SNMP, which allow for a standard
method of the transmission of queries and responses within the SNMP
protocol.*

*SNMP data objects are constructed from a simple set of atomic data
types (integer, octet string, and object identifier) and compounding
operators to allow arrays and records. Objects themselves are identified by a sequence of integers that correspond to selection of a
sequence of edges in a tree of data objects. The system description
object can be referred to as an SNMP object by the string*

an average packet size estimate, together with the output capacity. Of course,
such a measurement methodology assumes a simple FIFO queuing discipline
and a queue size that varies slowly over time. In a QoS environment, the first

1.3.6.1.2.1.1.1, or by the selection of the edges that lead to the sequence of labeled nodes iso.org.dod.internet.mgmt.mib.system.sysDescr. *The MIB used in SNMP, MIB-II, is an Internet-standard database of objects, together with an extension area, which enables equipment vendors to provide additional objects. Data objects are transported within a standard tag, length-value triplet.*

SNMP is a simple UDP-based protocol, using UDP port 162. SNMP allows for two query commands, GetRequest, *which retrieves the current value of the specified MIB object, and* GetNextRequest, *which allows the network management station to return the next MIB object in a depth-first traversal of the MIB-defined object tree. In addition, a* SetRequest *message allows the network management station to specify an MIB object and a new value for the object. All three messages elicit a* GetResponse *from the managed entity, which is used to return the specified value or an indication of why the requested action could not be performed by the entity. In addition to query-response interactions, SNMP allows for a trap message to be generated by the managed entity, in which the network management station can be informed of a number of standard events that include a power-up start, a restart, interface status change, SNMP access failure, or a vendor-defined event.*

As already noted, the security model for SNMP is relatively basic, allowing for communities of access. *These provide access to the managed entity using a simple password-based mechanism, which is sent in the clear with an accompanying query. Multiple communities can exist, each with a different access mode, whereby some communities have read-only access to the managed entity and others have read and write access.*

SNMPv2 adds the GetBulkRequest *message to the operator set to allow large blocks of data to be retrieved efficiently. Version 2 also adds the* InformRequest *operator to allow management station-to-management station notification. SNMPv2 also adds to the basic SNMP data set; 64-bit integer values complement the original 32-bit integer values. During the development of SNMPv2, the pressing need was for a more robust security architecture. This need remains unfulfilled until later work on SNMPv3.*

assumption is not necessarily the case. The queue size will oscillate with a relatively high frequency as a function of both the number and capacity of the input systems and of the capacity of the output system. In general, delay is not easily measured using network element polling.

While drop rates and fragmentation functions can be gathered from each network element, with the additional input of the current forwarding state of the network, it is possible to predict the path a packet will take through the network. This can be used to estimate the path probability of drop and the packet fragmentation limit. However, this information is still well short of service measurement.

One approach is to expand the functionality of the polling system by expanding the management model to include components of the service architectures. Efforts are underway at the IETF to standardize the MIBs relating to the DiffServ model and the operation of IntServ and RSVP. For the DiffServ MIB, it is first necessary to define an abstract model of a DiffServ admission router's operation, by looking at the major functional blocks of the router [ID-diffserv-model]. The first of these blocks is the definition of the supported behavior aggregates provided by the network. Within the network path, the initial active path element is the traffic classification module, which can be modeled as a set of filters and an associated set of output streams. The output stream is passed to the traffic-conditioning elements, which are the traffic meters and the associated action elements. A number of meter profiles can be used in the model: an average data rate, an exponential weighted moving average of one of a number of various traffic profiles that can be expressed by a set of token-bucket parameters using an average rate, a peak rate, and a burst size. More elaborate meter specifications can be constructed using a multilevel token-bucket specification. From the meter, the traffic is passed through an action filter, which may mark the packets and shape the traffic profile through queues or discard operations. Together, this sequence of components form a *traffic conditioning block*. The traffic is then passed into a queue through the use of a queuing discipline that applies the desired service behavior.

From this generic model it is possible to define instrumentation for SNMP polling, where each of these five components—the behavior aggregate, the classifier, the meter, profile actions, and the queuing discipline—correspond to a MIB table [ID-diffserv-mib]. With this structure it is possible to parameterize both the specific configuration of the DiffServ network element and its dynamic state. This MIB is intended to describe the configuration and operation of both edge and interior DiffServ network elements, the difference being that interior elements use just a behavior aggregate classifier and a queue manager within the management model while the edge elements use all components of the model.

A comparable MIB is defined for the Integrated Services architecture and an additional MIB for the operation of guaranteed services [RFC2213, RFC2214]. The Integrated Services MIB defines the per-element reservation table used to determine the current reservation state, an indication of whether or not the router can accept further flow reservations, and the reservation characteristics of each current flow. No performance polling parameters or accounting parameters are included in the MIB. The guaranteed services MIB adds to this definition with a per-interface definition of a backlog. This is a means of expressing *packet quantization delay*, a delay term, which is the packet propagation delay over the interface, and a *slack term*, which is the amount of slack in the reservation that can be used without redefining the reservation. Again, these are per-element status definitions and do not include performance or accounting data items.

The Integrated Services MIB is being further defined as an RSVP MIB for the operation of IntServ network elements [ID-rsvp-mib]. There are a larger number of objects within the MIB, including General Objects, Session Statistics Table, Session Sender Table, Reservation Requests Received Table, Reservation Requests Forwarded Table, RSVP Interface Attributes Table, and an RSVP Neighbor Table. Interestingly, the MIB proposes a writeable RSVP reservation table to allow the network manager to manually create a reservation state that can only be removed through a comparable manual operation. The MIB enables a management system to poll the IntServ network element to retrieve the status of every active IntServ reserved flow and the operational characteristics of the flow, as seen by the network element.

A complementary approach to instrumentation of network elements is active network probing. This requires the injection of marked packets into the data stream; collection of the packets at a later time; and correlation of the packets to infer some information regarding delay, drop, and fragmentation conditions for the path traversed by the packet. The most common probe tools in the network today are *ping* and *traceroute*.

THE POWER OF PING

The most common of these tools is the ICMP Echo Request and corresponding ICMP Echo Reply packets, which comprise the functionality of the ping utility. In its basic form, ping takes a target IP address as an argument, directs an ICMP Echo Request packet to that address, and awaits a matching ICMP Echo Reply response.

A ping response indicates that the target host is connected to the network, is reachable from the query agent, and is in a sufficiently functional state to respond to the ping packet. In itself, this response is useful information, indicating that a functional network path to the functioning target host exists. Failure to respond is not so informative because it cannot be absolutely inferred that the target host is not functional. The ping packet, or its response, may have been discarded within the network due to transient congestion, or the network may not have a path to the target host. In this latter case, because the ICMP does not allow the generation of ICMP host-unreachable messages in response to a nonroutable ICMP packet, no network error indication can be generated back to the query agent.

Further information can be inferred by ping with some basic modifications to the behavior of the ping algorithm. If a sequence of labeled ping packets is generated, the elapsed time for a response to be received for each packet can be recorded, along with the count of dropped packets, duplicated packets, and packets that have been reordered by the network. Careful interpretation of the response times and their variance can provide an indication of the load being experienced on the network path between the query agent and the target. Load will manifest a condition of increased delay and increased variance, due to the interaction of the router buffers with the traffic flows along the path elements as load increases. When a router buffer overflows, the router is forced to discard packets; and under such conditions, increased ping loss is observed. In addition to indications of network load, high erratic delay, and loss within a sequence of ping packets may be symptomatic of routing instability with the network path vacillating between a number of states.

A typical use of ping is to regularly test a number of paths to establish a baseline of path metrics. This enables a comparison of a specific ping result to these base metrics to give an indication of current path load within the network.

Of course, it is possible to infer too much from ping results, particularly when pinging routers within a network. Many router architectures use fast switching paths for data packets; the router's central processing unit is used to process ping requests. The ping response process may be given a low scheduling priority because the routing protocol operation is a more critical router function; therefore, it is possible that extended delays and loss may be related to the load of the target router processor rather than to the condition of the network path. Ping sequences do not necessarily mimic packet flow behavior of applications. Typical TCP flow behavior is prone to cluster bursts of packet transmissions on each epoch of the round-trip time. Routers may optimize their cache management, switching behavior, and queue management to take advantage of this behavior. Ping packets may not be clustered; instead, an evenly spaced pacing is used, meaning that the observed metrics of a sequence of ping packets may not exercise such router optimizations. Accordingly, the ping results may not necessarily reflect an anticipation of application performance along the same path.

It's important to note that packet delay variance and some small level of loss is not necessarily indicative of imminent congestion-induced collapse along the network path. Adaptive flow protocols, such as TCP, use a flow-control algorithm by which the TCP flow rate is increased to the point where a router buffer overflows and packet loss occurs. Therefore, some background level of packet delay variance and loss is not unusual within normal operational conditions of the network.

With these caveats in mind, monitoring a network through regular ping tests along the major network paths can yield useful information regarding the status of the network, indicating as a base the operational status of the network's routers and service platforms. Analysis of round-trip times and loss rates is also a basic indicator of network load, although care must be taken not to infer too much from such data.

A number of refinements to ping can extend its utility. Ping can use *loose source routing* to test the reachability of one host to another, directing the packet from the query host to the loose source routed host, then to the target host and back via the same path. However, many networks disable support for loose source routing, given that it can be exploited in some forms of security attacks. Consequently, the failure of a loose source routed ping may not be a conclusive indication of a network fault.

Ping also can be used in a rudimentary way to discover the provisioned capacity of network links. By varying the packet length and comparing the ping times of one router to the next-hop router on a path, the bandwidth of the link can be deduced with some degree of approximation required due to network jitter.

A more sophisticated variation of ping is to pace the transmission of packets from the received packets, mimicking the behavior of the TCP flow-control algorithm with slow start and subsequent congestion avoidance. TReno is such a tool [ID-ippm-treno]. In TReno, the transmission of ping packets is managed by the TCP Reno flow-control algorithm. Such a tool can indicate available flow rate-managed capacity on a chosen path—although this is a relatively intrusive network probe.

PATH DISCOVERY USING TRACEROUTE

The second common ICMP-based network management tool, traceroute, devised by Van Jacobson, is based on the ICMP Time Exceeded message. Here, a sequence of UDP packets are generated to the target host, each with an increased value of the TTL field in the IP header. This generates a sequence of ICMP Time Exceeded messages sourced from the router where the TTL expired. These source addresses are those of the routers, in turn, on the path from the source to the destination.

Like ping, traceroute measures the elapsed time between the packet transmission and the reception of the corresponding ICMP packet. In this way, the complete output of a traceroute execution exposes not only the elements of the path to the destination, but also the delay and loss characteristics of each partial path element. Traceroute also can be used with loose source route options to uncover the path between two remote hosts. The same caveats mentioned in the pin discussion relating to the deployment of support for loose source routing apply.

Traceroute is an excellent tool for reporting on the state of the routing system. It operates as an excellent "sanity check" of the match between the design intent of the routing system and the operational behavior.

The caveat to keep in mind when interpreting traceroute output has to do with asymmetric routes within the network. While the per-hop responses expose the routing path taken in the forward direction to the target host, the delay and loss metrics are measured across the forward and reverse paths. The reverse path is not explicitly visible to traceroute.

In a QoS Differentiated Services environment, ping and traceroute pose some interesting engineering issues. Ping is an ICMP packet. The network's QoS admission filters may choose a different classification for these packets from that chosen for TCP or UDP protocol packets; as a result, the probe packet may be scheduled differently or even take a completely different path to the network. In an IntServ network, the common classification condition for a flow is a combination of the IP header's source and destination addresses and the TCP or UDP header's source and destination port addresses. The ping probe packet cannot reproduce this complete flow description, and therefore cannot, by default, be inserted to the flow path that it is attempting to measure. With traceroute, the packet does have a UDP protocol address, but it uses the constant port address by default, causing a similar problem of attempting to be inserted to an IntServ flow. DiffServ encounters similar problems when attempting to pass the probe packet into the network via the DiffServ admission classification systems. Inside the network, it is possible to insert the probe packet into the network with the IP DSCP field set to the DiffServ behavior aggregate that is being measured.

The measurement of delay and loss taken by ping and traceroute is a cumulative value of both the forward and return path delay and loss. When attempting to measure unidirectional flow path behavior, such as an IntServ flow path, this measurement is of dubious value given the level of uncertainty as to which part of the path, forward or reverse, contributed to the ping or trace-route delay and loss reports.

There are techniques that perform a one-way delay and loss measurement, and they are better suited to measuring the service parameters of unidirectional flows [ID-ippm-delay, ID-ippm-loss]. A one-way approach does not use a single network management system, but relies on paired probe senders and receivers using synchronized clocks. For delay, the sender records the precise time a certain bit of the probe packet was transmitted into the network; the receiver records the precise time that same bit arrived at the receiver. Precisely synchronizing the clocks of the two systems is an interesting problem. Current experiments with this approach have used Global Positioning System receivers as a synchronized clock source. Consequent correlation of the data from repeated probes can reveal the one-way delay and loss patterns between sender and receiver. To correlate this to a service level requires the packets to travel along the same path as the service flow and with the same scheduling response from the network. In DiffServ networks, this can be done within the network, setting the DSCP field to the value of the service aggregate being monitored. Of course, from the customer's perspective, the DiffServ network service profile includes the admission traffic-conditioning block, and the interior one-way measurements are only part of the delivered service. In the IntServ network, the packets have to be structured to take the same path as the elevated service flows; they are classified by each element as part of the collection of such elevated service flows for the purposes of scheduling.

Measurement techniques using polling and modeling can track the performance of the network, on an element-by-element basis, but they cannot track per-path service levels across the network. Probe techniques, particularly one-way loss and delay, can perform such a complementary role of per-path service monitoring.

The Client Perspective

From the client's perspective, the measurement choices are more limited. A client does not normally enjoy the ability to poll network elements within a provider's network. One way for a client to measure service quality is to instigate probing of the network path, whereby a sender can pass a probe packet in to the network and measure the characteristics of the response. Of course, the problems of inserting probe packets into the service flow remain, as do the issues of unidirectional elevated service flows with bidirectional probes.

However, the client does have the advantage of being able to monitor and manipulate the characteristics of the service flow itself. For TCP sessions, the client can monitor the packet retransmission rate, the maximum burst capacity, the average throughput, the RTT, RTT variance, and misordered packets, by monitoring the state of the outbound data flow and relating it to the inbound ACK flow. For UDP sessions, there is no corresponding transport-level feedback information flow to the sender as a part of the transport protocol itself. The receiver can measure the service quality of the received data stream using RTP information—if RTP is being used for real-time data or as an application-related tool for other application types. If sender and receiver work in concert, the receiver can generate periodic quality reports and pass these summaries back to the sender. Such applications can confirm whether an application is receiving a specified level of service. This approach treats the network like a black box; no attempt is made to identify the precise nature or source of events that disrupt the delivered service quality. There are no standardized approaches to this activity, but there are a number of analysis tools available for host platforms that perform these measurements.

Though the client can measure and conform service quality on a per-application level of granularity, the second part of the client's motivation in measuring service quality is more difficult to address. The basic question is whether the service delivered in response to a premium service request is sufficiently differentiated from a best-effort service transaction. Without necessarily conducting the transaction a second time, the best approach is to use either one-way delay probes, for unidirectional traffic, or a bulk TCP capacity probe, to establish some indication of the relativity in performance.

Service Level Agreements

From a performance engineering marketing perspective, the obvious question to ask is whether it is possible to support service level agreements (SLAs) with clients of the service; and if so, in what form could these SLAs be phrased?

Some clients regard the service environment of best-effort delivery as too risky for them and not risky at all for the network provider. The most direct way for a network provider to increase the earning potential of the network service is to increase the traffic levels on the network without correspondingly augmenting carriage capacity. The certain outcome to the provider is a reduction in traffic-based unit costs of the network, as more traffic is being passed over a network infrastructure that has constant cost. The side effect on the client of this provider business strategy is declining levels of service; the increased load will raise the frequency and severity of congestion events occurring within the network.

The SLA is a means for the customer to impose some limit on the extent to which the network provider can indulge in over-subscription of carriage and switching capacity. The SLA establishes some minimum level of service level, or performance, which the provider must meet under the terms of the access contract. Although the intent is eminently sensible and prudent from the perspective of both parties, it presents implementation issues that are not easily addressed. These issues are related to QoS considerations.

The first consideration is how to phrase an acceptable service level or how to define a base level of quality within the service provider's network. Of course, an entire network does not exhibit a given level of quality at a particular moment in time. Rather, individual traffic flows encounter various quality conditions through the lifetime of the flow, and these quality conditions can be considered an expression of the quality of the network. Phrasing a quality constraint in terms of the complete absence of packet loss makes little sense in an environment where a background-level packet loss is a natural outcome of TCP flow behavior. However, specifying an average packet loss metric is open to widely varying interpretations. The underlying observation is that in an environment of opportunistic traffic flows, in which each TCP flow dynamically adjusts its behavior to occupy as much of the available network as possible, the outcome is variable-flow performance. In a variable-flow environment, each flow continually attempts to balance its performance with those flows, which are competing for the same resource. Gross limits can be expressed on packet loss, jitter, delay reliability, and path stability within an SLA, but it is likely that the SLA will be able to encompass only some instances of a large level of over-subscription, while the finer aspects of quality metrics may well prove elusive to any form of a realistic SLA.

The second consideration is how to measure an SLA. Once an SLA is specified in terms of a measurement technique, the ISP will divert attention from the overall delivery of a service quality level and concentrate on achieving those specifications in the SLA, even at the expense of quality of service. Accordingly, the measurement of the SLA must be generic enough so as not to create highly specific network tuning; at the same time, the measurement must not be so generic that the technique and the reported outcomes become the subject of dispute. The scope of the measurements is also relevant. For example, a provider would find the following SLA specification unacceptable: a set of end-to-end flow quality measurements in which the paths transited other party's networks and in which the end system platforms exhibited variable performance due to other load factors. If the measurement were via a mechanism that loaded a network with traffic, then the measurement itself would have a detrimental effect on available network capacity. On the other hand, network probe mechanisms, such as ping clusters, tend to produce outcomes in terms of reported round-trip loss and jitter, which are not well correlated to concurrent TCP and UDP flow performance.

Currently, the use of SLAs as a tool to enforce base levels of service quality within the Internet environment has questionable outcomes. Consistent poor service is still most effectively addressed by *client churn*, whereby clients move to other competitive providers in response to poor service experiences. Repeated effort to achieve high-quality service results in client loyalty to the service. So far, attempts to codify this into an SLA through financial penalty, to force an ISP into a certain mode of operation, have not always been as successful as either party would hope.

Service Accounting

The premium network services described here are intended to allocate network resources in a preferred manner. Such preferential allocation generates some level of benefit to the client of the premium service. To manage the level use of such a service, there is usually some associated accounting of the use of the service. In a commercial network service environment, this accounting is then associated with a higher tariff. In a corporate or campus environment, such a premium on access to a quality service may be expressed in terms other than pricing. Common to both environments is the use of some external process to place some cost on the premium service. Without such a cost transfer, the optimal behavior for all clients is to request a premium service for all transactions, and this results in a service outcome that is little different from an original best-effort service environment.

The objectives of the accounting system are to produce a collection of accounting entries that detail the nature of the service provided, the entity to which the service was provided, the volume of usage of the service, the profile of usage, the entity that requested the service, and the nature of the request. Note that in specifying an accounting system, no choice of a pricing model is inferred. The point of the accounting system is to provide sufficient detail to support a range of potential pricing models.

There are two generic models of accounting systems: the *transactional model* and the *continuous monitoring model*. Where there is a discrete start and stop event that defines a transaction, and where this event is visible to the network at some state change in a network element, then it is possible to support a transactional model of accounting. This is often described as a *Radius model*. To the dial-up accounting environment, which was the original impetus for Radius accounting, the transactional model is not all that different. For every long-held transaction, this model breaks down. The session information is only available to the accounting system on termination of the transaction. Intermediate checkpoint status accounting messages are a common response to this issue, where the accounting system is informed periodically of the current accounting for the transaction. Continuous monitoring is often associated with the Simple Network Management Protocol (SNMP) where an accounting server periodically polls a network element and retrieves the values of a number of defined SNMP variables. The change in value of these variables is passed to the accounting system as a continuous stream of data.

The major issues associated with service accounting are granularity of accounting and the definition of network events and locations that will trigger accounting activities. We will continue with our standard approach of looking at the issue first in conjunction with the IntServ state-based architecture, then examine what changes when we move to the aggregated service model of the DiffServ architecture.

The IntServ architecture relies on the establishment of a reservation *prior* to sending a distinguished service traffic flow. This RSVP Resv reservation message is the most likely trigger for an accounting entry where the network ingress point needs to create an accounting state in response to the message. The role of accounting is to allocate resource usage against a client. For this to occur with RSVP, it is necessary to include identity information in the RSVP message exchange. Within RSVP, the Policy Data Class object is defined as the appropriate vehicle to carry identity information that can authenticate a user. The policy object is further broken down into *policy elements*. The protocol element of interest to the accounting role is the Authentication Data policy element (Auth_Data PE). The Auth_Data PE can describe either the RSVP application or the RSVP user [ID-rap-rsvp-identity]. The authentication attributes that can be included in an Auth_Data PE include the policy locator, a directory locator pointer to the relevant admission policy; a user credential that may be the user identification or a digital certificate that can identify the user; and a digital signature that is used to sign the Auth_Data PE.

For the purposes of generality of accounting, an RSVP accounting process would need to match and store information generated at both network ingress and egress. At the point of network ingress of the RSVP data flow, the accounting information needs to identify the neighboring upstream network element, along with the policy element that permits this network element to send an

RSVP data stream into the network. The ingress element would need to collect per-flow volume and profile information, and pass this to the accounting server at the time of flow teardown. It is also appropriate that such per-flow information be pollable within an IntServ SNMP management information base (MIB). There is no such standard IntServ accounting MIB; the RSVP MIB does not include accounting entries, so for the moment, such a method of polling the state of usage of an RSVP reservation falls in the area of vendor-specific extensions to existing MIBs [ID-rsvp-mib]. A similar functionality is required on network egress, to identify the downstream next hop, the egress policy that allows the flow to be passed to this hop and the client identity as part of the accounting notification message. Here, too, there is a requirement for a pollable flow state and a teardown notification when the path is removed.

There is much yet to be defined in the area of accounting for network usage under the terms of an Integrated Services model. It is expected that some of these accounting issues will become clearer once the models of retail pricing of IntServ services use and interprovider financial settlement of cross-provider IntServ services become more stable. At this stage, there are no clear models for such pricing or settlement. For example, does the receiver of the IntServ flow bear the cost of the end-to-end flow? If so, then money is flowing in the opposite direction to the traffic flow, and each network would use egress accounting to generate information to be passed to the billing and settlement systems. In effect, each network provider sells the service to the next downstream provider, who adds further value in transiting the network, then sells the service at the sum of the purchase cost plus the value added by the local transit to the next transit network. The accumulated value is passed to the end receiver at a retail price. The receiver then is underwriting the costs of the entire IntServ-managed service transaction. Alternatively, does the sender of the flow bear the cost of the transaction? Here, money is flowing in the same direction as the traffic flow. As a network installs an RSVP reservation, the ingress accounting system must identify the adjacent upstream system, as well as generate accounting information to pass to the billing settlement systems. This model is one where the upstream provider purchases a service to the receiving client. It adds value by providing the local network transit, then invoices the total of these sums as the cost to reach the receiving client to the upstream network or sender. While both of these pricing and accounting models are relatively clear, problems arise when there is a mix of retail models and interprovider settlement models.

The DiffServ architecture imposes a different regime on the accounting system: There is no initial transaction between the network and the sender or the receiver to set up the service flow, nor is any binding undertaken on the part of the network that an individual service transaction will be honored. The commitment of a DiffServ network to its boundary is only that the network will honor aggregate traffic flows within the bounds of the network itself. There is no concept of an end-to-end traffic flow, because it is replaced by the concept of admission into a defined service class. The most logical view of the accounting system is of an admission meter that measures how much traffic is admitted into each aggregated service class.

The objective to be addressed by the accounting system is somewhat broader than accounting for the use of network resources by behavior aggregates. It includes the identification of the client of these services. The accounting issues are somewhat more involved, as there is now the requirement to undertake admission accounting at a level of granularity that is defined by the admission policies themselves, rather than at a level of a behavior aggregate or an individual micro-flow. If the admission system interfaces directly to end clients, the desired level of accounting may be by behavior aggregate per client. If the admission system interfaces to a transit network, the admission system may replicate a wholesale arrangement and work at a level of behavior aggregates alone. The ability to identify when a client's use of a DiffServ service commences and ceases is also a challenging problem. If there is an admission policy cache, where the binding of a traffic filter specification and associated admission policy are cached in a recently used policy set, then it is feasible to trigger the accounting start and stop points against the times when the binding is entered and removed from the cache.

Interprovider Issues

To clearly reveal the technical issues behind the support of multiservice IP networks, a deliberate effort has been made so far in this study to limit the scope of the network platform to that of a single provider. This approach is directly relevant to an enterprise service IP network or parts of the public Internet where the QoS service is limited in scope to that of a single Internet service provider (ISP). Now, however, it is appropriate to broaden the scope of this examination to include the issues that arise in providing QoS services transparently across multiprovider environments.

Internetwork issues are often different from intranetwork technical issues. For example, in the Internet routing system, a fundamental distinction is drawn between routing within a network and routing across a network administrative boundary. Intradomain routing generates explicit path information, while interdomain routing generates only network layer reachability information. We have yet to attain a comparable understanding of interdomain QoS requirements as compared to intradomain QoS mechanisms.

The policies relating to admission may be based on a number of factors. The two networks may have a bilateral peering policy that covers both the exchange of best-effort traffic and the exchange of traffic within a number of defined service levels. Alternatively, one network may be a service provider to the other, and the acceptance of service-qualified traffic may be based on a price premium. We will examine QoS and interprovider financial settlements in greater detail in the following section.

The major issue is that of the interdomain policy control framework and its impact on interdomain path selection. Within the uniform best-effort Internet environment, the interdomain policy framework is relatively straightforward: A network domain advertises reachability of a network prefix to an adjacent network domain, if the network is willing to carry the consequent traffic. The complementary action is that a network will accept and use a routing advertisement if the network's egress policies match the advertisement. Normally, such arrangements are bilateral so each network will accept the other's network prefix advertisements and pass traffic to each other. QoS changes this. The network may wish to advertise QoS-qualified reachability for a smaller set of network prefixes than the best-effort set. Most likely, the network would be passing on the original client service policies, so that if the network supported multiple levels of service, the interdomain reachability advertisements would be qualified by a service level that the network was willing to accept. The set of network address prefixes and their associated service levels that are visible to the adjacent network represent policies that determine admission to the network. The implication of this qualification of network reachability by an associated service level is that a standard method exists to encode service levels to interdomain routing advertisements. Such a standard encoding has yet to be defined. If such a set exists, it is then possible to tag an exterior routing advertisement with a set of attributes that describe the service levels that qualify the advertisement (see Figure 5.14).

Given that the network has a finite resource capacity, the network may also have an additional set of policies relating to the total level of service-qualified traffic that the network is willing to accept from its neighbor. Potentially, this means that even where a neighbor network advertises a network prefix and an associated service level, any service request may still be rejected due to the total service load policy.

In IntServ service models, the traffic flow itself is a consequence of the resource reservation negotiation, and the interprovider interaction is based on

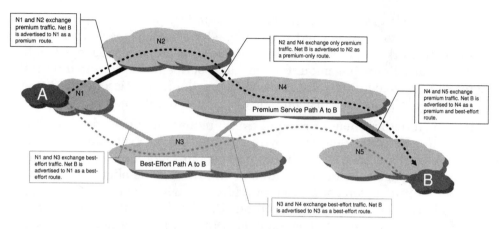

Figure 5.14 Service-qualified routing advertisements.

the negotiation of the RSVP messages. RSVP can span the network boundary quite easily. As the RSVP request passes across the network boundary, the change is in a policy framework, as each network will use local policies as the basis of their admission decisions. Across an interprovider boundary, the resource reservation messages remain unaltered, and the RSVP TSpec traffic profile specification has a constant interpretation across the network boundary. The advantage is that the IntServ architecture imposes uniform semantics on service requests, and end-to-end service connectivity is not affected by different service platform models. The TSpec service specification within the IntServ model specifies the parameters of the data flow itself. The universal assumption behind this specification is that of a limited service model comprising only guaranteed and control-load service classes.

The DiffServ model introduces the dimension of varying service offerings to allow for various efforts of service that may affect delay and jitter, packet loss, packet ordering, and available capacity. There is no requirement for one service provider to offer identical services to an adjacent service provider, nor is there a requirement to support a single standard set of DSCP encodings to describe such services. Still, two connected providers can agree on a boundary policy to admit traffic sent from the other, with either a common set of DSCPs and their mapping to PHBs or an agreed mapping of DSCP from one provider's service calls to its equivalent in the domain of the other. Such a boundary policy might, for example, allow traffic marked within the AF service group to be mapped across the service boundary using the default DSCP mappings, but explicitly map EF and CS class traffic to be remarked to the default best-effort DSCP. Furthermore, the boundary policy would provide for traffic profiles to be associated with mapped service classes so that the network-to-network admission systems enforce a resource management policy that explicitly limits the amount of elevated service traffic admitted from the network peer. Accordingly, for network peers, the interaction of the DiffServ boundary admission systems could be configured to interoperate.

How then does a client or an application request an elevated service to be applied to a network transaction with a remote party? Even though DiffServ can support the network peer interoperation, DiffServ alone is incapable of supporting such an interaction with a client. Even if the client has a service contract with its upstream service provider, allowing the client to mark the application's packets with the appropriate DSCP value for the desired service, this does not imply end-to-end service. If the path to the remote party traverses a network domain boundary, then this network boundary mapping function takes place. If the boundary mapping cannot map one provider's service profile to that of the neighboring network, or if the boundary policies do not match sufficiently to allow the transit, or if the boundary traffic load is already congested within the terms of the admission profile, then DiffServ end-to-end service cannot be provided. No mechanism is defined within the DiffServ architecture to allow control signaling back to the traffic sender about the state of the end-to-end DiffServ path. From a client's perspective, the DiffServ architecture, as currently defined, cannot provide the necessary signaling to client to assure that the requested service extends from end to end. This leads to the network-centric model of DiffServ interprovider interaction whereby two networks can create and implement a set of boundary policies that allow the exchange of DiffServ traffic. This model lacks the capability to translate this local interaction into a cohesive end-to-end service architecture that creates reliable end-to-end services. As it stands, this is a critical weakness in the current DiffServ architectural model.

Interprovider Settlements and QoS

To date, no pressure is evident to change the technology base of the current ISP service model to accommodate more sophisticated settlement structures than the simple sender-keep-all model. The fundamental observation is that any financial settlement structure is robust only where a retail model exists that is relatively uniform in both its nature and deployment, and encompasses the provision of services on an end-to-end basis. Where a broad diversity of partial-service retail mechanisms exists within a multiprovider environment, the stability of any form of interprovider financial settlement structure will remain dubious at best.

Perhaps the strongest factor driving change here is the shift toward an end-to-end service model associated with the QoS mechanisms. Where a client signals the requirement for some level of preemption or reservation of resources to support an Internet transaction or flow, the signal must be implemented on an end-to-end basis if the service request is to have any meaning or value. The public Internet business model to support practical use of such QoS technologies will shift to that of the QoS signal initiator bearing the cost of the entire end-to-end traffic flow associated with the QoS signal. This is a retail model in which the application initiator funds the entire cost of data transit associated with the application (see Figure 5.15). It is analogous to the end-to-end retail models of the telephony, postal, and freight industries, in which the participating agents are compensated for the

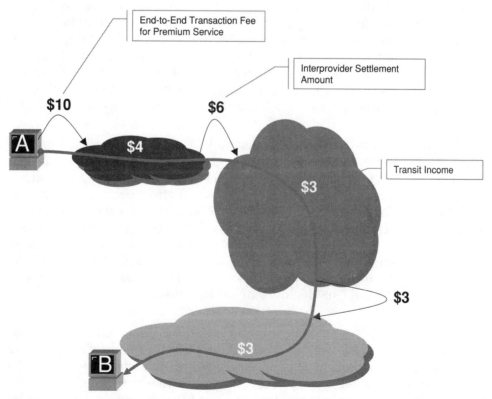

Figure 5.15 QoS Settlement Flows.

use of their services through a financial distribution of the original end-to-end revenue with a logical base for interagent financial settlements the outcome. Therefore, meaningful interprovider financial settlements within the Internet industry are highly dependent on the introduction of end-to-end service retail models. They in turn are dependent on a shift from universal deployment of a best-effort service regime with partial path funding to the introduction of layered end-to-end service regimes that feature both end-to-end service-level undertakings and end-to-end tariffs applied to the initiating party.

The number of conditions in this argument is significant. If QoS technologies are developed that scale to the size of the public Internet, provide sufficiently robust service models to allow the imposition of service level agreements with service clients, and are standardized so that QoS service models are consistent across all vendor platforms, then this area of interprovider settlements will need to change as a consequence. The pressure to change will come from emerging market opportunities to introduce interprovider QoS interconnection mechanisms and the associated requirement to introduce end-to-end retail QoS services. Subsequently, there will be pressure to support this with interprovider financial settlements whereby the originating provider will apportion the revenue gathered from the QoS signal initiator with all other providers that are along the associated end-to-end QoS flow path.

This type of end-to-end QoS settlement model assumes significant proportions that may in themselves impact the QoS signaling technologies. It is conceivable that each provider along a potential QoS path may need to signal not only that it can support the QoS profile of the potential flow, but also the unit settlement cost that will apply to the flow. The end user may then use this cost feedback to determine whether to proceed with the flow, given the indication of total transit costs, or to request alternate viable paths in order to choose between alternative provider paths so as to optimize both the cost and the resulting QoS service profile. The technology and business challenges posed by such an end-to-end QoS deployment model represent an impressive quantum change to today's best-effort Internet.

With this in mind, one potential future is that the public Internet environment will adopt a QoS-mediated service model that is capable of supporting a diverse competitive industry through interprovider financial settlements. The alternative is the current uniform best-effort environment with no logical role for interprovider settlements and the associated strong pressures for provider aggregation. The reliance on Internet QoS technologies to achieve not only Internet service outcomes, but also to achieve desired public policy outcomes in terms of competitive pressures, is evident within this perspective. It is unclear whether the current state of emerging QoS technologies and QoS interconnection agreements will be able to mature and be deployed in time to forge a new chapter in the story of the Internet interconnection environment. The prognosis is not good.

References

[Bakre 1995] *Handoff and Systems Support for Indirect TCP/IP*, A. Bakre and B. Badrinath, Proceedings of the Second USENIX Symposium on Mobile and Location-Independent Computing, Ann Arbor, Michigan, April 1995.

A description of an approach to low-speed adaptation using two TCP sessions and an explicit adaptation function to pass data between the two sessions.

[Floyd 1999] *Promoting the Use of End-To-End Congestion Control in the Internet*, S. Floyd and K. Fall, submitted to IEEE/ACM Transactions on Networking, www-nrg.ee.lbl.gov/floyd/end2end-paper.html, May 1999.

This paper examines the impacts of increasing use of non-congestion-management traffic within an internet environment, noting the potential for congestion collapse under such circumstances. The paper proposes a number of possible router mechanisms that can identify such flows in order to apply some form of corrective response.

[Huston 1999] *An Introduction to Web Caching*, G. Huston, *Internet Protocol Journal*, vol. 2, no, 3., September 1999.

An overview describing the role of Web proxy caches and the engineering issues that arise in the deployment of caches in Internet service networks.

RFCs

Request for Comments documents (RFCs) are published by the RFC editor. They are available online at www.rfc.editor.org

[RFC1144] *Compressing TCP/IP Headers for Low-Speed Serial Links*, V. Jacobson, RFC1144, Proposed Standard, February 1990.

This document describes a method for compressing the headers of TCP/IP datagrams to improve performance over low-speed serial links. The motivation, implementation, and performance of the method are described.

[RFC2213] *Integrated Services Management Information Base Using SMIv2*, F. Baker, J. Krawczyk, and A. Sastry, RFC2213, Proposed Standard, September 1997.

This document defines objects for managing the interface attributes defined in the Integrated Services model.

[RFC2214] *Integrated Services Management Information Base Guaranteed Service Extensions Using SMIv2*, F. Baker, J. Krawczyk, and A. Sastry, RFC2214, Proposed Standard, September 1997.

This document defines objects for managing the interface attributes defined in the guaranteed service of the Integrated Services model.

[RFC2340] *Nortel's Virtual Network Switching (VNS) Overview*, B. Jamoussi, D. Jamieson, D. Williston, and S. Gabe, RFC2340, Informational RFC, May 1998.

This document provides an overview of Virtual Network Switching (VNS), a multiprotocol switching architecture that provides class of service-sensitive packet switching; provides logical networks and traffic segregation for virtual private networks, security, and traffic engineering; enables efficient WAN broadcasting and multicasting; and reduces address space requirements.

[RFC2393] *IP Payload Compression Protocol*, A. Sacham, R. Monsour, R. Pereria, and M. Thomas, RFC2393, Proposed Standard, December 1998.

This document describes a protocol intended to provide lossless compression for Internet Protocol datagrams in an Internet environment. IP payload compression is a protocol to reduce the size of IP datagrams. This protocol will increase the overall communication performance between a pair of communicating hosts/gateways by compressing the datagrams—provided that the nodes have sufficient computation power, through either CPU capacity or a compression coprocessor, and that the communication is over slow or congested links.

[RFC2481] *A Proposal to add Explicit Congestion Notification (ECN) to IP*, K. Ramakrishnan and S. Floyd, RFC2481, Experimental RFC, January 1999.

This experimental RFC proposes the addition of explicit congestion notification (ECN) to IP. With the addition of active queue management, such as RED, to the Internet infrastructure, where routers detect congestion before the queue over-flows, routers are no longer limited to packet drops to indicate congestion. In this proposal, routers set a Congestion Experienced (CE) bit in the header of packets from ECN-capable transport protocols (such as TCP). Upon receipt of a CE signal, the transport protocol enters a congestion experienced response state and adjusts its flow rate accordingly.

[RFC2508] *Compressing IP/UDP/RTP Headers for Low-Speed Serial Links*, S. Casner and V. Jacobson, RFC2508, Proposed Standard, February 1999.

This document describes a method for compressing the headers of IP/UDP/RTP datagrams to reduce overhead on low-speed serial links. In many cases, all three headers can be compressed to 2 to 4 bytes.

IETF Internet Drafts

IETF Internet drafts are work-in-progress documents. They are valid within the IETF process for a maximum period of six months from the date of submission. Internet drafts are not normally referenced, but those cited here are the best pointers to current research and developmental efforts, and so have relevance to this topic of quality of service and network performance. Typically, these documents follow the Internet Standards process and are published as RFCs sometime in the future. The current collection of Internet drafts, along with pointers to the RFC documents, can be found at www.ietf.org references.

[ID-ackspace] *ACK Spacing for High-Delay Bandwidth Paths with Insufficient Buffering*, C. Partridge, Internet Draft, draft-rfced-info-partridge-01.txt, September 1998.

This document informally describes an approach to addressing buffering shortages in routers on high delay-bandwidth paths through the technique of spacing out the TCP ACK packets.

[ID-cm] *The Congestion Manager*, S. Seshan and H. Balakrishnan, Internet Draft, draft-balakrishnan-cm-00.txt, June 1999.

This document describes the Congestion Manager (CM), an end system module that enables an ensemble of multiple concurrent flows that share the same receiver and congestion behavior to display proper congestion behavior; it also allows applications to easily adapt to network congestion. This framework integrates congestion management across all applications and transport protocols. The CM maintains congestion parameters (available aggregate and per-flow bandwidth, per-receiver round-trip times, etc.) and exports an API that enables applications to learn about network characteristics; obtain information from and pass information to the CM; share congestion information; and schedule data transmissions. This document focuses on applications and transport protocols with their independent per-byte or per-packet sequence number information. It does not address networks with reservations or service discrimination.

[ID-diameter-reliable] *DIAMETER-Reliable Transport Extensions*, A. Rubens and P. Calhoun, Internet Draft, draft-calhoun-diameter-reliable-01.txt, March 1999.

Many services that require DIAMETER need retransmission and time-out faster than TCP can provide. This DIAMETER specification defines the extensions necessary for the base-reliable protocol to operate over a nonreliable transport, such as UDP.

[ID-diffserv-mib] *Management Information Base for the Differentiated Services Architecture*, F. Baker, Internet Draft, draft-ietf-diffserv-mib-00.txt, July 1999.

A proposed MIB for the DiffServ architecture, developing the DiffServ conceptual model into a set of parameterization points and pollable instrumentation points.

[ID-diffserv-model] *A Conceptual Model for DiffServ Routers*, Y. Bernet, A. Smith, and S. Blake, Internet Draft, draft-ietf-diffserv-model-00.txt, June 1999.

This draft proposes a conceptual model for use in the management of DiffServ routers. It describes the fundamental packet classification principles that allow traffic streams to be differentiated, and the fundamental traffic conditioning elements that comprise the traffic conditioning functionality of Diffserv routers. The draft also addresses the fundamental queue elements that comprise the per-hop behavior (PHB) functionality of Diffserv routers, and proposes a formal model for them. Finally, the document identifies parameters and variables that may be used to monitor the operation of these network elements.

[ID-ippm-delay] *A One-way Delay Metric for IPPM*, G. Almes, S. Kalidini, and M. Zekauskas, Internet Draft, draft-ietf-ippm-delay-07.txt, May 1999.

This document defines a metric for one-way delay of packets across Internet paths. It builds on ideas introduced and discussed in the IPPM framework document, RFC2330. This document is parallel in structure to one for packet loss. The interesting challenge posed by this approach is to use precisely synchronized clocks that operate at the sender and receiver.

[ID-ippm-loss] *A One-way Packet Loss Metric for IPPM*, G. Almes, S. Kalidini, and M. Zekauskas, Internet Draft, draft-ietf-ippm-loss-07.txt, May 1999.

This document defines a metric for one-way loss of packets across Internet paths. It builds on ideas introduced and discussed in the IPPM framework document, RFC2330. The document is parallel in structure to one for packet delay.

[ID-ippm-treno] *Treno Bulk Transfer Capacity*, M. Mathis, Internet
    Draft, draft-ietf-ippm-treno-btc-03.txt, February 1999.

TReno is a tool to measure bulk transport capacity (BTC), as defined in an accompanying IPPM draft. This document specifies details of the TReno algorithm as
required by the BTC framework document. The basic approach is to emulate an
end-to-end TCP session using probe echo request packets, where the sending rate
of the probe requests is governed by a TCP rate control algorithm.

[ID-isis-omp] *IS-IS Optimized Multipath*, C. Villamizar and T. Li, Internet Draft,
    draft-ietf-isis-omp-01.txt, February 1999.

IS-IS may form multiple equal cost paths between points. This is true of any link-
state protocol. In the absence of any explicit support to take advantage of this, a
path may be chosen arbitrarily. Techniques, referred to as Equal Cost Multipath
(ECMP), have been utilized to divide traffic somewhat evenly among the available paths. An unequal division of traffic among the available paths is generally
preferable. Routers usually have no knowledge of traffic loading on distant links,
and therefore have no basis on which to optimize the allocation of traffic. Equal
Cost Multipath is an extension to IS-IS, utilizing additional type/length/value
(TLV) tuples to distribute loading information.

[ID-megaco-tudp] *Transaction UDP Plus Timers (TUDP) for MEGACO*, F.
    Cuervo and A. Rayhan, Internet Draft, draft-cuervo-megaco-tudp-00.txt, June
    1999.

This document describes transaction-oriented reliable UDP with timers (TUDP).
It can be used to provide reliable delivery of transactions for Media Gateway
Controller (MEGACO) applications. The objective of the protocol is to deliver
transactions and messages within the time constraints of the application, that is,
the application timer. Transaction sequences are ordered per-stream, and super-
seding of transactions is allowed.

[ID-mpls-diffppp] *MPLS Support of Differentiated Services over PPP Links*, F.
    Le Faucheur, S. Davari, R. Krishnan, P. Vaananen, and B. Davie, Internet
    Draft, draft-lefauchuer-mpls-diff-ppp-00.txt, June 1999.

This document proposes a mechanism for MPLS to support Differentiated
Services (DiffServ) over PPP links. This solution allows the service provider to
flexibly define how DiffServ behavior aggregates (BAs) are mapped to LSPs, so
that they can best match the DiffServ and traffic engineering objectives within a
network. This mechanism can use LSPs where the behavior aggregate's scheduling treatment is inferred by the LSR from the packet's label value; the drop precedence is indicated in the EXP field of the MPLS PPP header. This mechanism can

also use LSPs where both the behavior aggregate's scheduling treatment and drop precedence are conveyed to the LSR in the EXP field of the MPLS PPP Header.

[ID-mpls-rsvp] *Use of Label Switching with RSVP*, B. Davie, Y. Rekhter, E. Rosen, A. Viswanathan, V. Srinivasan, and S. Blake, Internet Draft, draft-ietf-mpls-rsvp-00.txt, March 1998.

Multiprotocol Label Switching (MPLS) allows labels to be bound to various granularities of forwarding information, including application flows. This document presents a specification for allocating and binding labels to RSVP flows and for distributing the appropriate binding information using RSVP messages.

[ID-ospf-omp] *OSPF Optimized Multipath*, C. Villamizar, Internet Draft, draft-ietf-ospf-omp-02.txt, February 1999.

OSPF may form multiple equal cost paths between points. This is true of any link-state protocol. In the absence of any explicit support to take advantage of this, a path may be chosen arbitrarily. Techniques, referred to as Equal Cost Multipath (ECMP), have been utilized to divide traffic somewhat evenly among the available paths. An unequal division of traffic among the available paths is generally preferable. Routers typically have no knowledge of traffic loading on distant links, and therefore have no basis to optimize the allocation of traffic. Optimized Multipath is a compatible extension to OSPF, utilizing the Opaque LSA to distribute loading information, proposing a means to adjust forwarding, and providing an algorithm to make the adjustments gradually enough to ensure stability yet provide reasonably fast adjustment when needed.

[ID-pilc-error] *Performance Implications of Link-Layer Characteristics: Links with Errors*, S. Dawkins, G. Montenegro, M. Kojo, V. Magret, and N. Vaidya, Internet Draft, draft-pilc-error-00.txt, June 1999.

Because TCP is still the major protocol for reliable data transport on the Internet, and because TCP congestion-avoidance procedures interact badly with high uncorrected error rates, this document is focused on TCP over high error rate links. The definition of "high error rate" isn't a formal one: The sender spends an excessive amount of time waiting on acknowledgments that aren't forthcoming, whether due to data losses in the forward path or acknowledgment losses in the return path, and these losses are not due to congestion-related buffer exhaustion. The sender then transmits at substantially reduced traffic levels while it probes the network to determine "safe" traffic levels.

[ID-pilc-ltn] *Long Thin Networks*, S. Dawkins, G. Montenegro, M. Kojo, V. Magret, and N. Vaidya, Internet Draft, draft-montenegro-pilc-ltn-02.txt, May 1998.

In view of the unpredictable and problematic nature of long thin networks (for example, wireless WANs), arriving at an optimized transport is a daunting task. This document reviews existing proposals, along with future research items, and recommends a set of mechanisms for implementation in long thin networks. The goal is to identify a TCP that works for all users, including users of long thin networks.

[ID-pilc-slow] *Performance Implications of Link-Layer Characteristics: Slow Links*, S. Dawkins, G. Montenegro, M. Kojo, and V. Magret, Internet Draft, draft-ietf-pilc-slow-00.txt, May 1998.

This document makes a series of recommendations for improved protocol performance in network paths that traverse "extreme" link conditions, focusing on network paths that traverse very low bitrate links. Very low bitrate implies "slower than we would like." This recommendation may be used in any network where hosts can saturate available bandwidth.

[ID-purdet] *PURDET Reliable Transport Extensions on UDP*, K. Toney, Internet Draft, draft-toney-purdet-00.txt, March 1999.

This PURDET specification defines a very lightweight protocol with the minimum transport functions necessary to provide reliability and flow control over nonreliable UDP transport. PURDET supports sequencing, flow control, multiplexing/protocol identification (using an optional field), error retransmission, and link loss detection. PURDET is a work-in-progress presented for comment.

[ID-rap-rsvp-identity] *Identity Representation for RSVP*, S. Yadav, R. Yavatkar, R. Pabbati, P. Ford, T. Moore, and S. Herzog, Internet Draft, draft-rap-rsvp-identity-04.txt, July 1999.

This document describes the representation of identity information in the POLICY_DATA object for supporting policy-based admission control in RSVP. The goal of identity representation is to allow a process on a system to securely identify the owner and the application of the communicating process, and convey this information in RSVP messages (Path or Resv) in a secure manner. The document describes the use of this identity information in an operational setting.

[ID-rsgp] *Reliable Signaling Gateway Protocol (RSGP)*, M. Holdrege and L. Ong, Internet Draft, draft-ong-rsgp-ss7-info.txt, August 1998.

This document describes a combined message and function set for the Control Protocol used between a network access server (NAS) and a signaling gateway (SG). The Control Protocol supports Call Control, Circuit Maintenance, and Resource Management.

[ID-rsvp-mib] *RSVP Management Information Base*, F. Baker, J. Krawczyk, and A. Sastry, Internet Draft, draft-ietf-rsvp-mib-v2-00.txt, April 1998.

This document defines objects for managing the Resource Reservation Protocol (RSVP) within the interface attributes defined in the Integrated Services model.

[ID-sigtran-rudp] *Reliable UDP Protocol*, R. Stewart and S. Hussain, Internet Draft, draft-ietf-sigtran-reliable-udp-00.txt, February 1999.

This document discusses Reliable UDP (RUDP), a simple packet-based transport protocol. RUDP is based on RFC1151 and RFC908, Reliable Data Protocol. RUDP is layered on UDP/IP and provides reliable in-order delivery (up to a maximum number of retransmissions) for virtual connections. RUDP has a very flexible design, making it suitable for a variety of transport uses, for example, to transport telecommunication signaling protocols.

chapter six

What Is Quality of Service Anyway?

How could he say whether Quality was mind or matter when there was no logical clarity as to what was mind and what was matter in the first place? And so: he rejected the left horn. quality is not objective, he said. It doesn't reside in the material world.

Then: he rejected the right horn. Quality is not subjective, he said. It doesn't reside merely in the mind.

And finally: Phaedrus, following a path that to his knowledge had never been taken before in the history of Western thought, went straight between the horns of the subjectivity-objectivity dilemma and said Quality is neigher a part of mind, nor is it a part of matter. It is a third entity which is independent of the two...

The world now, according to Phaedrus, was composed of three things: mind, matter and Quality.

—Robert M. Pirsig
Zen and the Art of Motorcycle Maintenance

At the heart of supporting the multiservice IP network is the requirement for quality of service within the network. Quality of service continues to be a topic of considerable interest among vendors, service providers, and Internet consumers today. One reason for the continuing fascination with this topic is that, to date, it has not been adequately addressed. And despite marketing hyperbole, the QoS continues to present a number of engineering challenges that remain unmet.

To service an ever increasing range of applications and an ever divergent set of values attached to such applications, the network platform is being placed in the position of having to make critical choices. The network itself is being assigned the role of actively differentiating its response by providing high-value applications with the service profile they desire, and ensuring that best-effort traffic is not entirely compromised in so doing. This is a problem easy to state, but very difficult to solve. It is necessary to recognize that the tools available today form only part of the picture; much yet needs to be constructed to produce the desired outcomes.

The first step to achieve quality of service is to establish whether the service objective is one best addressed by a small number of simple improvements within the best-effort network or one that requires engineering measures to construct active differentiation of service levels to various classes of traffic. Accordingly, to begin our examination of service responses from Internet networks, we'll look first at how to address base service quality. From that platform we can better determine what needs to be done to achieve differentiated service responses from the network.

Achieving Service Quality

It may seem a contradiction of sorts, but it is extremely difficult to engineer differentiated service responses on top of a network that is performing poorly in the first place. There is no substitute for proper network engineering of the base service platform. This applies equally to the network itself as to the host systems that drive traffic into the network. Here we will look at a number of steps to take to achieve a basic level of service quality.

Service Quality and the Network

Network loads generally exhibit peak load conditions at certain hours of every day and if the network cannot handle these loads, the consequent overload conditions created by bandwidth exhaustion, queue exhaustion, and switch saturation cannot be readily ameliorated by QoS measures. The underlying resource-starvation issue must be addressed if any level of service is to be delivered to the network's clients. Additionally, the stability of the routing environment is of paramount importance to ensure that the network platform behaves predictably. Therefore, two primary prerequisites exist for effective network management:

> **Provision adequate network resources to handle normal load conditions.** This refers to both a need for adequate bandwidth to manage the imposed traffic without excessive queue buildup and for adequate queue sizes in the routers to cope with the inherent requirements for buffering under normal load. But this is not entirely a bandwidth and queue matter, because a requirement also exists to deploy a processing and switching capability commensurate with the packet loads each router faces. Because any imposition of QoS services imposes addi-

tional load on the processing and switching functions, this capacity requirement is a critical one. The processing function is exercised most strenuously when the network is under peak load, so a processing and switching capability should be deployed so that it can comfortably manage the full extent of the peak network loads it will face. This advice applies to both the edge and the core of the network. In the Internet service provider market, frequently the client does not purchase a sufficiently large access circuit to cope with the client's peak traffic levels; the typical result is a significant congestion point at the boundary of the ISP network. Of course, this may be beyond the client's control, as is the case with modem-based access.

Ensure network stability. This refers to the stability of the underlying transport substrate and the routing system over which the network directs traffic. Without an environment where the routing system can converge rapidly, then operate for an extended period without further need to recompute the internal forwarding tables, the network quickly degenerates and degrades in such a way that no incremental introduction of QoS structures can salvage it.

Major performance and efficiency gains can be made by allowing the network to signal the end systems regarding the likely onset of congestion conditions so that the end systems can take action to reduce the traffic rate long before the network is forced into queue tail-drop behavior. The most effective steps a network operator can take to improve network efficiency and end user flow performance are as follows:

Implement random early detection (RED). This is to ensure that the initial congestion back-off signals are sent to those TCP senders who are pushing hardest at the network congestion points before comprehensive packet discard occurs. RED is statistically likely to signal those stacks that are operating with large transmission windows. The effect of this discard mechanism is to signal a reduction in the transmit window size. It attempts to prevent signaling loss to those small-scale flows that are not causing the overall congestion problem.

Support explicit congestion notification (ECN). Explicit congestion notification allows a host to be signaled about the potential onset of congestion without requiring the host to detect certain forms of small-scale packet loss and distinguish them from more significant incidents of packet loss. This is a measure intended to improve the efficiency of the network through improvement in the signaling between the network and the host. This does not imply that a host can then interpret all loss as packet corruption loss. Queue exhaustion will still occur, as will congestion-induced packet loss. The intent of ECN is to make the signaling of the onset of congestion more explicit to the host, thereby allowing for an unambiguous signal with an unambiguous response to slow down the sending rate on the part of the host.

Separate TCP and UDP queues within the router. This step is a more general recommendation, to separate the queue handling of congestion-managed data from other data flows. Congestion-managed flows, generically labeled here as TCP flows, tend to derive their behavior from a *network clock,* where the sending rate is timed against the network's load condition feedback. The consequent sending pattern is to send a packet burst at a speed somewhat higher than the return signal rate being received, and rely on the queues within the network to absorb the burst. This calls for large queues within the routers if network efficiency is to be maintained. Non-congestion-managed flows, generically labeled here as UDP flows, are typically used in supporting applications that use external conditions for clocking data rates. Such systems tend to rely on short network queues to reduce the levels of network jitter, in order to preserve the integrity of the implicit data clocking rate. One mechanism to separate these two traffic behaviors is to implement profile filters to limit the level of queue resources given to nonflow-controlled UDP traffic. This allows the TCP queues to behave more predictably. The most direct way to implement this is to place UDP traffic and TCP traffic in different output queues and use a weighted scheduling algorithm to select packets from each queue according to a network-imposed policy constraint of relative resource allocation. This method allows the TCP end system stacks to oscillate faster in order to estimate the amount of total end-to-end traffic capacity based on the behavior of the flow-controlled traffic passing through the queues. The outcome of this recommendation is to limit the extent of impact a nonnetwork-clocked UDP flow can generate. By making UDP queues relatively short and TCP queues longer in the router, there is a greater probability that the TCP queues can behave in a way that attempts to avoid tail-drop congestion, thereby increasing network-clocked throughput efficiency.

Administer admission traffic shaping using traffic conditioners at the network edges, to reduce the burstiness of the data traffic rates. The effect of this measure is to act as a surrogate method of data burst-rate limiting using the queues at the periphery of the network to reduce the level of queue load in the more critical high-capacity interior of the network.

Measure and monitor the service performance of the network. The network operator must understand the load conditions imposed on the network and the way those conditions vary over time. Without base-line information on the service performance of the network, it is impossible to determine the requirement for any form of differentiated service performance measures.

By taking these actions alone you can gain marked increases in efficiency of bandwidth usage in terms of actual data transfers. RED provides for the enhanced efficiency of bandwidth resource sharing, making best-effort service more uniformly available to all flows at any given moment. On the other hand, queue segregation attempts to limit the damage that high-volume real-time (externally clocked) flows may inflict on network-clocked reliable data flows.

Service Quality and the Host

The second part of this "Good Housekeeping Guide" list is intended to aid the host in working within reasonable operating parameters. The host and the host application also have the capability to support a quality service platform, particularly in relation to the management of TCP traffic.

Using host TCP buffers and window advertisements commensurate with the delay-bandwidth product of the end-to-end traffic path helps the end system to effectively utilize available bandwidth. Because a data packet may need to be retransmitted, the buffer cannot remove data until it has been positively acknowledged. It takes one round-trip (RTT) time to receive the acknowledgment. To drive the network at maximal efficiency, the buffer must contain at least the amount of data equal to the path bandwidth multiplied by one round-trip time. If the host buffer is smaller than this amount, the sender will transition to an idle state even though there may be available network capacity, because no more data can be sent until a signal of receipt of earlier packets is received. Therefore, the buffer size must be greater than the bandwidth-RTT product to drive data transfer to the point of maximal performance. Thus, the only limiting factor becomes the true bandwidth of the network, not inadequate host buffering.

For very high bandwidth systems with bulk data transfer applications, TCP itself becomes the limiting factor. The use of a TCP implementation that supports TCP window-scaling options, along with very large buffers, can offer significant improvements in end-to-end in such situations. This is especially true where very long propagation delay is evident, as in paths that use high-bandwidth geostationary satellites.

Even with a model where the host attempts to synchronize itself with the network, a complementary level of network buffers is necessary. The sender has no clear view of the precise state of the path bandwidth at every moment in time. A TCP session can be thought of as a quasi-constant delay loop. When a sender begins in slow-start mode, it transmits one packet; and following a delay of one RTT, it will receive the corresponding ACK packet. The sender will transmit two packets in response to this ACK. This cycle of immediately transmitting two data packets upon receipt of each ACK packet continues through the slow-start mode. If, at any place in the traffic path there are slower links than at the sender's end of the traffic path, or the presence of other traffic, queuing of the second packet occurs. The steady state of slow start is an exponential increase of volume, placed into the end-to-end path using ACK pacing to double the data in each ACK "slot." The resulting behavior is naturally bursty traffic. At each RTT interval, the packet train contains successive packet trains of length 2, 4, 8, and so on, where the pacing of the doubling is based on the bit rate of the slowest link in the end-to-end traffic path. Within the network, this sequencing of packets at a rate of twice the previous rate within each RTT interval is smoothed out to the available bandwidth of the end-to-end path by the use of queuing within the routers. This queuing requirement can grow to half the delay-bandwidth product. For one half of this RTT delay interval, the queuing requirement is to store one segment for every segment it can transmit.

This queuing burst can be mitigated in this way: The sender implements bandwidth estimates using sender packet back-off, where the initial RTT is used to pace the emission of the second of the two packets at one-half the RTT, and for the second iteration at one-fourth the RTT, and so on. Alternatively, ACK spacing can be done at the receiver or in the interior of the network, although interior ACK pacing may require symmetric paths to allow automatic initiation of ACK spacing by the router. The slow-start phase of doubling the data rate per RTT continues to the onset of data loss; thereafter, further increases of data rate are linear. This next phase, the congestion-avoidance algorithm, operates differently. Instead of doubling the amount of data in the network in each RTT interval, TCP congestion-avoidance uses an algorithm intended to sustain a linear rate increase of one MTU per RTT.

How can the host optimize its behavior in such an environment? Again, there are a number of recommendations to consider. The following suggestions are a combination of those measures that have been well studied and are known to improve performance and those that appear to be highly productive areas of further research and investigation.

Use a good TCP protocol stack. Many of the performance pathologies that exist in the network today are not necessarily the by-product of oversubscribed networks and consequent congestion. Many of these performance pathologies exist because of poor implementations of TCP flow-control algorithms; inadequate buffers within the receiver; poor, or no, use of path-MTU discovery; no support for fast-retransmit flow recovery; and imprecise use of protocol-required timers. It is unclear whether network ingress-imposed QoS structures will adequately compensate for such implementation deficiencies. The conclusion is that attempting to address the symptoms is not the same as curing the disease. A good protocol stack can produce even better results in the right environment.

Implement a TCP selective acknowledgment (SACK) mechanism. SACK, combined with a selective repeat-transmission policy, can help overcome the limitation that traditional TCP experiences when a sender can learn only about a single lost packet per RTT.

Implement larger buffers with TCP window-scaling options. The TCP flow algorithm attempts to work at a data rate that is the minimum of the delay-bandwidth product of the end-to-end network path and the available buffer space of the sender. Larger buffers at the sender assist the sender in adapting more efficiently to a wider diversity of network paths by permitting a larger volume of traffic to be placed in flight across the end-to-end path.

Support TCP explicit congestion notification (ECN) negotiation. ECN enables the host to be explicitly informed of conditions relating to the onset of congestion without having to infer such a condition from the reserve stream of ACK packets from the receiver. The host can react to such a condition promptly and effectively with a data flow-control response without having to invoke packet retransmission.

Use a higher initial TCP slow-start rate than the current 1 MSS (maximum segment size) per RTT. A size that seems feasible is an initial burst of 2 MSS segments. The assumption is that there will be adequate queuing capability to manage this initial packet burst; the provision to back off the send window to 1 MSS segment should remain intact to allow stable operation if the initial choice was too large for the path. A conservative initial choice is 2 segments, although simulations have indicated that 4 initial segments is also highly effective in many situations.

Implement sender data pacing or receiver ACK spacing so that the burst nature of slow-start increase is avoided. This is perhaps a more controversial recommendation. The intent of such pacing is to reduce the inherent burst nature of the slow-start TCP algorithm, and, in so doing, to relieve the queuing pressure placed on the network where the end-to-end path traverses a relatively slower hop. However, modern routers

Recently, a number of better-than-TCP protocol stacks have appeared on the market, most commonly in conjunction with Web server systems, where the performance claim is that these protocol stacks can interoperate with standard TCP clients, but offer superior download performance to a standard TCP protocol implementation. This level of performance is achieved by modifying the standard TCP flow control systems in a number of ways. The modified implementation may use a lower initial RTT estimate to provide a more aggressive startup rate, and a more finely grained RTT timer system to allow the sender to react more quickly to network state changes. Other modifications may include using a larger initial congestion window size or may use an even faster version of slow start, where the sending rate is tripled, or more, every round-trip time interval. The same technique of incremental modification can be applied to the congestion avoidance state, where the linear rate increase of one segment size per round-trip time interval can be increased to some multiple of the segment size, or use a timebase other than the round-trip time for linear expansion of the congestion window. The back-off algorithm can also be altered such that the congestion window is reduced by less than half during congestion back-off. Resetting the TCP session to slow start mode following the ACK time-out can also be avoided in such modified protocol implementations.

These techniques are all intended to force the sender to behave more aggressively in its transmission of packets into the network, thereby increasing the pressure on the network's buffers. The network

can use small, fast caches to detect and, optimally, switch such burst packet trains. Packet pacing breaks apart such trains, and may cause slightly slower router switching performance. The advantage of such an approach is to allow the network flow to quickly find the available end-to-end flow speed without receiving transient load signals that may confuse the availability calculation being performed.

Use a host platform that has sufficient processor and memory capacity to drive the network. The highest-quality service network and optimally provisioned access circuits cannot compensate for a host system that does not have sufficient capacity to drive the service load. This is a condition that can be observed in large or very popular public Web servers, where the peak application load on the server drives the platform into a state of memory and processor exhaustion, even though the network itself has adequate resources to manage the traffic load.

is not the only subject of this increased sending pressure; such modified protocol systems tend to impose a significant performance penalty on other concurrent TCP sessions that share the path with these servers. The aggressive behavior of the modified systems in filling the network's queues tends to cause the other concurrent standard TCP sessions to reduce their sending rate. This in turn opens additional space in the network for the modified TCP session to increase its transmission rate. In an environment where the overall network resource-sharing algorithm is the outcome of dynamic equilibration between cooperative sending systems, such aggressive flow control modification can be considered to be extremely antisocial behavior at the network level. Paradoxically, such systems can also be less efficient than a standard TCP implementation. TCP server systems modified in this way tend to operate with higher levels of packet loss because their efforts to saturate the network with their own data packets make them less sensitive to the signals of network congestion. Consequently, when delivering large volumes of traffic, or where there are moderately low levels of competitive pressure for network resources, the modified TCP stack may often perform less efficiently than a standard TCP implementation. Accordingly, these modified better-than-TCP implementations remain in the experimental domain as potential alternatives to the conventional TCP models of congestion management. Within the production environment, their potential to impose undue performance penalties on concurrent TCP sessions and their potential to reduce overall network efficiency are reasonable indicators that such modified stacks should be used with considerable care and discretion, if at all.

All these actions have one feature in common: They can be deployed incrementally at the edge of the network and can be deployed individually, allowing end systems to obtain superior service performance even in the absence of the network provider attempting to distinguish traffic by class distinction.

The conclusion is that if the performance of end-to-end TCP is the perceived problem, the most effective answer is not necessarily to add service differentiation in the network. Often, the greatest performance improvement can be made by upgrading the way that hosts and the network interact through the flow-control protocol. That said, there are many cases where some form of service-specific network response is necessary and essential.

Providing QoS

Before describing how to provide QoS, it's important to review the relevance of the underlying model of the network itself. To quote a report from the Internet Research Task Force, "The advantages of [the Internet Protocol's] connectionless design, flexibility, and robustness have been amply demonstrated. However, these advantages are not without cost: careful design is required to provide good service under heavy load" [RFC2309]. Careful design is not exclusively the domain of the end system's protocol stack, although good end system stacks are of significant benefit. Careful design also includes consideration of the mechanisms within the routers that are intended to prevent congestion collapse. Differentiation of services places additional demands on this design, because in attempting to allocate additional resources to certain classes of traffic, it is essential to ensure that the use of resources remains efficient and that no class of traffic is so starved of resources that it suffers throughput and efficiency collapse.

QoS and the Internet Router

A router has a very limited set of actions at its disposal after it receives an IP packet. The router can select an outbound interface, then queue the packet on the associated outbound queue; or the router can discard the packet. If the packet is queued, the output driver will subsequently select the datagram to transmit, once the packet reaches the head of the queue. Within this conceptual model, the router performs only two actions: forwarding, and queue management.

Forwarding

In a traditional unicast environment, the router uses the destination IP address of the datagram when determining to which outbound interface to pass the packet for transmission. One course of possible action is to make a forwarding decision based on some QoS-related setting. Quality of service routing can take a variety of factors into consideration, as discussed earlier. One action is to overlay a number of distinct topologies based on differing results of calculating end-to-end (or hop-by-hop) metrics. These metric calculations might include such factors as estimates of propagation delay or configured link capacity. Certainly, it is theoretically possible to also use current idle capacity as a metric. However, a very tight relationship is required between forwarding decisions and the subsequent need to redefine the routing protocol, making this option an unstable choice, at least for the short-term future. In fact, QoS routing technologies still are in the conceptual stages, and nothing appears to be viable in this area for the near term, especially when considering this as a candidate routing mechanism for large-scale Internet environments. Alternatively, the forwarding decision can be based on an imposed state, whereby the flow identification of the packet is matched against a table of maintained state information. Such constructs are used within RSVP mechanisms, and a variant of this approach is used in MPLS networks.

At present, knowledge is scant regarding the stability and scalability of QoS mechanisms directed at altering the choice of the outbound interface as the mechanism of service differentiation. Indeed, given the succession of routing protocols that have been deployed in the Internet over the past decade, it also may be relevant to observe that knowledge of the stability and scalability of large-scale routing protocols themselves is still somewhat inadequate. Adding complexity in terms of overlaying multiple QoS-related logical topologies should be done with extreme caution. As with a dancing circus bear, the fact that it can dance is amazing enough; getting it to sing is possibly asking too much.

Queue Management

Conventionally, FIFO queues are used in routers. This preserves the ordering of packets within the router, so that sequential packets of the same flow remain in order along a particular end-to-end path within the network. Given the finite length of an output queue, the router performs packet discard once the queue is exhausted, creating what commonly is referred to as a tail-drop behavior. In this state, discard is undertaken as an alternative to queuing. If space is available on the queue, the packet is added to the tail end of the queue. Otherwise, the packet is discarded.

In general, tail-drop queue behavior should be avoided. Tail drop is when a sequence of packets are discarded because the queue is saturated and cannot accept any other packets into the queue. This may cause those TCP sessions that traverse the saturated queue to lose part, or all, of an entire packet train. Tail drop often causes a TCP session to lose multiple trailing packets from bursts within a single flow. This form of packet loss causes those TCP sessions to restart their congestion management state using slow start mode, which in turn induces a large-scale efficiency drop in overall network throughput. This basic queue-admission policy can be modified by establishing admission criteria that direct which packets may be discarded even though space is available on the queue. The Random Early Detection (RED) queue management policy is an important component of active queue management. Within flow-controlled end-to-end environments, this approach is easily interpreted by the TCP sender as a signal to throttle back the rate expansion instead of doing extensive packet drop and retransmits.

QoS criteria can be admitted into the queue-admission policy through the selective behavior of RED. The technique is to weight the preference for packet discard to lower-precedence packets, which causes a rate-damp signal to be sent to lower-precedence streams (assuming that all packets within a stream have identical precedence) before the signal is sent to the higher-precedence streams. The intrinsic behavior of RED is a relatively subtle form of service differentiation. Short-lived TCP flows and noncongestion-managed UDP streams are somewhat immune from the effects of RED; while the packet drop may cause retransmission, the total flow rate is not substantially altered. This is either because the flow is of very short duration or the end-to-end application protocol is not drop-sensitive. Studies of network flow patterns in the Internet indicate that the majority of traffic is part of short TCP sessions; only a small proportion of traffic is part of a high-volume data transfer. Thus, though RED can promote more efficient use of network resources by long-held, well-behaved TCP flows, and though weighting of RED by some form of QoS precedence can allow some level of QoS flow differentiation within the network, overall, service-level differentiation is difficult to discern within today's traffic profiles.

The efficient operation of congestion-managed flows normally calls for extensive queue sizes within the router, as the router's queues need to absorb the sender's burst characteristics. But longer queues cause jitter distortion to data flows, and present a service quality issue to real-time data flows. The queue-admission policy can be altered further by modifying the basic FIFO operation to one in which QoS packets may be inserted at the head of the queue or in an intermediate position that maintains an overall queue ordering by some QoS form of precedence indicator.

Uncontrolled User Datagram Protocol (UDP) flows are a relevant consideration here, and with the increasing deployment of UDP traffic flows under the multimedia umbrella, this issue will become critical in the near future. A long-lived UDP flow, operating at a rate in excess of an intermediate hop's resource capacity, exerts growth pressure on the associated output queue, ultimately causing queue saturation and, subsequently, a queue tail-drop condition that affects all flows across this path. Controlling the level of nonflow-controlled UDP can be a difficult ingress policy to effectively police. An alternative is to use distinct queuing disciplines for TCP and UDP traffic, thereby attempting to minimize the interaction between flow-controlled and nonflow-controlled data. A weighted fair-queuing technique can be used across the two queues, to provide a bandwidth allocation regime. This queuing discipline can be refined to use one queue per service class and apply different queuing disciplines to each queue. Such a technique allows the per-queue service to be defined by the queue length, the queue management discipline, and the relative weighting of the service queue.

The capability of the QoS application is an outcome of the queue structure, the queue-admission criteria, and the packet-scheduling mechanism. QoS weighting on the queues, where higher-precedence packets are scheduled at a higher priority, does, however, introduce grossly visible differentiation of service levels under load. As the network load increases, the incremental congestion load is expressed in increased queue delay in queues that surround the network load point. This queuing delay is visible in terms of extended RTT estimates for traffic flows, which in turn reduces the TCP traffic rates. Both slow-start and congestion-avoidance TCP algorithms are RTT-sensitive. By using QoS scheduling, the higher-precedence traffic consumes a greater share of the congested segments, providing the high-precedence TCP control systems with stable RTT estimates for the associated traffic flows. The difference between a relatively stable RTT and one that exhibits relatively high variability and higher average value is eminently visible in performance terms, both for short and long TCP traffic flows. It also is highly visible to simple diagnostic probes such as ping and traceroute. It is admittedly a relatively coarse differentiator of traffic, and one that can be used to cause high levels of differentiation, even under relatively light levels of load. Of course, if QoS is tied to a financial premium, one of the key market attributes of QoS will need to be highly visible QoS differentiation. Weighting the queue scheduling is the mechanism that offers the greatest promise here.

QoS Signaling

For QoS to be functional, it may be necessary for all the nodes in a given path to behave in a similar fashion with respect to QoS parameters, or at the very least, not to impose additional QoS penalties other than conventional best-effort into the end-to-end traffic environment. The sender or network ingress point must be able to create some form of signal associated with the data that can be used by down-flow routers, potentially to modify their default outbound interface selection, queuing behavior, and discard behavior.

The insidious issue here is that the host is attempting to exert *control at a distance*. The objective in this QoS methodology is for an end system to generate a packet that can trigger a differentiated handling of the packet by each node in the traffic path, so that the end-to-end behavior exhibits performance levels in line with the end-user's expectations and, potentially, in conformance to a contracted service level agreement.

This control-at-a-distance model can take the form of a "guarantee" between the user and the network. This guarantee is one in which, if the ingress traffic conforms to a certain profile, the egress traffic maintains that profile state, and the network does not distort the desired characteristics of the end-to-end traffic beyond specific allowed parameters. To provide such guarantees, the network must maintain some form of transitive state along a determined path, where the first router commits resources to honor the traffic profile, and passes this commitment along to a neighboring router that is closer to the nominated destination (or nominated sender) and that is also capable of committing to honor the same traffic profile. This is done on a hop-by-hop basis along the transit path between the sender and receiver, or from receiver to sender. To ensure that the commitment continues to be honored, the path is repeatedly refreshed with the service profile. Any changes in the topology of the network or in service load that impinge on this reservation can be discovered by such refresh operations.

Each activation of a service-requesting application causes an end-to-end state reservation to be created across the network. This type of state maintenance is viable within small-scale networks; but in the heart of large-scale public networks, the cost of state maintenance is overwhelming. This mode of RSVP operation presents some serious scaling considerations, and is inappropriate for deployment in the core of large networks. RSVP scaling considerations bring up another important point, too. RSVP's deployment constraints are not limited simply to the amount of resources it might consume on each network node as per-flow state maintenance is performed. It is easy to understand that as the number of discrete flows increases in the network, the more resources it will consume. Of course, this can be somewhat limited by defining how much of the network's resources are available to RSVP. Everything in excess of this value then is treated as best-effort. More subtle, is that when all available RSVP resources are consumed, all further requests for QoS are rejected until RSVP-allocated resources are released. This is similar in functionality to the way the telephone system works, where the network's response to a flow request is commitment or denial. Reservation denial is an option for the network by using a simple time order of requests to determine reservation priority. Alternatively, policy control of reservations can be exerted; and, as determined by the policy settings, existing reservations can be preempted by higher-precedence reservations.

The alternative to state maintenance and resource reservation schemes is the use of mechanisms for preferential allocation of resources, essentially creating varying levels of best-effort service. With the absence of end-to-end guarantees of traffic flows, this removes the criteria for absolute state maintenance, so that better-than-best-effort traffic with classes of distinction can be constructed inside larger networks. Currently, the most promising direction for these better-than-best-effort systems appears to be to modify the queuing and discard algorithms based on packet-based signals. This is the essence of the Differentiated Services architecture. These stateless mechanisms rely on an attribute value within the packet's header so that queuing and discard preferences can be signaled at each intermediate node. The initial step for the network provider is to define a set of supported services, then select associated packet codepoints that will trigger the service on an element-by-element basis. The network elements then need to be configured to react to the packet codepoints in a manner that supports the defined service. This may entail some local resource reservation to adequately support the service profile. The network admission points then need to apply the code values to the packet using some form of policy-driven classification, in conjunction with a profile meter that limits the amount of traffic admitted within each service profile to some maximum load. To complete the picture, a number of signal systems are necessary. The interior of the network needs to exchange signals with the network admission systems to ensure that the admission systems do not oversubscribe the network by admitting too much service traffic, thereby potentially compromising delivered service levels. In addition, the application or the host must be able to exchange signals with the admission system to ensure that its service needs are being achieved by the network.

The cumulative behavior of such stateless, local-context algorithms can result in distinguished service levels, and it hold the promise of excellent scalability. You still can mix best-effort and QoS-enabled better-than-best-effort nodes, but all nodes in the latter class should conform to the entire QoS-selected profile or a compatible subset. This is an example of the principle that it is better to do nothing than to do damage.

QoS and the Network

It is unreasonable to believe that QoS deployment can fully compensate for poor network engineering or poor protocol implementations in hosts. The network platform and the connected hosts must already exhibit a relatively favorable level of service quality for the following measures to be effective.

The first step is to determine which services are to be supported by the network. The service may take the form of an emulation of a point-to-point synchronously clocked circuit, where the network should impose no additional jitter or loss, nor reorder any packets, as long as the packets remain within the emulated circuit's profile. The service may take the form of a delay-bounded point-to-point service, where there is a jitter budget, and any packets exceeding this budget will not be delivered. In addition to a number of path-specific services, are a number of generic premium services not associated with any particular path. This may allow the end host to mark packets to be delivered with a higher scheduling priority or a lower loss probability, without respect to a particular path that the packets may take through the network, nor with respect to a particular destination. The reverse may also be the case, where packets destined to a client are marked with such premium service indicators. The loss probability can be manipulated by active queue management systems, allowing the network to bias the effective TCP throughput of concurrent streams by selectively weighting loss probabilities according to some service class definition.

What is required in network devices to support such services? The initial response is *more of everything*. The processing capability of the router needs to be sufficient to assemble, classify, forward, schedule, and pass every packet to the output driver in a highly efficient manner. For the router to be able to drive the network efficiently, the router must be able to generate output packets at a rate no slower than the input rate. This infers that the per-packet processing interval is no greater than the time it takes to pass a minimal packet along the wire. For high-speed communications systems, this is not a viable option using serial processing techniques. Instead, the router must use pipelining and parallel processing techniques to increase the number of packets being processed by the router at any one time. And the basic equation must hold if the router is going to perform efficiently; then, the length of time to process a packet, divided by the level of parallelism within the processing architecture, must be no greater than the peak packet arrival rate. The requirements for increased processing loads for QoS tend to require both increased processor speeds and increased levels of parallelism within the router architecture.

The number of queues within the router must be at least as large as the number of output drivers, multiplied by the number of discrete service classes. And it does not stop there, as there are a number of additional queue structures that may be required, including those to support traffic profile shaping. A number of recent router architectures propose the use of a dedicated queue per active micro-flow using up to 64,000 queues per output interface. The schedulers associated with the queues also require significant processing resources, particularly if a discipline such as weighted fair queuing is used on a per-flow basis. Of course, consideration of network scale leads to the conclusion that in the core of many large public service provider networks, 64,000 per-flow queues are not enough to manage the number of active microflows at any point in time; therefore, this resource management approach is directed more to the enterprise service environment domain.

Moreover, the devices require significant local storage to allow the unit to actively manage bandwidth resources and to honor reservation commitments for particular services. The device must track the available resource levels, ensure that the schedulers work within the relevant operational parameters, and ensure that resource requests are managed accurately.

Routers are also differentiated according to their roles within the network. Edge routers have a higher relative processing capability in comparison to the packet throughput levels, and support the signaling interface between the network and the network's client. The edge routers also perform the admission control function, ensuring that admitted traffic conforms to both an admission policy and to an associated admission traffic profile. Interior, or core, routers normally have a significantly larger transmission capacity to manage, and do not have the same level of processor capacity on a per-packet basis. The interior router normally performs forwarding and scheduling actions using the information in the packet header to determine the appropriate action in each case, rather than undertaking a comprehensive traffic classification and traffic profiling function.

Deploying RSVP

One additional comment must be made regarding RSVP deployment by the user and the network. End-to-end deployment of RSVP could be quite expensive, and often the results may be no better than a properly designed network. After implementing the mechanisms and principles discussed here, you can cut down queue length, reduce extraneous congestion signaling, and control misbehaving flows so that propagation times more accurately reflect physical signal-hop propagation.

There is a tight synergy between the end-system behavior of flow-controlled protocols and the router's queuing behavior. The interaction of many such systems results in a somewhat chaotic behavior. When you then introduce nonflow-controlled UDP traffic, the result is a system in which predictive traffic management is virtually impossible. When a diverse collection of differentiation mechanisms is introduced, not even an experienced protocol engineer, network architect, or network engineer would be able to determine how to make the network behave efficiently. Thus, it is unclear that RSVP has a role as a universally deployed resource reservation mechanism. What is clear is that RSVP can serve as an effective end-to-end signaling protocol, and as such, there are logical roles RSVP can play in a service differentiated network.

The model used by the Internet is one of distributed intelligence, whereby the functionality of data flow control is passed over the network fence to the end-user's platform. The interior of the network is reduced to the bare essentials of basic transmission elements and simple switches. The result is a network of unsurpassed cost-efficiency. Such a network is vastly different from the circuit-switching models of previous communications systems in which significant functionality and cost were preserved within the network in a centralized functionality model. The long-term challenge is scaling this model efficiently in terms of switching and in terms of consumer models, so that the cost of the transmission and switching elements is fairly distributed to the consumer. QoS is an intermediate step and, to a degree, represents intermediate failure to converge on this path.

QoS Observations

A number of contradictory questions tend to interfere with efforts to engineer possible solutions to the quality of service requirement on the Internet. Is meeting the QoS requirement addressing an economic or a technical problem? Is QoS an economic or technical solution to a poorly specified problem? Is QoS the imposition of state onto a stateless network? By attempting to differentiate degradation, do you increase total degradation of service? Is QoS a mechanism imposed by the network on a customer-by-customer basis or activated by the user on a per-transaction basis? Is the need to provide QoS the promotion of a network service provider that wants to oversell a limited network resource? Are some users being forced to compensate for a poor network model? Is QoS a product of the marketing department or a consumer-driven approach to the marketing problem of over-subscription? Is QoS a solution borrowed from another networking environment, such as circuit-switched networks, and placed into the Internet without a complete understanding of the nature of the problem to be solved? Let's tackle some of these questions by making some observations about the QoS landscape.

So far, QoS has been viewed as a wide-ranging solution set that can be applied to a very broad problem area. This fact often can be a liability. Ongoing efforts to provide "perfect" solutions have demonstrated that attempts to solve all possible problems result in technologies that are far too complex, have poor scaling properties, or simply do not integrate well with the diversity of the Internet. Nevertheless, close scrutiny of the issues and technologies available reveal that some very clever mechanisms have been developed. Determining the usefulness of these mechanisms is perhaps the most challenging aspect of assessing the merit of a particular QoS approach. So, as we make these observations, we'll try to assess their practical utility as well.

There's No Such Thing as a Free Lunch

Before we can begin to come to conclusions on how to practically implement a QoS methodology, we must revisit one basic observation. We also must quantify several criteria before launching into a discussion about which approach might be appropriate in which environment.

The observation we are revisiting is this: When a network is under severe load stress, the network operator can increase the available bandwidth (and scale up a simple switching environment to match the additional bandwidth) or leave the base capacity levels constant and attempt to ration access to the bandwidth according to some predetermined policy framework. There is no magic in engineering. Additional network capacity cannot be created simply by imposing a different queuing discipline or by using a different signaling protocol. QoS engineering cannot generate more network capacity out of thin air. Or, there is no such thing as a free lunch.

An easily supportable second conclusion is that bandwidth is not unlimited in every location where there is a need for data. Remember that a majority of the public still accesses the Internet via modems, where a large and diverse collection of high-speed, high-volume data servers are being accessed through small-capacity edge modem circuits. Within this framework, the need to differentiate services becomes a technical argument, not necessarily one of introducing new revenue-generating products and services. Although the latter certainly may be a by-product of the former, unfortunately, new business services rarely are driven by technical requirements.

Dumb Networks and Smart Networks

Quality of service engineering is also an attempt to create a major shift in the basic architecture of the Internet. The end-to-end model is used to support reliable and efficient data transfer over a diverse concatenation of subnetworks. It makes remarkably few assumptions about the behavior of features of the underlying network. At the most basic level, the assumption is that a reasonable proportion of packets passed into the network reach their destinations.

TCP makes a small number of additional assumptions about the network. The most critical is that the network has a finite carriage capacity, and attempting to exceed its capacity results in packet loss. In addition TCP assumes a level of predictability of network behavior, that there is a high probability that the elapsed time to deliver a packet is relatively similar for successive packets. Beyond these basic behavioral assumptions, all remaining functions are under the control of the end hosts.

In this architecture, the network is viewed as a passive device from this perspective. Under equivalent input conditions, the network will deliver consistent responses to traffic. The functionality that is strictly required of the network's elements is limited to the requirement to switch the packet into a path hop that leads closer to the packet's destination. With an essentially passive network, the responsibility for both efficiency and reliability of data transfer is passed to the host as part of an end-to-end host function. Both efficiency and reliability of operation rely on a simple feedback control loop, in which the feedback flow allows the sender to infer lost packets that require retransmission and whether the network can support the sender increasing its transmission rate.

This approach leads to the architectural characterization of the best-effort Internet as one of *smart hosts, dumb network*. The network is a passive device, and hosts dynamically adjust their behavior in a manner that equilibrates with all other concurrent flows to produce an outcome of maximal network efficiency and fairness of allocation to each flow.

One of the major consequences of this model is of cost. Reducing the functionality of the network to that of a switched packet stream, and reducing the complexity of the operation of each element of the network, has implications not only to the resulting capital cost of network hardware, but also to reductions in operating costs of the network. This reduced network cost, coupled with a host behavior that attempts to drive the network to the point of maximal operating efficiency, produces a cost outcome that allows Internet-based services to reach the market with very low price points. Much of this outcome stems from the basic choice of a dumb network.

The QoS approach dispels some of these dumb network assumptions. The architectural change is that the network becomes an active agent within the service delivery architecture by virtue of the network's role in classifying traffic according to some defined policy and then offering certain traffic classes access to a managed level of network switching and transmission resource. The network can then be characterized, perhaps optimistically, as a *smart network*. In a best-case scenario, the network and the host are both configured with a similar service objective, so that the host is aware of the network's augmented set of functions and is capable of adjusting its behavior to make best use of this expanded functionality. In the worst case, the network and the host work to opposite objectives, potentially with the result that is worse than a conventional best-effort service.

Adding to the value of the service through supporting service quality functionality within the network is not without cost implications. Not only is there a requirement for more complexity to be placed into the network elements, there is also the increased scope of the role of operational management, where now there are a number of dimensions to the resource being managed, including policy management. Cost allocation is also more involved in that there is a rational role for resource usage accounting and cost apportionment in a manner directly related to actual resource consumption. All of these tasks add to the cost of the service.

A Matter of Marketing?

The current model of the Internet marketplace is an undifferentiated one. Every subscriber sees a constant best-effort Internet where, at any point within the network, packets are treated uniformly. No distinction is made in this model for the source or destination of the packet. Nor is a distinction made in relation to any other packet header field or in regard to an attempt to impose a state-based contextual "memory" within the network.

The base best-effort model is remarkably simple. A single routing-generated topology is imposed on the network, which is used by the switches to implement simple destination-based, hop-by-hop forwarding. At each hop, the packet may be queued in the next hop's FIFO buffer; or if the buffer is full, the packet is discarded. If queuing delay or buffer exhaustion occurs, discarded packets are affected equally. Thus, although the Internet service model is highly variable across a large network, the elements of the service model are imposed uniformly on all traffic flows.

Predictability and Quality

The marketing question is: Does a uniform variable-service model cause unacceptable uncertainties in transaction times; and if this is the case, is there a secondary market for a more consistent service model within the same environment? Is what is being requested here the requirement for predictability of transaction performance? Or is this to be combined with the requirement to define the performance in advance? This is no longer *best* service; it is *predictable* service.

Predictability and opportunism do not go well together. Opportunism tends to be a dominant factor; and, in general, opportunism is a highly seductive approach to marketing. In this light, clients of a predictable constant response service doubtless would want to avail themselves of superior service levels when the resources are available to other clients of the service. Their expectation would be that when the service levels degrade for other clients, their service levels are maintained at a *base* or contracted level. The bottom line for this approach raises a number of other questions: Can an absolute base-service QoS level be specified for general connectivity within an Internet environment? Is a QoS contract a comparative contract in which the service level for QoS-differentiated clients is specified as different from traditional best-effort services?

Is QoS Financially Viable?

The second facet of marketing a QoS service is addressing whether it will provide a financial return to the operator and yield results for the client that can translate into justification of the presumably increased subscription fee. In very general terms, it is reasonable to state that QoS services may cause the network to operate at a lower level of overall efficiency within periods of peak load; consequently, the QoS service must operate at a premium. This price premium can offset the reduced base revenue stream due to the lower peak carriage capacity for such base normal service traffic. Working through service and price options is a critical exercise in marketing this service.

The challenge is to come up with a comparison of fixed price and variable service. Although the prices will be fixed within the fee schedule, the service level differential from the base best-effort service will be both variable and extremely difficult to determine from the customer's perspective. In an uncongested network, there may be no visible service difference, unless the network operator undertakes traffic degradation for non-QoS traffic during unloaded periods, which, in a competitive market, appears to be a very short-sighted move. As the network load increases, the difference will become greater as the QoS traffic consumes a proportionately larger share of the congested resources. Of course, there is no such thing as an infinite resource, and at some load point, the QoS traffic will congest within the QoS category. If this happens, the service differential will then start to decline.

Subscription versus Packet-Option Models

Is QoS a subscription service or a user-specified per-packet option? From a marketing perspective, the QoS subscription model offers many advantages, including some stability of the QoS revenue stream, some capability to plan the QoS traffic levels and consequent engineering load, and a resulting capability to deliver a stable QoS service from an engineering standpoint. In many ways, a subscription service represents an insurance policy. Outside of the peak load periods, there may be little measurable difference between the quality service and an equivalent best-effort service. However, during the peak load periods, when the load levels reach the available network capacity, the QoS service agreement is intended to insulate the client from any adverse conditions, providing a service that remains within the contracted service level agreement. The subscription service can be regarded as a form of service-level insurance, in that the continual price premium will return a benefit to the client only in those intermittent periods when the normal best-effort service-response levels fall.

Packet-option, or transaction-based, models are equivalent to a spot market in service quality. When the base best-effort service starts to decline under load, there is a resulting instant market for a premium service level. A packet-option model, where the client marks the packet with a codepoint to request elevated service levels for the packet, allows the client to purchase a premium service only when required. Packet-option models create a more variable service environment with highly dynamic QoS loads that are visible only at periods of intense network use. A per-packet option also entails extensive packet-level accounting and verification at the entry to the network, because the consumer anticipates that any fee premium will apply only to packets marked with precedence directives. Of course, this also implies that a well-informed client would need to use some dynamic service-level discovery probe as a diagnostic tool to determine when to activate a request for a premium service response. Unfortunately, no such probes exist within the QoS environment.

From a customer's perspective, there is no fixed notion as to which of these options is preferable. If the subscription model offers access to the quality service at a low premium, then it may represent an attractive value proposition to the customer. The packet option model will inevitably attract a much larger relative price premium, as this model assumes that the network operator expects that such service requests will be activated only when best-effort traffic is already filling the network. Accordingly, honoring the service request will disrupt concurrent best-effort service profiles. Such disruption is not without a considerable compensatory price premium.

Effect on Non-QoS Clients

The other aspect to the marketing service is the negative message it sends to non-QoS clients, that base quality is unacceptable and that a QoS service level agreement is required to make the service acceptable. In the highly competitive commercial retail market for Internet access, the negative message—that a client must pay a premium to obtain an acceptable Quality of Service—is a difficult marketing message to sustain. Nevertheless, there is an element of truth in this negative message. The immediate assumption on the part of the customer is that when the network biases available resources toward QoS traffic, non-QoS traffic will receive a lower resource level, which in turn will lead to an average profile of lower levels of service.

This is an instance of a more general issue with adaptive consumer patterns with a fixed supply resource. When launched in the market, the first consumers see an essentially idle resource that presumably is capable of handling their entire traffic load. As the marketing and sales effort picks up, followed by an increase in the number of consumers, the share of the network available to each client declines. The addition of a QoS service would exacerbate the decline of service levels for best-effort traffic. Presumably, the best-effort service fee would remain constant. The result? Constant pricing with declining expectation of service levels. In a competitive market, this is not a scenario that results in customer loyalty. The point is: Although it is possible to implement traffic precedence within the network, the marketing approach should attempt to limit the negative impact of QoS on the network. This can be done by deploying entry-level traffic profilers to create an environment of bounded levels of QoS network preemption in an attempt to preserve average levels of service performance for non-QoS customers.

Looking at QoS solely from the perspective of the transaction between the service provider and the consumer, and taking the approach that QoS is a matter of marketing, does, however, tend to ignore the larger issue of market economics—which are very critical here. This raises the question: Is QoS a matter of market and supply-side economics?

A Matter of Economics?

Economics is sometimes defined as a means of describing the allocation, flow, and value of resources within a society. QoS can be described as a means of allocating network resources within a group of competing demands. In this light, it should not be surprising that there are some similarities in the two areas of study. In fact, as mentioned earlier, it has been suggested that Internet service quality becomes nothing more than a pricing argument from such a perspective.

QoS is not a uniform requirement in any network. If networks are engineered to the point where there is never any congestion, the argument for QoS weakens considerably. A line of argument asserts that it is in the provider's own economic best interest to ensure that engineering an abundant best-effort network is undertaken, in that the added operational complexity and associated costs of operating QoS mechanisms within a network are greater than simply purchasing additional capacity at those points where network congestion is being experienced [Odlyzko 1998]. This is not to say that the argument for service quality is any weaker; this is clearly not the case. However, the differentiation of services is somewhat of a non-sequitur to this objective of service quality. Network administrators will continue to face the task of provisioning networks that meet customer demands for availability, speed, latency, and so on. Consequently, service quality is a constant demand. The line of reasoning within the economics of abundance is that it is cheaper to raise the service quality for all customers than it is to selectively raise the service quality for a subset of the customer base.

Other lines of thought tend to lead to a different conclusion regarding the potential abundance of network infrastructure. There has been some indication that the initial wave of consumers wanting to connect to the Internet has been carried through on the margins of over-supply in, and subsequent build-out of, the traditional telephone system infrastructure. Now that Internet traffic has effectively consumed that resource, there is no high-margin cross-subsidization agent for further build-out of the Internet infrastructure. Surprisingly, this line of thought leads to the conclusion that Internet service prices could rise. That said, service providers cannot raise prices uniformly in a highly competitive market. Accordingly, QoS may be the levering mechanism for price escalation. QoS allows the service provider to generate selective price escalation, by offering differentiation during periods of congestion. The conclusion to this line of argument is that current flat-rate pricing mechanisms that dominate much of the

Internet landscape today cannot rise, and QoS is one lever used to meet the financial challenge of funding the additional communications capacity necessary to ride the continuing aggressive expansion of the Internet. In this light, QoS is seen as an add-in across a base, flat-rate commodity Internet market.

The alternative course of action is to increase the level of differentiation of delivered services and couple this with service-based pricing for access to such a range of services. The fervent hope is that this will enable network operators to break free of single-fee, single-quality-level market structures and let the operator market differentiated service levels at differentiated prices. Perhaps the concomitant unspoken hope is that the entry of premium pricing structures will correct the potential risks of ever decreasing quality levels associated with a steadily decreasing unit-pricing model, and also improve the operating margins of Internet service providers at the same time.

Supply and Demand

As in economics, the basic issue in quality of service is one of supply and demand. When all demands are met because of over-supply, there is no inherent need for resource management. Each component of demand can be met from the pool of available resource. But when there is a shortfall of supply, the demand must be moderated; then the allocation of available resources against the demands becomes the required resource allocation task. In this section, we will take a look at the issues of supply and demand for network capacity, to determine whether the drive for QoS services is based on a fundamental scarcity of network capacity to service a rapacious market demand for Internet services.

One place to examine these issues is the long-distance international carriage market. In this market, undersea cables and satellite communication systems present a delicate balance between aspects of network engineering, network architectural design, and scales of economy, so, we will look as this market as an example of the broader carriage market. With the changes introduced by the wave of Internet demand, some of the new balances are still not well understood across the telecommunications industry. In our example market, the brokering of transcontinental bandwidth is a complex and convoluted exercise. The entrenched players are global telecommunications industry giants that have become accustomed to a conservative growth model in terms of demand. The investment process for these international projects was historically intended to reflect the high investment levels required, and to account for the high risks of investing in these projects, as well as the relatively slow growth of the traditional consumer of the product—the voice market. The industry geared itself to an artificial constraint of supply to ensure stability of pricing, which in turn was intended to ensure that return on investment remained high. (Those familiar with the history of the diamond industry no doubt will see some parallels in this situation.) Thus, international cable systems have been historically geared to ensuring that at any stage, the supply of capacity to the market just matches the current level of demand, so that wholesale pricing does not crash and so that the investment profile remains attractive in the face of its associated risk. The relatively complex investment vehicles to underwrite the costs of undersea cables, and the careful construction of matching half-circuit orders from respective carriers on either side of the cable in order to commission end-to-end circuits, together with a relatively constrained process of releasing capacity for use from the common shared cable, were all components of an environment intended to ensure that supply never outstripped demand within the market, thus preserving the value of the initial investment.

Now that recent data demands are changing the model from a well-tempered growth model to an "all you can eat" version of capacity requirement, the balance is shifting. Historically, well-understood forward-capacity plans have been scrapped as the data networks ravenously chewed through all available inventory. The same observation holds true for local exchange carriers (LECs) in a competitive domestic telecommunications market. The LECs are just now beginning to understand that architecting "data-friendly" networks deviates somewhat from their traditional method of architecting circuit-switched voice networks.

Traditionally, Internet service providers (ISPs) simply have bought or leased circuits from the telcos (the local telecommunications entity or telephone company) with which they constructed their networks. This worked reasonably well in the earlier days of the Internet, when traffic volumes were relatively low and circuits were relatively low-speed by today's standards. The circuit orders were provisioned from the margins of the over-supply of a vastly greater voice network, whose supply model was one of advance provisioning—up to two decades in advance of consumption. The supply of carriage for data leases can be seen as reducing the point of capacity augmentation by some small number of months two decades hence. In the intervening period, the telcos have a revenue stream for otherwise idle capacity.

A paradigm shift has occurred over the course of the past decade, resulting in many telcos taking up an active presence in the ISP market, competing directly with the entrepreneurial ISP businesses to which they were also selling capacity. In entering this market the telcos have been playing dual role as wholesale services supplier and retail competitor. Their provisioning systems, and the associated capital investment structures, are in many cases still well entrenched within a traditional slow linear growth market model heavily influenced by the traditional demands of the voice market. Given the finite nature of the telcos' carriage inventory, the relatively complex provisioning systems used to support data services, and the unprecedented level of both internal and external demand for basic telco carriage services, some signs of stress are appearing in terms of availability of carriage services. In many instances, this almost vertical growth in data services demand has caused a staggering lag in provisioning new carriage services. Coupled with the fact that the number of ISPs has grown phenomenally during the same time frame, as has the demand for bandwidth, capacity is not readily available in every location, even where there may be a relative abundance of underlying fiber. It will take the traditional telco some time to meet the capital and logistical demands required to service this recent explosive wave of data. Simply stated, radical changes to the capital investment programs and provisioning processes of such capacity providers cannot occur overnight.

Another significant reason for this paradigm shift is that the telcos are beginning to appreciate an apparent conflict of interest in meeting the entire set of supply demands from a rapacious ISP sector. The traditional high barriers for competitive entry into the voice market have been eroded both by deregulation and technological advances. Telcos may perceive their current ownership of an extensive cable inventory as the last bulwark of protection in their voice revenues. Moreover, as the telcos are now becoming active players in the "value-added" data business, either through direct business development or by acquisition, they now can adopt a more reluctant view of releasing significant levels of capacity outside their immediate in-house interests.

Cost of Additional Capacity

The potential outcome of these factors is an artificial environment where, regardless of the levels of installed fiber system inventory, data carriage capacity is not always readily available. Alternatively, data carriage services may be tariffed at a premium to reflect both the limited ability and some degree of business level reluctance to service the market to the entire extent of its capacity demands. Under these circumstances, the non-telco ISP must decide whether to pay such tariffs for additional capacity, find alternative methods for connectivity, or simply find methods to mitigate network degradation in the face of inadequate network resources. This leads to the observation that in some circumstances, even when an abundance of underlying fiber optic cable capacity appears to exist, there is still an economic driver for QoS services that is fueled by a lack of data carriage bandwidth. Possibly, as market outcomes from the current phase of deregulation are examined, this issue will become increasingly important from a regulatory standpoint. Then the question will be whether the deployment of QoS is an acceptable outcome of the current market-based methods of bandwidth distribution to the telecommunications industry. It poses the somewhat curious question of whether, in some markets, the requirement level for Internet QoS services is based to some degree on market failure, or as an unintended by-product of a particular regulatory regime.

Of course, in some geographic locations, such as parts of Africa or Central Asia, bandwidth cannot be purchased for any price because it simply does not exist. In this case, the deployment of QoS, in response to bandwidth scarcity, is based on local infrastructure investment conditions instead of being an artifact of the state of the telecommunications industry.

We find ourselves asking some familiar questions—most of them technical, but some economic. Is the desire to implement QoS introduced by fundamental economic imbalances in the evolution of the telecommunications industry? If it were possible to bring abundant carriage capacity to bear on the levels of demand, and bring it at an affordable price, would all demand be completely satiated? If the answer is yes, then there would appear to be a much weaker rationale for the deployment of QoS.

In those markets where the telecommunications industry is fully deregulated, the response to demand exceeding supply is to introduce new providers, as one would expect in an open market. While open markets are often observed to equilibrate supply and demand, the sudden appearance of new demand or supply factors often leads to short-term market distortions and over-compensation. In the case of the Internet, the combined factors of an explosive uptake in demand, a relatively rapid deregulation of the supply industry, and a very fortuitous technology shift in fiber optic cable systems is leading to an almost instantaneous switch from scarcity to abundance in some markets. This can be observed in the long-distance data carriage market in the United States; and similar changes appear to be taking place in the trans-continental undersea fiber cable markets in both the trans-Atlantic and trans-Pacific sectors.

A question left unanswered is whether the level of available transmission can be raised to the point at which it is far in excess of sustainable demand from all the clients of the network. If such a scenario comes to bear, the engineering of the Internet will inevitably shift from the model of adaptive behavior to maximize cost efficiency of transmission to a model of engineering for abundance.

Economics of Abundance

A number of factors that may restore the historical situation of large margins of over-supply of transmission capacity are becoming evident in the area of fiber optic cables. The distance between signal regenerators on long-distance cables, multiplied by the bandwidth of the cable, is a metric that is seeing a tenfold increase every four years. Additionally, new strands of fiber optic cable are being installed across the world at a rate of some 100 meters per second, or at an equivalent speed of Mach 2. The bandwidth per strand is also increasing; dense wave division multiplexing allowing, in late 1999, 320 wavelengths, each carrying an OC-192 (10Gbps) bearer, or some 3.2Tbps of total capacity per strand. No doubt this figure will increase in the very near future. Expensive synchronous digital hierarchy switching systems are now being complemented by alternative optical switching solutions and data clocking solutions, resulting in a decrease in the cost of trunk switching networks. The cost of carriage, when measured in the cost of a megabit per second of installed carriage capacity per kilometer distance, continues to shrink rapidly. In this model, the control function is being pushed to the edge of the network, while the interior of the network performs only basic switching functions.

The model of abundance points to the current areas of shortfall of supply of carriage systems, and the rapid industry escalation to respond to this. The response is twofold, with both technology, in the form of greater per-strand carriage capacities and increased levels of installed fiber cable, and in the associated shrinking unit cost of fiber-based carriage. The prediction is that the industry response will over-compensate, and that again there will be significant amounts of over-supply of carriage capacity. The critical difference is that this over-supply is predicted to occur within a highly competitive commodity supply market, as the model also requires one additional assumption: that the supply of capacity to the consumer market not be artificially constrained by monopolistic or cartel trading behaviors on the part of the carriage providers. Here the model points to the deregulation of the communications industry and the number of new entrants in the form of Internet service providers in order to make the point that in such a diverse competitive industry, the formation of an effective cartel is unlikely.

The combination of technology advances in per-strand carrying capacity and the introduction of a deregulated market in fiber provisioning leads to the prediction that many network transmission markets will soon see supply exceeding demand levels, with the economic consequence of rapidly decreasing trunk transmission costs for network providers by some orders of magnitude. However, it does not necessarily follow that lower transmission costs will lead to higher-quality networks; the current pattern of high utilization rates could persist, leading to a continuing requirement for QoS service deployment.

The question then becomes whether, within an abundant low-cost transmission market, Internet service networks will continue to operate at high utilization levels, or whether the network operators will be able to reduce their transmission capacity utilization levels to the point at which QoS mechanisms would create only marginal, if any, performance difference. Bear in mind that a low utilization network model is one in which the capacity is sufficiently large compared to the sum of the access channel loads that the network can manage. In such a network model, the residual service quality issue is one of packet quantization jitter, because queuing jitter and congestion loss are effectively eliminated. There is some reason to predict that network utilization levels will drop in an environment of transmission abundance, due to consumer preference to purchase service from networks where the average performance level is superior. Given that transmission costs are a much lower proportion of total service costs in this type of environment, there is little opportunity to further reduce costs through a high network load factor. In addition, the consumer preference is not necessarily for higher data volumes but, rather, toward faster transactions. The prediction from such an analysis is that significantly lower transmission prices will prompt a market shift toward lower utilization rates and uniformly improved best-effort service quality. An evolutionary path to higher network quality levels without the need for active differentiation mechanisms should not be surprising. Similar evolutionary paths of uniform quality improvements have been seen in many other commodity markets throughout the twentieth century.

An additional factor reinforces this view of uniformly improved service quality as an outcome of market factors in an environment of abundant network capacity. It is the issue of retail pricing and interprovider financial settlements. Active quality differentiation within networks will imply some form of associated differentiated pricing as a necessary control factor to ensure that the entire traffic load does not simply shift into the premium service class, thereby creating an exact replica of the original best-effort service class. Such a pricing model will need to have either a tariff premium or a usage-sensitive tariff component. In a highly competitive retail market, consumer preference tends to favor usage-insensitive pricing; this is coupled with the observation that there is a commodity price threshold below which the costs of usage-based accounting cannot be sustained. This implies that a service provider either can attempt to keep prices high and best-effort service levels low, to allow for the accounting margin on differentiated services usage-sensitive retail models, or to simply drop the accounting requirement by lifting the service quality of the base best-effort service and operate a flat-rate retail service. In a competitive market where the base commodity price is dropping due to abundant supply, the latter course of action is the logical market outcome.

Consideration of interprovider financial settlements reinforces this conclusion. It would appear that any quality differentiation at the retail level would have a somewhat different intent to the best-effort retail service model, as a quality service offering will need to encompass a single point of tariff for an end-to-end service. In a quality-differentiated environment the quality initiator pays the premium for the entire end-to-end service, whereas in the current best-effort Internet, both parties to a transaction fund a part of the transaction costs. In a multiprovider environment, this quality tariff implies that the originating service provider has collected the revenue for a service supplied by a group of providers along the end-to-end service path, and that some form of revenue distribution is necessary to realign the revenue to approximately match the source of cost. This shift in the retail and interprovider financial settlement model represents a truly significant shift in the nature of the Internet industry. Given the marginal returns from such a shift, and the relatively large organizational and marketing effort to implement such a change to the existing industry model, it is unlikely that there will be any driving motivation within the market to force such changes to occur. This leads to the conclusion that in a highly competitive market, with abundant resource availability the optimal market approach lies in adopting a position of low network utilization levels, with a high-quality level provided as a uniform best-effort service class, sold into the retail market through fixed-price retail structures.

In such a network model, engineering priorities must be raised. Is QoS differentiation the most critical engineering problem, or are service availability and rapid restoration more critical? In a very large network, each trunk bearer may be carrying the traffic of millions of end customers. Cable systems are vulnerable to physical damage, and network elements have failure rates. Very rapid service restoration is the most critical service factor, with engineering concentrating on carriage and element resiliency rather than on differentiated service levels. In this model of abundance, where there is no resource contention, there is no natural need for QoS engineering, and other engineering priorities tend to surface in their place.

From this economic perspective, there are various segments of the market where, due to abundant supply of carriage services, the prognosis for QoS deployment is not good. Where abundance is not necessarily evident, the question is whether the remaining market segments have sufficient market influence to impose QoS solutions upon the entire network, or whether the spread of abundant capacity will be broad enough and fast enough to obviate any need whatsoever for QoS mechanisms within the network. To continue this line of thought, we need to look at these other market segments where the prospect of abundance is not necessarily in evidence.

A Uniform Market?

The preceding line of argument would tend to conclude that there is scant benefit in deploying QoS within the network. However, a number of additional considerations could modify this into a more conditional conclusion. The major caveat appears to be that this conclusion is most relevant in the trunk capacity market. Other parts of the network, such as the wireless network, the copper loop access network, and the broadband cable access network, are affected by various forms of technology constraints that limit the network's capacity in some form or other. In the case of wireless services, this is further limited by the finite spectrum space and the presence of alternative high-value uses of the radio spectrum, ensuring that the base price of the resource will remain relatively high and that the supply will remain constrained to some extent. The importance of nonuniformity of network availability and cost is that an end-to-end network service may traverse a number of distinct underlying carriage markets; and while some of these markets may see little need for QoS services, other markets, particularly in the access area, see value in deploying QoS resource management mechanisms. However as we have seen, QoS is not a mechanism that can be deployed piecemeal. QoS is a *service*, and as such it operates in the services domain on an end-to-end basis. To support this end-to-end service model, a relatively ubiquitous model of QoS response is called for from the network. The question raised is whether the QoS demands from the more constrained access environments will force QoS mechanisms to be deployed within the trunk networks to support full end-to-end QoS functionality.

At a more general level, bandwidth, congestion, and quality are interwoven topics. There is a certain clash between the capability of the technology and the emerging expectation of the user community over the Internet. The Internet model gains its significant cost efficiencies—that is, a cheap Internet—through a deliberate effort by the Internet protocols to completely fill the bottleneck wire segment with data. The critical point is that adaptive data protocols use a mechanism of rate adaptation to the point of queue overload to ensure that the end hosts are making the most efficient use of available network resources. Some level of packet loss is a natural outcome of this operational model—the end host systems always attempt to overload the network. The mutual resolution of this condition by the end hosts is the resource arbitration system. So, how much network capacity is too little? How much is too much? Given the adaptive load profile of the Internet, these questions do not have objective answers.

If the entire network load conformed to an adaptive load model, this would be the end of the answer: The network is a collection of passive switches, and the host systems themselves negotiate their collective behavior. This can be likened to a highway system in which the roads are passive, and the cars both contribute to and mediate the resultant load. In this model, there are few levers, if any, for the network to mediate the load profile.

Other factors impinge on this: The introduction of nonadaptive, real-time traffic, such as voice and video streams, leads to an expectation of a somewhat different network model, where the network operates with sufficient margin to be able to accommodate such applications without imposing distortion in the form of jitter and loss on the associated traffic streams.

The crux of the performance issue is that adaptive systems allow and expect over-subscription of the critical bottleneck component of the network, thereby sacrificing precise predictable performance for the benefit of cost economies of operation. The systems cooperate to collectively push as much data into the network as possible. Certainly, this allows the network operator to defray costs over the maximum volume of traffic, to make a cheap network. Nonadaptive systems use a different model, as their requirement for predictable network response forms a large part of the value of the traffic. Here the network operator must dimension the network and its associated admission control systems to admit only as much nonadaptive traffic as there is capacity within the constrained sector of the network.

Historically, the long-distance carriage market has been the most expensive and most constrained component of the network, and the engineering of the network has inevitably been directed toward maximizing the use efficiency of this resource. It is possible to imagine a near-term environment where this shifts completely, wherein the last mile is the most constrained component of the network. This is already the case for modem-based access; but perhaps the so-called high-speed access systems of DSL and broadband cable will fall into the same service category. In this type of access environment, the same factors of maximizing the utilization of this constrained resource still apply. While this does imply a continuing value in using QoS mechanisms to control the contention within this resource, the focus of QoS changes radically. Rather than attempting to manage the aggregation of separate service demands from many users in their access to a common resource in a technically rational and economically feasible fashion, the focus shifts to managing a resource on behalf on a number of concurrent demands from a single user across a dedicated access system. Here the objective of QoS shifts to allow the user to express a service preference as to how to allocate access service to each of these concurrent demands.

A Matter of Technology?

An alternative line of thought is that QoS is just a case of isolating the correct technical approach, followed quickly by deployment. To date, three directions have been pursued in the attempt to isolate the appropriate technology:

◆ The host or the application signals the desired per-packet service characteristic in the header of the packet. The network is requested to honor this request without further negotiation. This is the Type of Service selector.

◆ The application signals the intended traffic profile to the network, and the network responds with a resource commitment that will honor the request. This is the Integrated Services architecture.

◆ The ingress of the network determines—via some policy mechanism—both the service that will be applied to packets and the amount of traffic admitted within each service class. This is the Differentiated Services architecture.

All three approaches are only partial solutions to a common, and as yet unspecified, architecture. But though the entire architecture has not been specified as yet, certain features of this broader service architecture can be identified from these three efforts.

Resource Negotiation

Service quality requires some level of negotiation between the network and the end host regarding the level of availability of resource within the network to meet the service demands. If the amount of traffic requiring some premium level of service exceeds the resource capacity of the network, then the network will be unable to meet such service commitments.

The network and the host must be able to negotiate the host's service request and the capability of the network to devote resources to meet the request. This negotiation can take a number of forms. If the host is able to predict the load profile of the application, then it may be possible for this profile to be passed to the network and for the network to respond with a reservation of resources necessary to meet the request. A network is a dynamic system, so it is also necessary to periodically refresh the reservation to ensure that the service levels continue to be met, or that the host is informed when this is not possible. If the host is able to moderate its behavior dynamically, then the negotiation may be a process of continual feedback. The feedback would allow the network to regularly signal the service state of its admission system to the host, which in turn could moderate its traffic profile to meet the requirements of the admission system.

Whatever the exact form of negotiation, one feature is common to this process: feedback control. The host must interact with the network to the extent that the host is aware at all times that the network is able to meet its service requirements. To achieve this, the network must engage in some form of information feedback back to the host system. The standards-track technology choice for this negotiation appears to be the RSVP protocol. The extensibility of the objects carried within RSVP messages offers sufficient flexibility for most forms of negotiation.

Policy Negotiation

Assuming a finite amount of dedicated resource to devote to the support of premium services, there must be some form of policy management to determine how the service is to be allocated to clients. This results in the network having to make decisions from among conflicting resource requests. The means by which such conflicting requests are resolved involves the use of policy.

Initial technology efforts in this area have been directed to the architectural model of policy determination, along with the protocol required to support the interaction of network elements and policy-based decision makers. The associated effort is to define, in a sufficiently general manner, a ruleset for traffic filters that can isolate the desired traffic class to which a policy refers. The policy itself is relatively straightforward, given that its intent is to create a priority of decisions. A typical priority is simply a precedence value of the request used to compare with other values of concurrent requests.

Management and Measurement

Any approach to QoS service provision will have to ensure that the performance of the service can be monitored and measured. It is an act of misplaced faith to assume that configuration alone will result in a robust service platform. The network operator needs more direct assurance that the network is working within the bounds defined by the network configuration. This is usually achieved through polling each network element to retrieve its operational status, coupled with a systematic methodology of network probes intended to discover the service characteristics of paths that traverse the network.

The feedback from a measurement system is not necessarily restricted to a network management station. In the case of a Differentiated Services boundary admission architecture, the admission systems must be able to perform a dynamic adjustment of their admission meters based on the load levels within the network. If there is a transient load condition within the network interior that causes overload, or if there is some operational failure that results in a large shift of traffic to under-dimensioned bearers, the carriage capacity of the network may be seriously reduced. The risk if the admission systems continue to admit premium traffic at a constant rate is twofold: denial of service for the base best-effort traffic and service breakdown for the premium-level service. One way to avoid these situations is to associate the admission profiles with the current state of the network. To achieve this, the admission systems need feedback from the interior of the network regarding the available service-qualified capacity on each forwarding path from this admission system. This enables the admission system to alter its admission profiles on a path-by-path basis in the event of dynamic changes in network load and consequent changes in the service capacity of the network. For this to happen, a service measurement system must be deployed, one that produces information that can be used immediately by an admission system to configure the admission profile meter.

The service environment also needs mechanisms that let the client measure the service quality delivered by the network. Unfortunately, standard mechanisms to perform such measurements may still be some time coming, for without measurement tools the client has no trusted method to evaluate the service quality and to compare the service against the best-effort base service.

Service Accounting

It may be possible to create premium service levels using no accounting function at all, but it is unlikely. Understandably, client and provider alike expect that the use of a premium service will also incur some incremental charge within the price structure. The consequence is to meter the usage of the service and record the associated quality of the service as part of the data collection requirements in constructing an accounting system.

A number of efforts have attempted to standardize an architecture for accounting measurements within the Internet environment. The confounding issue for most of these efforts has been the standardization of the client model within the scope of the accounting system. Some accounting models work at the micro-flow level of granularity, generating an accounting record for each application activation that makes use of network services. In large network environments, this type of accounting model generates an overwhelming amount of data to analyze. Consequently, the accounting model may move to a coarser level of granularity, to generate periodic interval records based on a source address and some defined service level.

Scale

According to Paul Vixie of the Internet Software Consortium, the two relevant engineering questions for Internet deployment are: "What if *everybody* did that?" and "What if there were 6 million *times* as many as there are now?" The point here is that any proof of concept within a controlled environment does not have that much relevance to determining its suitability for the Internet, both now and in the near term future. Therefore, the major technical challenge for all approaches to QoS is to create a technology environment that can withstand the pressures of growth.

As the circumference of the network grows, the subsequent growth in the core of the system rapidly outstrips the capacity of available hardware control systems. And the competitive nature of the service provider market means that there is little room for over-engineered solutions. Thus, the technical requirement is to place as much (but no more) processing as is adequate for the required task in the core of the network. This implies that detailed, computationally intensive tasks are best managed at the edge of the network, while more basic, easily aggregated functions are best performed in the core of the network.

This leads to the observation that the technology to support differentiated service levels in the network will not be uniformly deployed across all network elements. Edge network elements will perform specific flow management, signaling, and service management functions. Internet network elements will operate at a more aggregated level, using coarsely defined traffic-conditioning controls. The essential glue in this approach of role specialization of network elements is that of an appropriate signaling feedback system to allow the edge devices to work with a clear model of the current capacity available within the network core. A standard feedback technology has yet to be adopted by the industry.

The conclusion is that QoS technology has not been comprehensively defined. Numerous tools still require definition. No doubt these gaps will be filled, given sufficient marketing and economic impetus.

What's Wrong with This Picture?

It's not finished yet.

If QoS is so effective in addressing a range of service requirements, and if QoS can address client needs for any particular service, then why aren't all Internet service networks enabled to support QoS today?

There are many answers to this question. We list a few here to illustrate the problems that the emerging QoS architectures will need to solve before they can gain wide acceptance.

> **There is no clear model of how to add QoS to an application.** Is QoS an application option or a transport-layer option? As a transport layer option, it could be envisaged that any application could use some QoS-enabled network service by changing the host configuration, or by changing the configuration at some other network control point, without making any explicit changes to the application itself. The strength of the transport layer approach is that there is no requirement to substantially alter application behavior. In the case of the Integrated Services architecture, this transport level control does not appear to be an available option. The application does require some alteration to function correctly as an IntServ requestor. The application must be able to provide to the reservation module a profile of its anticipated traffic, or in other words the application must be able to predict its traffic load. In addition the application must be able to share the reserva-

tion state with the network, so that if the network state fails, the application can be informed of the failure. In the case of the Differentiated Services architecture there is no explicit provision for the application to communicate with the network regarding service levels. In this case, if the Differentiated Services boundary traffic conditioners enter a load shedding state, the application is not signaled of this condition, and is not explicitly aware that the requested service response is not being provided by the network. While there is no explicit need to alter application behavior in this architecture, as the basic DiffServ mechanism is one that is managed within the network itself, the consequence is that an application may not be aware that a particular service state is being delivered to the application, or not.

There is no comprehensive service environment. The maintained reservation state of the Integrated Services architecture and the end-to-end signaling function of RSVP are part of the picture; however, it is not cost-effective, or even feasible, to operate a per-application reservation and classification state across the high-speed core of a network. While the aggregated behavior state of the Differentiated Services architecture does offer excellent scaling properties, its lack of signaling facilities means it cannot be operated in isolation. The Differentiated Services architecture can be characterized as a boundary-centric operational model. It requires both state signaling from the core of the network to the DiffServ boundary and signaling from the boundary to the client. A service architecture that spans end-to-end service paths will need to integrate these two approaches into a larger single architecture.

There is no robust mechanism for path discovery with service performance attributes. How can a service path to an arbitrary destination be discovered? Assuming that the deployment of a service-differentiating infrastructure will be piecemeal, even if only in the initial stages of service rollout, how can a host application determine whether there is a distinguished service path to the destination? Much refinement must be done to the model of service discovery and the associated task of resource reservation.

There is no integration of interdomain routing and service attributes. How does path discovery with service attributes impact the interdomain routing state? Taking piecemeal deployment of a service differentiation network infrastructure one step further, it is reasonable to suggest that we will require the qualification of routing advertisements with some form of service quality description. This implies that we will also require some form of quality vector-based forwarding function, at least in the interdomain space, and some associated routing protocol can pass a quality of service vector in an operationally stable fashion.

High performance TCP requires service symmetry. How is TCP bidirectionality supported? Is it necessary to support it? A congestion-managed flow uses the feedback from the ACK packet stream to time subsequent data transmissions. If the ACK stream is treated by the network as a different service profile from the outgoing data packets, to what extent will the data-forwarding service be compromised in terms of achievable throughput? High rates of jitter on the ACK stream can cause ACK compression, which in turn will cause high burst rates on the subsequent data send. These bursts will stress the service capacity of the network and will compromise throughput rates. If symmetric service profiles are important for TCP sessions, how can this be structured so that it does not incorrectly account for service usage? In other words, how can both directions of a TCP flow be accurately accounted to one party?

Traffic Conditioning TCP is yet to be defined. Current traffic conditioning models use a token bucket as the basic mechanism to condition traffic into a defined profile. Token buckets operate in a manner similar to a FIFO queue, and impose similar tail drop behaviors on the traffic flow. Such tail drop behaviors result in the TCP session recycling through retransmission timeouts and slow start mode. Where the traffic flow is based on TCP traffic, token bucket traffic conditioning elements are best described as being TCP hostile.

There are too many service models. The Differentiated Services architecture does not restrict the number of potential services, and many industry players have recognized this as an opportunity. True, the network operator may be limited to a choice of up to 64 discrete services in terms of the 6-bit service codepoint in the IP header, but because the mapping from service to codepoint can be defined by each network operator, there can be any number of potential services. Of course, there can be too much of a good thing, and a large number of services will undoubtedly raise questions revolving around end-to-end service coherency. A small set of distinguished services can be supported across a large set of service providers by equipment vendors and by application designers alike, whereas an ill-defined large set of potential services has little productive purpose.

There is no service measurement architecture. It is reasonable to anticipate that premium services will attract a premium tariff. How can the network operator provide objective measurements to substantiate the claim that the delivered service quality conformed to the service specifications? Furthermore, how can the client measure the delivered service quality so that the additional expense can be justified in terms of superior application performance.

There is no service accounting architecture. If there are to be tariff premiums, then some accounting of the use of the premium service would appear to be necessary, to relate use of the service to a particular client. So far, no such accounting model has been defined, nor has any method been developed to gather the data to support the resource accounting function.

There is no standard method for cost apportionment and settlement. How does the network operator apportion the service operational costs? What data is required for a network operator to determine whether supporting a particular service is cost-effective? How can the network operator determine whether supporting a service is a profitable proposition? Structural cross-subsidies tend to weaken an operator's market position in a competitive unbundled market, and accurate cost apportionment is an essential component of a stable and well-founded business operation. Currently, few tools are available within the network to support cost apportionment across a set of services. The common processing and switching elements used within the network tend to make such an exercise difficult.

There is no transition plan. How does the deployment of service-aware networks commence? Which gets built first? Host applications or network infrastructure? This is another instance of the chicken and egg problem. No network operator will make the significant investment in distinguished service infrastructure unless clients and applications are available to make immediate use of such facilities. No application designer will attempt to integrate service quality features into the application unless a model of operation is supported by widespread deployment, to make the additional investment in application complexity worthwhile. With both sides in this scenario waiting for the other to move, deployment of distinguished services may take some time.

Diversity is always a factor. This means that it is highly improbable that any single form of service differentiation technology will be rolled out across the Internet and across all enterprise networks. Some networks will deploy a form of service differentiation technology while others will not. Some of these service platforms will interoperate seamlessly and others less so. To expect all applications, host systems, network routers, network policies, and interprovider arrangements to coalesce into a single homogenous service environment that can support a broad range of service responses is unrealistic. More likely, we will see a number of small-scale deployment of service differentiation mechanisms and some efforts to bridge these environments. In this heterogeneous service environment, the task of service capability discovery is as critical as being able to invoke service responses and measure the service outcomes. We will need to further develop protocol capabilities in supporting service discovery mechanisms. In addition, such a heterogeneous deployment environment will exert greater scaling pressure on the operational network, because now there is an additional dimension to the size of the network. Each potential path to each host is potentially qualified by the service capabilities of the path. While one may be considered as best-effort path candidate, another may offer a more precise match between the desired service attributes and the capabilities of the path to sustain the service. The brunt of such scaling pressures will be felt in the interdomain and intradomain routing domains, where the pressure is on to increase the number of attributes of a routing entry, and to use the routing protocol in some form of service signaling role.

None of these, either singly or as a group, form sufficient reason to condemn a multiservice internet network as impossible, or to believe that the efforts being made to define quality of service architectures will not come to fruition. This list is intended to illustrate that before we see widespread deployment and use of such multiservice networks a number of pieces need to be found to complete the picture. These "pieces" will form the engineering agenda over the coming months, as we move from fundamental service architecture definitions to the details of deployment and operation.

CRCRCR

Remember, things are not bad all over; it just pays to do your homework. Operating a multiservice internet platform is possible today, and there are many examples of networks that perform well in meeting their service objectives. The efforts of the evolving QoS technologies are intended to improve the generality and scalability of this effort. Arguably, network operators who want to implement a multiservice platform within a single administrative domain will have much better success in providing differentiated services than anyone who attempts to provide similar services across multiple administrative domains in the public Internet—at least for the foreseeable future. Nevertheless, the same technical principles hold true in the broader environment, and similar considerations must be given to network performance, stability, scale, management, and control. Differentiating the service response to support multiservice networks will not be a free ride. Whether you are trying to implement QoS in a private network or within a segment of the global Internet, service differentiation will come at a cost. There's no magic here.

Caveat emptor.

ଅଅଅ

References

[Odlyzko 1998] *The Economics of the Internet: Utility, utilization, pricing, and Quality of Service*, A. M. Odlyzko, 1998. Available at www.research.att.com/~amo.

This paper argues that over-provisioning data networks is a viable and economically sustainable response to the demands for service quality within data networks, and that such a response is technically and economically superior to implementing QoS responses within the network.

RFCs

Request for Comments documents (RFCs) are published by the RFC editor. They are available online at www.rfc.editor.org.

[RFC2309] *Recommendations on Queue Management and Congestion Avoidance in the Internet*, B. Braden, D. Clark, J. Crowcroft, B. Davie, S. Deering, D. Estrin, S. Floyd, V. Jacobson, G. Minshall, C. Partridge, L. Peterson, K. Ramakrishnan, S. Shenker, J. Wroclawski, and L. Zhang, RFC2309, Informational RFC, April 1998.

This presents two recommendations to the Internet community concerning measures to improve and preserve Internet performance. The first recommends testing, standardization, and widespread deployment of active queue management in routers to improve the performance of today's Internet. It also urges a concerted effort of research, measurement, and ultimate deployment of router mechanisms to protect the Internet from flows that are not sufficiently responsive to congestion notification. It is a fitting final reference for further reading for the study of service quality.

glossary

AAL (ATM Adaptation Layer) A connection of protocols that takes data traffic and frames it into a sequence of 48-byte payloads for transmission over an Asynchronous Transfer Mode (ATM) network. Currently, four AAL types are defined that support various service categories. AAL1 supports constant bit-rate connection-orientated traffic. AAL2 supports time-dependant variable bit-rate traffic. AAL3/4 supports connectionless and connection-oriented variable bit-rate traffic. AAL5 supports connection-oriented variable bit-rate traffic.

ABR (Available Bit Rate) One of the service categories defined by the ATM Forum. ABR supports variable bit-rate traffic with flow control. The ABR service category supports a minimum guaranteed transmission rate and peak data rates.

ACF/VTAM (Advanced Communications Facility/Virtual Telecommunications Access Method) In traditional legacy SNA networks, ACF/VTAM on the IBM mainframe is responsible for session establishment and activation of network resources.

ACK (Acknowledgment) A message that indicates the reception of a transmitted packet.

ANSI (American National Standards Institute) One of the American technology standards organizations.

API (Application Programming Interface) A defined interface between an application and a software service module or operating system component. Conventionally, an API is defined as a subroutine library with a common definition set that extends across multiple computer platforms and operating systems.

APPN (Advanced Peer-to-Peer Networking) APPN represents IBM's second-generation SNA networking architecture, which accommodates peer-to-peer communications, directory services, and dynamic data routing between SNA subdomains.

ARP (Address Resolution Protocol) The discovery protocol used by host computer systems to establish the correct mapping of Internet layer addresses, also known as IP addresses, to Media Access Control (MAC) layer addresses.

ARPA (Advanced Research and Projects Agency) A U.S. federal research funding agency credited with initially deploying the network now known as the Internet. The agency was referred to as DARPA (Defense Advanced Research and Projects Agency) in the past, indicating its administrative position as an agency of the U.S. Department of Defense.

AS (Autonomous System) The term used to describe a collection of networks administered by a common network management organization. The most common use of this term is in interdomain routing, where an autonomous system is used to describe a self-connected set of networks that share a common external policy, with respect to connectivity; in other words, networks that generally are operated within the same administrative domain.

ASIC (Application-Specific Integrated Circuit) An integrated circuit that is an implementation of a specific software application or algorithm within a silicon engine.

ATM (Asynchronous Transfer Mode) A data-framing and transmission architecture that features fixed-length data cells of 53 bytes, consisting of a fixed format of a 5-byte cell header and a 48-byte cell payload. The small cell size is intended to support high-speed switching of multiple traffic types with low end-to-end jitter. The architecture is asynchronous, so there is no requirement for clock control of switching and transmission.

Authentication A process to determine the identity of a party, system, or application, securely and uniquely.

B Channel (Bearer Channel) Traditionally refers to a single, full-duplex logical Integrated Services Digital Network (ISDN) interface that operates at 64Kbps.

BECN (Backward-Explicit Congestion Notification) A notification signal passed to the originator of traffic to indicate that the path to the destination exceeds a threshold load level. This signal is defined explicitly in the Frame Relay frame header.

Best Effort The default service level that has no associated undertaking of delay, loss, or jitter levels.

BGP (Border Gateway Protocol) An Internet routing protocol used to pass routing information between different administrative routing domains or ASes. The BGP routing protocol does not pass explicit topology information; it passes a summary of reachability between ASes. BGP is most commonly deployed as an inter-AS routing protocol.

Border Router Generally describes routers on the edge of an AS. Uses BGP to exchange routing information with another administrative routing domain. The term also can describe any router that sits on the edge of a routing subarea, such as an Open Shortest Path First (OSPF) area border router. *See also* Edge Device.

BRI (Basic Rate Interface) A user interface to an Integrated Services Digital Network (ISDN) that consists of two 64Kbps data channels (B channels) and one 16Kbps signaling channel (D channel) sharing a common physical access circuit.

Bridging The process of forwarding traffic based on address information contained at the data-link framing layer. Bridging allows a device to flood, forward, or filter frames based on the Media Access Control (MAC) address. Contrast with routing.

Broadcast The process of sending a single message to all reachable receivers simultaneously.

CBQ (Class-Based Queuing) A queuing methodology by which traffic is divided into separate classes and queued according to its assigned class in an effort to provide differential forwarding behavior for certain types of network traffic.

CBR (Constant Bit Rate) An ATM service category that corresponds to a constant bandwidth allocation for a traffic flow. The Cell Loss Priority (CLP) bit is set to 0 in all cells to ensure that they are not discard eligible in the event of switch congestion. The service supports circuit emulation, as well as continuous bitstream traffic sources (such as uncompressed voice or video signals).

CBT (Core-Based Trees) A method for multicast routing that uses a core node as the root of a spanning tree that distributes the multicast data.

CDV (Cell Delay Variation) An ATM QoS parameter that measures the variation in transit time of a cell over a virtual connection (VC). For service classes that are jitter-sensitive, this is a critical service parameter.

CIDR (Classless Interdomain Routing) An Internet routing paradigm that passes both the network prefix and a mask of significant bits in the prefix within the routing exchange. This supercedes the earlier paradigm of classful routing, where the mask of significant bits was inferred by the value of the prefix (where Class A network prefixes inferred a mask of 8 bits, Class B network prefixes inferred a mask of 16 bits, and Class C network prefixes inferred a mask of 24 bits). CIDR commonly is used to denote an Internet environment in which no implicit assumption exists of the Class A, B, and C network addresses. BGP version 4 is used as the de facto method of providing CIDR support in the Internet today.

CIR (Committed Information Rate) A Frame Relay term describing a minimum access rate at which the service provider commits to provide the customer for any given permanent virtual circuit (PVC).

Circuit A physical or logical state configured into a network to allow sender and receiver to communicate without further qualification of the data with address information.

CLP (Cell Loss Priority) A single-bit field in the ATM cell header to indicate the discard priority. A CLP value of 1 indicates that an ATM switch can discard this cell in a congestion condition.

CLR (Cell Loss Ratio) An ATM QoS metric defined as the ratio of lost cells to the number of transmitted cells.

Codec (Coder/Decoder) A hardware or software mechanism to encode a signal into a different format, with a matching mechanism to decode the format back to the original signal.

Connectionless A method of communication whereby each component of the communication is handled separately by the network. The network does not require the instantiation of a circuit state to support connectionless communication.

Connection-Oriented A method of communication that uses a circuit to support the transmission of data from the sender to the receiver.

Controlled-Load Service A service profile defined within the Integrated Services architecture. The profile is described as offering a service level equivalent to an unloaded network operating within the parameters of a specified level of bandwidth.

COPS (Common Open Policy Service) A TCP-based request-response protocol that can be used by a network element to refer decisions to a remote policy decision point. This allows a network manager to implement networkwide policies from a single configuration point.

CoS (Class of Service or Services) A categorical method of dividing traffic into separate classes to provide differentiated service to each class within the network.

CPE (Customer Premise Equipment) The equipment deployed on the customer's site when the customer subscribes (or simply connects) to a carrier's service.

CPU (Central Processing Unit) The arithmetic, logic, and control unit of a computer that executes instructions.

CRC (Cyclic Redundancy Check) An error-detection algorithm used to verify the integrity of data. CRCs are used within the Internet Protocol on IP, TCP, and UDP headers.

CSU/DSU (Channel Service Unit/Data Service Unit) A customer premise equipment (CPE) device that provides the telephony interface for circuit data services, including the physical framing, clocking, and channelization of the circuit.

CTD (Cell Transfer Delay) An ATM QoS metric that measures the transit time for a cell to traverse a virtual connection (VC). The time is measured from source UNI to destination UNI.

Datagram A data packet that includes sufficient address information to allow the packet to be delivered to its destination without reference to any other similarly addressed packets. There is no guarantee of delivery.

DCE (Data Communications Equipment) A device on the network side of a user-to-network interface (UNI). Typically, this is the customer premise equipment (CPE), such as a modem or channel service unit/data service unit (CSU/DSU).

D Channel (Data Channel) A full-duplex control and signaling channel on an ISDN Basic Rate Interface (BRI) or Primary Rate Interface (PRI). The D channel is 16Kbps on an ISDN BRI and 64Kbps on a PRI.

DE (Discard Eligible) A bit field defined within the Frame Relay header to indicate that a frame can be discarded within the Frame Relay switch when the local queuing load exceeds a configured threshold.

Delay The amount of time it takes for a data packet to traverse the network from its source to its destination.

DHCP (Dynamic Host Configuration Protocol) A protocol that is beginning to see widespread use on end-system computers to automatically obtain an IP host address, subnet mask, and local gateway information. A DHCP server dynamically supplies this information in response to end-system broadcast requests.

Differentiated Services A service architecture that applies a per-hop service response to a packet, based on the marking of the Differentiated Services field of the IP packet header.

Dijkstra Algorithm Also commonly referred to as Shortest Path First (SPF), it is a single-source, shortest-path algorithm that computes all shortest paths from a single point of reference based on a collection of link metrics. This algorithm is used to compute path preferences in both Open Shortest Path First (OSPF) and Intermediate System to Intermediate System IS-IS)). *See also* SPF.

DLC (Data Link Control) Refers to IBM data-link layer support, which supports various types of media, including mainframe channels, Synchronous Data Link Control (SDLC), X.25, and Token Ring.

DLCI (Data Link Connection Identifier) A numerical identifier given to the local end of a Frame Relay virtual circuit (VC). The local nature of the DLCI is that it spans only the distance between the first-hop Frame Relay switch and the router, whereas a VC spans the entire distance of an end-to-end connection between two routers that use the Frame Relay network for link-layer connectivity.

DLSw (Data Link Switching) Provides a standards-based method for forwarding Systems Network Architecture (SNA) traffic over TCP/IP networks using encapsulation. DLSw provides enhancements to traditional Remote Source-Route Bridging (RSRB) encapsulation by eliminating hop-count limitations; it removes unnecessary broadcasts and acknowledgments, and provides flow control.

Downstream The direction toward the destination, corresponding to the direction of data flow.

DS0 (Digital Signal Level 0) A circuit-framing specification for transmitting digital signals over a single channel at 64Kbps on a T1 facility.

DS1 (Digital Signal Level 1) A circuit-framing specification for transmitting digital signals at 1.544Mbps on a T1 facility in the United States, or at 2.108 Mbps on an E1 facility elsewhere.

DS3 (Digital Signal Level 3) A circuit-framing specification for transmitting digital signals at 44.736Mbps on a T3 facility.

DSBM (Designated Subnet Bandwidth Manager) A device on a managed subnetwork that acts as the subnet bandwidth manager (SBM) for the subnetwork to which it is attached. This is done through a complicated election process specified in the SBM protocol specification. The SBM protocol is an IETF proposal for handling resource reservations on shared and switched IEEE 802-style local area media. *See also* SBM.

DS Field An 8-bit field of the IP header used to specify the requested per-hop service within the scope of the Differentiated Services architecture. This field was originally specified as the Type of Service (TOS) field.

DTE (Data Terminal Equipment) A device on the user side of a user-to-network interface (UNI). Typically, this is a computer or a router.

E1 A carrier transmission circuit that carries data at a rate of 2.048Mbps. Predominantly used outside the United States.

E3 A carrier transmission circuit that carries data at a rate of 34.368Mbps. Predominantly used outside the United States.

Edge Device Any device on the edge or periphery of an administrative boundary. Traditionally used to describe an ATM-attached host or router that interfaces with an ATM network switch. *See also* Border Router.

Encapsulation The mechanism of enclosing one session's transport protocol packet within the data payload of another transport session. This technique is used by tunneling.

End System Any device that terminates an end-to-end communications relationship. Traditionally used to describe a host computer. May also include intermediate network nodes where a particular end-to-end communications substrate relationship terminates on an intermediate device (e.g., a router and an ATM VC).

EPD (Early Packet Discard) A congestion-avoidance mechanism generally found in ATM networks. EPD uses a method to preemptively drop entire ATM Adaptation Layer 5 (AAL5) frames instead of individual cells in an effort to anticipate congestion situations and make the most economic use of explicit signaling within the ATM network.

Ethernet A common bus LAN specification invented by the Xerox Corporation, then jointly developed by Xerox, Intel, and Digital Equipment Corporation. Ethernet uses Carrier Sense Multiple Access/Collision Detection (CSMA/CD) and operates on various media types. It is similar to the IEEE 802.3 series of protocols.

FAQ (Frequently Asked Questions) Compiled lists of the most frequent questions and their answers on a particular topic. A FAQ generally can be found in various formats, such as HTML Web pages, as well as in printed matter.

FDDI (Fiber-Distributed Data Interface) A LAN standard defined in American National Standards Institute (ANSI) Standard X3T9.5 that operates at 100Mbps, uses a token-passing technology and fiber-optic cabling for physical connectivity. FDDI has a base transmission distance of up to 2 kilometers, and uses a dual-ring architecture for redundancy.

FECN (Forward Explicit Congestion Notification) A notification signal passed to the receiver of traffic to indicate that the path to the originator exceeds a threshold load level. This signal is defined explicitly within the Frame Relay frame header.

FIFO (First In, First Out) FIFO queuing is a method of transmitting packets that are presented to a device for subsequent transmission. Packets are transmitted in the precise order in which they are received.

FIN (FINish Flag) Used in the TCP header to signal the end of TCP data.

Flow A sequence of packets generated by an application, all addressed to a common remote application. If the application uses TCP or UDP, a flow is characterized by a common source and destination IP address and a common source and destination port address.

FRAD (Frame Relay Access Device or Frame Relay Assembler/Disassembler) A device that operates natively at the Frame Relay data-link layer; it is less robust than multiprotocol routers (and in fact, usually does not provide network-layer routing). A FRAD simply frames and transmits traffic over a Frame Relay network; and on the opposite side of a Frame Relay network unframes the traffic and places it on the local media.

Fragmentation Where an IP packet is larger than the maximum transmission unit size of a link, the IP packet may be broken into a number of smaller packets, or fragments, with each fragment header sharing a common portion of the original IP header.

FTP (File Transfer Protocol) A TCP-based, transaction-oriented file transfer protocol used in TCP/IP networks.

Full Duplex Simultaneous bidirectional communication over a common transmission path.

Gbps (Gigabits Per Second) The data world avoided using the term billion, which is interpreted variably as either one thousand million or one million million, in favor of the term giga as one thousand million. Some confusion between the telecommunications and data-storage worlds still exists as to whether a giga is really the value 10^9 or 2^{30}. The communications industry tends to use giga to mean 10^9.

GCRA (Generic Cell Rate Algorithm) A specification for implementing cell-rate conformance for ATM variable bit rate (VBR) virtual connections (VC). The GCRA algorithm uses parameters to characterize traffic that is conformant to administratively defined admission criteria. The GCRA implementation commonly is referred to as a leaky bucket.

Guaranteed Service An Integrated Services architecture service class that provides an absolute limit on delay and bandwidth, given a traffic load that conforms to a specified traffic profile.

Half Duplex Bidirectional communication over a common transmission path; only one party can send at any time.

HDLC (High-Level Data Link Control) A bit-oriented, synchronous data-link layer transport protocol developed by the International Standards Organization (ISO). HDLC provides an encapsulation mechanism for transporting data on synchronous serial links using framing characters and checksums. HDLC was derived from Synchronous Data Link Control (SDLC).

Header The protocol-specific information associated with a data payload to form a protocol unit.

Hop A packet transaction where the data packet is passed through a transmission circuit to the next network switching element.

Host A terminating network element capable of originating and terminating transmission packets. This can be some form of workstation computer, or it can encompass any device that can interface with the network.

HSSI (High-Speed Serial Interface) The networking standard for high-speed serial connections for wide area networks (WANs), accommodating link speeds up to 52Mbps.

HTML (HyperText Markup Language) A simple hypertext document-formatting language used to format content that is presented on the World Wide Web (WWW) and read using one of the many popular Web browsers.

HTTP (HyperText Transfer Protocol) A TCP-based application-layer protocol used for passing information between Web servers and Web clients, also known as Web browsers.

IAB (Internet Architecture Board) The group concerned with the ongoing architecture of the Internet. IAB members are appointed by the trustees of the Internet Society (ISOC). The IAB also appoints members to several other organizations, such as the Internet Engineering Steering Group (IESG).

iBGP (Internal BGP or Interior BGP) A method to carry exterior routing information along the backbone of a single administrative routing domain, obviating the need to redistribute exterior routing to interior routing. iBGP is a unique implementation of BGP, not a separate protocol.

ICMP (Internet Control Message Protocol) A network-layer protocol that provides feedback on errors and other information specifically pertinent to IP packet handling.

I-D (Internet Draft) A proposal submitted to the IETF by the members of a particular working group or by individual contributors. I-Ds may or may not be subsequently published as IETF Requests for Comments (RFCs).

IEEE (Institute of Electrical and Electronics Engineers) A professional organization that develops communications and network standards, traditionally, link-layer LAN signaling standards.

IESG (Internet Engineering Steering Group) IESG members are appointed by the Internet Architecture Board (IAB) and manage the operation of the IETF.

IETF (Internet Engineering Task Force) An engineering and protocol standards body that develops and specifies protocols and Internet standards, generally in the network layer and above. These include routing, transport, application, and, occasionally, session-layer protocols. The IETF works under the auspices of the Internet Society (ISOC).

Integrated Services In a broad sense, encompasses the transport of audio, video, real-time, and classical data traffic within a single network infrastructure. More narrowly refers to the Integrated Services architecture, which consists of five key components: QoS requirements, resource-sharing requirements, allowances for packet dropping, provisions for usage feedback, and a resource-reservation model (RSVP).

Interface The point of termination of a physical or logical circuit.

Internet The global Internet. Commonly used as a reference for the loosely administered collection of interconnected networks around the world that share a common addressing structure for the interchange of traffic.

Intranet Generally used to refer to the interior of a private network, which either is not connected to the global Internet or is partitioned so that access to some network resources is limited to users within the administrative boundaries of the domain.

I/O (Input/Output) The process of receiving and transmitting data, as opposed to the actual processing of the data.

IP (Internet Protocol) The network-layer protocol in the TCP/IP stack used in the Internet. IP is a connectionless protocol that provides extensibility for host and subnetwork addressing, routing, security, fragmentation and reassembly; and as far as QoS is concerned, it offers a method to differentiate packets with information carried in the IP packet header.

IPng (IP Next Generation) A vernacular reference to the follow-on technology for IP version 4, otherwise known as IP version 6 (IPv6).

IP precedence A bit value that can be indicated in the IP packet header and used to designate the relative priority with which a particular packet should be handled.

IPv4 (Internet Protocol Version 4) The version of the Internet protocol in wide use today. This version number is encoded in the first 4 bits of the IP packet header, and is used to verify that the sender, receiver, and routers all agree on the precise format of the packet and the semantics of the formatted fields.

IPv6 (Internet Protocol Version 6) The version number of the IETF standardized next-generation Internet protocol (IPng), proposed as a successor to IPv4.

IRTF (Internet Research Task Force) An organization composed of a number of focused and long-term research groups, working on topics related to Internet protocols, applications, architecture, and technology. The chair of the IRTF is appointed by the IAB. The IRTF is described more fully in RFC2014.

ISDN (Integrated Services Digital Network) An early adopted protocol model; currently offered by many telephone companies for digital end-to-end connectivity for voice, video, and data.

IS-IS (Intermediate System to Intermediate System) A link-state routing protocol for connectionless Open Systems Interconnection (OSI) networks, similar to Open Shortest Path First (OSPF). The protocol specification for IS-IS is documented in ISO 10589.

ISO (International Standards Organization) The complete name for this body is the International Organization for Standardization and International Electrotechnical Committee. Its members are the national standards bodies, such as ANSI in the United States and the British Standards Institution (BSI) in the United Kingdom. The documents produced by the ISO are termed International Standards.

ISOC (Internet Society) An international society of Internet users and professionals who share a common interest in the development of the Internet.

ISP (Internet Service Provider) A vendor that provides external transit for a client network or individual user, providing connectivity and associated services to access the Internet.

ISSLL (Integrated Services over Specific Link Layers) An IETF working group that defines specifications and techniques needed to implement Internet Integrated Services capabilities within specific subnetwork technologies, such as ATM or IEEE 802.3z Gigabit Ethernet.

Jitter The distortion of a signal as it is propagated through the network; the signal varies from its original reference timing. In packet-switched networks, jitter is a distortion of the interpacket arrival times compared to the interpacket times of the original signal transmission. Also known as delay variance.

Kbps (Kilobits Per Second) A measure of data-transfer speed. Some confusion exists as to whether this refers to a rate of 10^3 bits per second or 2^{10} bits per second. The telecommunications industry typically uses this term to refer to a rate of 10^3 bits per second.

L2TP (Layer 2 Tunneling Protocol) A proposed mechanism whereby discrete virtual tunnels can be created for each dial-up client in the network, each of which may terminate at different points upstream from the access server. This allows individual dial-up clients to do interesting things, such as use discrete addressing schemes and have their traffic forwarded via the tunneling mechanisms along completely different traffic paths. At the time of this writing, the L2TP protocol specification is still being developed by the IETF.

LAN (Local Area Network) A local communications environment, typically constructed with privately operated wiring and communications facilities. The strict interpretation of this term is a broadcast medium in which any connected host system can contact any other connected system without the explicit assistance of a routing protocol.

LANE (LAN Emulation) A technique and ATM forum specification that defines how to provide LAN-based communications across an ATM subnetwork. LANE specifies the communications facilities that allow ATM to be interoperable with traditional LAN-based protocols, so that among other things, address resolution and broadcast services will function properly.

Latency The amount of time it takes for a data packet to traverse the network from its source to its destination. Also referred to as delay.

Layer 1 Commonly used to describe the physical layer in the OSI reference model. Examples include the copper wiring or fiber-optic cabling that interconnects electronic devices.

Layer 2 Commonly used to describe the data-link layer in the OSI reference model. Examples include Ethernet and ATM.

Layer 3 Commonly used to describe the network layer in the OSI reference model. Examples include IP and Internet Packet eXchange (IPX).

Leaky Bucket Generally, a traffic-shaping mechanism in which the input side is an arbitrary size and the output side is of a smaller, fixed size. This implementation has a smoothing effect on bursty traffic, because traffic is "leaked" into the network at a fixed rate. Contrast with token bucket.

LIJ (Leaf-Initiated Join) A feature introduced in the ATM Forum Traffic Management 4.0 specification. Any remote node in an ATM network can connect arbitrarily to a point-to-multipoint VC without explicitly signaling the VC originator.

LIS (Logical IP Subnetwork) An IP subnetwork in which all devices have a direct communication path to other devices sharing the same LIS, such as on a shared LAN or point-to-point circuit. In a Non-Broadcast Multi-Access (NBMA) ATM network, where all devices are attached to the network via VCs, the LIS is a method by which attached devices can communicate at the IP layer so that the IP believes all devices are connected directly to a local network medium, although they are not.

LLC (Link Layer Control) The higher of the two sublayers of the data-link layer defined by the IEEE. The LLC sublayer handles flow control, error correction, framing, and MAC-sublayer addressing. *See also* MAC.

LPM (Local Policy Module) A component of a policy architecture describing a module that resides within the network element itself. The local module may make policy decisions or may refer the decisions to a remote policy decision point.

LSA (Link State Advertisement) A packet-forwarding link-state routing process to neighboring nodes that includes information concerning the local node, the link state of attached interfaces, or the topology of the network. LSAs are generated by link-state routing protocols such as OSPF and IS-IS.

MAC (Media Access Control) The lower of the two sublayers of the data-link layer defined by the IEEE. The MAC sublayer handles access to shared media—for example, Ethernet and Token Ring—and decides whether methods such as media contention or token passing are used. *See also* LLC.

MARS (Multicast Address Resolution Server) A mechanism for supporting multicast in ATM networks. MARS serves a collection of nodes by providing a point-to-multipoint overlay for multicast traffic.

maxCTD (Maximum Cell Transfer Delay) An ATM QoS metric that measures the transit time for a cell to traverse a VC. The time is measured from the source UNI to the destination UNI.

Mbps (Megabits Per Second) A unit of data transfer. The communications industry typically refers to a mega as the value 106, whereas the data-storage industry uses the same term to refer to the value 220.

MBS (Maximum Burst Size) An ATM QoS metric that describes the number of cells that may be transmitted at the peak rate while remaining within the Generic Cell Rate Algorithm (GCRA) threshold of the service contract.

MCR (Minimum Cell Rate) An ATM service parameter related to the ATM ABR service. The allowed cell rate can vary between the MCR and the PCR to remain in conformance with the service.

Merging In the context of RSVP, describes the joining of two or more resource reservations that are part of the same multicast flow, so that a single reservation is passed upstream.

MIB (Management Information Base) A database of network-management information used by the network-management Simple Network Management Protocol (SNMP). Network-managed objects implement relevant MIBs to allow remote-management operations. *See also* SNMP.

MPLS (Multi-Protocol Label Switching) An emerging technology in which forwarding decisions are based on fixed-length labels inserted between the data-link and network layer headers to increase forwarding performance and path-selection flexibility.

MPOA (Multi-Protocol-over-ATM) An ATM Forum standard specifying how multiple network-layer protocols can operate over an ATM substrate.

MSS (Maximum Segment Size) A TCP option in the initial TCP Synchronize Sequence Numbers (SSN) three-way handshake that specifies the maximum size of a TCP data packet that the remote end can send to the receiver. The resultant TCP data-packet size is normally 40 bytes larger than the MSS: 20 bytes of IP header and 20 bytes of TCP header.

MTU (Maximum Transmission Unit) The maximum size of a data frame that can be carried across a data-link layer. Every host and router interface has an associated MTU related to the physical medium to which the interface is connected; an end-to-end network path has an associated MTU that is the minimum of the individual-hop MTUs within the path.

Multicast A form of communication whereby a single information flow can be directed to multiple receivers simultaneously.

NBMA (Non-Broadcast Multi-Access) A multiaccess network that does not support broadcasting, or on which broadcasting is not feasible.

NHOP (Next Hop) The next downstream switching point referenced as an object within the Integrated Services architecture protocol specifications.

NHRP (Next-Hop Resolution Protocol) A protocol used by systems in an NBMA network to dynamically discover the MAC address of other connected systems.

NLRI (Network-Layer Reachability Information) Information carried within BGP updates that includes network-layer information about the routing-table entries and associated previous hops, annotated as prefixes (IP addresses).

NMS (Network Management System) A computer system that understands the network so well that it can warn the operator of impending disaster—the distant dream of many a network operations manager.

NNI (Network-to-Network Interface) An ATM Forum standard that defines the interface between two ATM switches operated by the same public or private network operator. Also used within Frame Relay to define the interface between two switches in a common public or private network.

NOC (Network Operations Center) Where you try to call to inform a network operator that the network has failed.

nrt-VBR (Non-real-Time Variable Bit Rate) One of two variable bit-rate ATM service categories in which timing information is not crucial. Generally used for delay-tolerant applications with bursty characteristics.

OSI (Open Systems Interconnection) A network architecture developed under the auspices of the ISO throughout the 1980s as a standards-based technology suite to allow multivendor interoperability. Now primarily of historical interest.

OSPF (Open Shortest Path First) An interior gateway routing protocol that uses a link-state protocol coupled with a shortest path first (SPF) path-selection algorithm. The OSPF protocol is widely deployed as an interior routing protocol within administratively discrete routing domains.

Outgoing Interface The interface selected for the next hop.

Packet A synonym for datagram.

Packet Classifier A traffic management module that inspects parts of the packet header to allow the packet to be associated with a local state or a locally supported service class.

Packet Scheduler A packet management process that schedules packets for service based on the local operating state and the packet's service class.

Packet Switching A mechanism of information transfer whereby information is segmented into a series of discrete elements, and each element is switched to the destination as a series of independent actions.

PAP (PPP Authentication Protocol) A protocol that allows peers connected by a PPP link to authenticate each other using the simple exchange of a username and password.

Path Message An RSVP message generated by the data source that describes the characteristics of the data flow.

PCR (Peak Cell Rate) An ATM service parameter. PCR is the maximum value of the transmission rate of traffic on an ABR service category virtual connection.

PDP (Policy Decision Point) A network element that is able to make a network admission decision.

PEP (Policy Enforcement Point) A network element that enforces the policy decisions made by a PDP.

PHOP (Previous Hop) The preceding network switching element, as referenced by an object within the Integrated Services architecture protocol specifications.

PNNI (Private Network-to-Network Interface) The ATM Forum specification for distribution of topology information among switches in an ATM network, to allow the computation of end-to-end paths. The specification is based on similar link-state routing protocols. Otherwise known as the ATM routing protocol.

PPP (Point-to-Point Protocol) A data-link framing protocol used to frame data packets on point-to-point links. PPP is a variant of the HDLC data-link framing protocol. The PPP specification also includes remote-end identification and authentication (PAP and CHAP), a link-control protocol (to establish, configure, and test the integrity of data transmitted on the link), and a family of network-control protocols specific to different network-layer protocols.

PRA (Primary Rate Access) Commonly used as an off-hand reference for ISDN PRI network access.

PRI (Primary Rate Interface) An ISDN user-interface specification. In North America, a PRI is a single 64Kbps D channel used for signaling, and 23 64Kbps B channels used for voice or data (using a T1 access bearer). Elsewhere, the specification is two 64Kbps D channels and 30 64Kbps B channels (using an E1 access bearer).

PSTN (Public-Switched Telephone Network) A generic term referring to the public telephone network architecture.

PTSP (PNNI Topology State Packet) A link-state advertisement distributed between adjacent ATM switches that contain node and topology information. Analogous to an OSPF link state advertisement (LSA).

PVC (Permanent Virtual Connection or Permanent Virtual Circuit) An end-to-end VC that is established permanently.

QoS (Quality of Service) Read this book and find out.

QoSR (Quality of Service Routing) A dynamic routing protocol that has expanded its path-selection criteria to consider issues such as available bandwidth, link and end-to-end path utilization, node-resource consumption, delay and latency, and induced jitter.

RAPI (RSVP Application Programming Interface) An RSVP-specific API that enables applications to interface explicitly with the Resource ReSerVation Setup Protocol (RSVP) resource-reservation process.

RED (Random Early Detection) A congestion-avoidance algorithm developed by Van Jacobson and Sally Floyd at the Lawrence Berkeley National Laboratories in the early 1990s. When queue depth begins to fill on a router to a predetermined threshold, RED begins to randomly select packets from traffic flows that are discarded, in an effort to implicitly signal the TCP senders to throttle back their transmission rate. The success of RED is dependent on the basic TCP behavior, where packet loss is an implicit feedback signal to the originator of a flow to slow down its transmission rate. The ultimate outcome of RED is the avoidance of congestion collapse.

RFC (Request for Comments) RFCs are documents produced by the IETF for the purpose of documenting IETF protocols, operational procedures, and similarly related technologies.

RIP (Routing Information Protocol) RIP is a classful, distance-vector, hop-count-based, interior-routing protocol. RIP has been moved to "historical" status within the IETF, and is widely considered to have outlived its usefulness.

Routing The process of calculating network topology and path information based on the network-layer information contained in packets. Contrast with bridging.

RSVP (Resource ReSerVation Setup Protocol) An IP-based protocol used for communicating application QoS requirements to intermediate transit nodes in a network. RSVP uses a soft-state mechanism to maintain path and reservation state in each node in the reservation path.

RTT (Round-Trip Time) The time required for data traffic to travel from its origin to its destination and back again.

rt-VBR (Real-Time Variable Bit Rate) One of the two variable bit-rate ATM service categories in which timing information is critical. Generally used for delay-intolerant applications with bursty transmission characteristics.

SBM (Subnet Bandwidth Manager) A proposal of the IETF for handling resource reservations on shared and switched IEEE 802-style local area media. *See also* DSBM.

SCR (Sustained Cell Rate) An ATM traffic parameter that specifies the average rate at which ATM cells may be transmitted over a given virtual connection.

SDH (Synchronous Digital Hierarchy) The European standard that defines a set of transmission and framing standards for transmitting optical signals over fiber-optic cabling. Similar to the SONET standards developed by Bellcore.

SDLC (Synchronous Data-Link Control) A serial, bit-oriented, full-duplex, SNA data-link layer communications protocol. Precursor to several similar protocols, including HDLC.

SECBR (Severely Errored Cell Block Rate) An ATM error parameter used to measure the ratio of badly formatted cell blocks (or AAL frames) to blocks that have been received error-free.

SLA (Service Level Agreement) Generally, a service contract between a network service provider and a subscriber that guarantees a particular service's quality characteristics. SLAs vary from one provider to another and usually are concerned with network availability and data-delivery reliability. Violations of an SLA by a service provider may result in a pro-rated service rate for the next billing period for the subscriber, as compensation for the service provider not meeting the terms of the SLA.

SNMP (Simple Network Management Protocol) A User Datagram Protocol (UDP)-based network-management protocol used predominantly in TCP/IP networks. SNMP can be used to monitor, poll, and control network devices. SNMP traditionally is used to manage device configurations, gather statistics, and monitor performance thresholds.

Soft State A self-administered state that will time out and remove itself unless it is periodically refreshed.

SONET (Synchronous Optical Network) A high-speed synchronous network specification for transmitting optical signals over fiber-optic cable. Developed by Bellcore. SONET is the North American functional equivalent of the European Synchronous Digital Hierarchy (SDH) optical standards. SONET transmission speeds range from 155Mbps to 2.5Gbps.

SPF (Shortest Path First) Also commonly referred as the Dijkstra algorithm. SPF is a single-source, shortest-path algorithm that computes all shortest paths from a single point of reference based on a collection of link metrics. This algorithm is used to compute path preferences in both OSPF and IS-IS. *See also* Dijkstra algorithm.

SRB (Source-Route Bridging) A method of bridging developed by IBM and used in Token Ring networks, where the entire route to the destination is determined prior to the transmission of the data. Contrast with transparent bridging.

SVC (Switched Virtual Connection or Switched Virtual Circuit) A virtual circuit dynamically established in response to UNI signaling and torn down in the same fashion.

SYN (SYNchronize Sequence Numbers Flag) A bit field in the TCP header used to negotiate TCP session establishment.

T1 A WAN transmission circuit that carries DS1-formatted data at a rate of 1.544Mbps. Predominantly used within the United States.

T3 A WAN transmission circuit that carries DS3-formatted data at a rate of 44.736Mbps. Predominantly used within the United States.

Tbps (Terabits Per Second) The data world avoided using the term billion, which is interpreted variably as either one thousand million or one million million, in favor of the term tera as one thousand million. The communications industry tends to use tera to mean 10^{12}.

TCP (Transmission Control Protocol) A reliable, connection- and byte-oriented transport layer protocol within the TCP/IP protocol suite. TCP packetizes data into segments, provides for packet sequencing, and provides end-to-end flow control. TCP is used by many of the popular application-layer protocols, such as HTTP, Telnet, and FTP.

TDM (Time Division Multiplexing) A multiplexing method popular in telephony networks. TDM works by combining several signal sources on a single circuit, allowing each source to transmit during a specific timing interval.

Telnet A TCP-based terminal-emulation protocol used in TCP/IP networks predominantly for connecting to and logging in to remote systems.

TLV (Type, Length, Value) A standard IETF format for protocol packet formats, whereby individual fields are allocated to indicate the type and length of a particular packet, as determined by a specific value expressed in each field.

Token Bucket A traffic-shaping model in which the capability to transmit packets from any given flow is controlled by the presence of tokens. For packets belonging to a specific flow to be transmitted, for example, a token must be available in the bucket; otherwise, the packet is queued or dropped. This particular implementation controls the transmit rate and accommodates bursty traffic. The operation of a token bucket is equivalent to a smoothing function, coupled with a high-frequency filter, so that rapid oscillation of traffic load levels are damped down to a more constant load by the shaping function.

TOS (Type of Service) A bit field in the IP packet header designed to contain values that indicate how each packet should be handled in the network. This particular field has never been used much.

Transparent Bridging A method of bridging used in Ethernet and IEEE 802.3 networks. Frames are forwarded one hop at a time, based on forwarding information at each hop. Transparent bridging gets its name from the fact that the bridges themselves are transparent to the end systems. Contrast with SRB.

TTL (Time To Live) A field in an IP packet header that indicates how long the packet is valid. The TTL value is decremented at each hop; and when the TTL equals 0, the packet no longer is considered valid, because it has exceeded its maximum hop count.

UBR (Unspecified Bit Rate) An ATM service category used for best-effort traffic. The UBR service category provides no QoS controls, and all cells are marked with the CLP bit set, to indicate that all cells may be dropped in case of network congestion.

UDP (User Datagram Protocol) A connectionless transport-layer protocol in the TCP/IP protocol suite. UDP is a simplistic protocol that does not provide for congestion management, packet loss notification feedback, or error correction; UDP assumes these will be handled by a higher-layer protocol.

UNI (User-to-Network Interface) Commonly used to refer to the ATM Forum specification for ATM signaling between a user-based device, such as a router or similar end system, and the ATM switch.

UPC (Usage Parameter Control) A reference to the traffic policing done on ATM traffic at the ingress ATM switch. UPC is performed at the ATM UNI level and in conjunction with the GCRA implementation.

Upstream The direction pointing to the data source. This is the reverse of the direction of the flow of data.

VBR (Variable Bit Rate) An ATM service characterization for traffic that is bursty by nature or is variable in the average, peak, and minimum rates at which data is transmitted. There are two service categories for VBR traffic: real-time and non-real-time VBR. See also rt-VBR and nrt-VBR.

VC (Virtual Connection or Virtual Circuit) An end-to-end connection between two devices that spans a Layer 2 switching fabric (e.g., ATM or Frame Relay). A VC may be permanent (PVC) or temporary (SVC), and is wholly dependent on the implementation and architecture of the network. Contrast with VP.

VCI (Virtual Connection Identifier or Virtual Circuit Identifier) A numeric used to identify the local end of an ATM VC. The local nature of the VCI is that it spans only the distance between the first-hop ATM switch and the end system (e.g., router), whereas a VC spans the entire distance of an end-to-end connection between two routers that use the ATM network for link-layer connectivity.

VLAN (Virtual Local Area Network or Virtual LAN) A networking architecture that allows end systems on topological disconnected subnetworks to appear to be connected on the same LAN. Predominantly used in reference to ATM networking. Similar in functionality to bridging.

VP (Virtual Path) A connectivity path between two end systems across an ATM switching fabric. Similar to a VC; however, a VP can carry several VCs within it. Contrast with VC.

VPDN (Virtual Private Dial Network) A VPN tailored specifically for dial-up access. A more recent example of this is L2TP, where tunnels are created dynamically when subscribers dial into the network, and the subscriber's initial Layer 3 connectivity is terminated on an arbitrary tunnel end-point device that is predetermined by the network administrator.

VPI (Virtual Path Identifier) A numeric used to identify the local end of an ATM VP. The local nature of the VPI is that it spans only the distance between the first-hop ATM switch and the end system (e.g., router), whereas a VP spans the entire distance of an end-to-end connection between two routers that use the ATM network for link-layer connectivity.

VPN (Virtual Private Network) A network that can exist discretely on a physical infrastructure consisting of multiple VPNs; similar to the "ships in the night" paradigm. There are many ways to accomplish this, but the basic approach is that many individual, discrete networks may exist on the same infrastructure without knowledge of the others' existence.

WAN (Wide Area Network) A network environment where the elements of the network are located at significant distances from each other, and the communications facilities typically use carrier facilities rather than private wiring. Typically, the assistance of a routing protocol is required to support communications between two distant host systems on a WAN.

WDM (Wave Division Multiplexing) A mechanism used to allow multiple signals to be encoded into multiple wavelengths, so that the light signals can be transmitted on a single strand of fiber-optic cable.

WFQ (Weighted Fair Queuing) A combination of two distinct concepts: fair queuing and preferential weighting. WFQ allows multiple queues to be defined for arbitrary traffic flows, so that no one flow can "starve" other, lesser flows of network resources. The weighting component in WFQ enables the administrator to create the queue size and to delegate which traffic is identified for a particular size queue.

WRED (Weighted Random Early Detection, or Weighted RED) A variant of the standard RED mechanism for routers, in which the threshold for random packet discard varies according to the service precedence level of the packet. The weighting is such that RED is activated at higher queue-threshold levels for higher-precedence packets.

WWW (World Wide Web) The global collection of Web servers, interconnected by the Internet, that use the HyperText Transfer Protocol.

index